MAN AND ATOM

MAN AND ATOM

BUILDING A NEW WORLD
THROUGH NUCLEAR TECHNOLOGY

by Glenn T. Seaborg
and William R. Corliss

E. P. DUTTON & CO., INC.   NEW YORK   1971

Grateful acknowledgment is made for permission to reprint the following copyright material:

Extract from the poem "Cape Hatteras" by Hart Crane, from the collection *The Complete Poems and Selected Letters and Prose of Hart Crane.* Copyright © renewed 1966 by Liveright Publishing Corp., New York. Reprinted by permission of publisher.

Table on page 42 and diagrams on pages 39 and 40 (adapted) from G. T. Seaborg and J. L. Bloom, "Fast Breeder Reactors," *Scientific American,* Vol. 223, November 1970. Copyright © 1970 by *Scientific American.* All rights reserved.

Diagram on page 163 adapted from A. Wolman, "The Metabolism of Cities," *Scientific American,* Vol. 213, September 1965. Copyright © 1965 by *Scientific American.* All rights reserved.

Diagram on page 169 reprinted from *Science and Economic Development* by R. L. Meier, by permission of The M I T Press, Cambridge, Massachusetts. Copyright © 1966 by The Massachusetts Institute of Technology.

Diagram on page 251 reprinted by permission of *Nuclear News* from its December 1970 issue.

Map on page 348 from *Science,* Vol. 157, August 4, 1967. Copyright © 1967 by the American Association for the Advancement of Science. Reprinted by permission of *Science* and Dr. P. M. Hurley.

Contents

PART II. APPLYING THE TOOLS

Chapter 4. More Food and Water

8   Contents

PART III. THE ATOM AND SOCIETY

Illustrations

Preface

Man and Atom, conceived partly as a "labor of love," became during the course of its writing a work of necessity, the completion and publication of which seemed to become more urgent each day.

This book, written during a period of both great progress and controversy in atomic affairs, has dwelled in the mind of the senior author for many years prior to its actual writing. He has spent the better part of his life engaged in investigating and working with this new source of energy and with colleagues involved with nuclear matters in the United States and throughout the world. During this time he became increasingly aware that as our knowledge of the atom and our ability to use it grew, so also did there grow a gap of understanding between those engaged in nuclear affairs and the general public.

Broadly speaking, a deficiency of public understanding of science poses a problem in a democratic society—especially one that is also a technological society so dependent for its human progress on scientific progress. But in recent years we have become impressed with the fact that public understanding of the atom specifically is an even more urgent problem, as to a growing extent our very future may hinge on how wisely we manage this great new source of energy and its myriad applications.

ᚲ Human civilization is rapidly approaching a series of crises that can be managed only through some radical departures in man's dealings with the relationship between energy and matter. Nuclear energy holds one key—a crucial one—to the successful resolution of these crises. Without it there is no doubt that civilization, as we know it, would slowly grind to a halt. With it not only will we be able to raise a greater part of the world's people to a decent standard of living, but we will be able to move all mankind ahead into an era of new human advancement—human advancement which takes place in harmony with the natural environment that must support it.

It is understandable that a source of energy introduced to the world through the destructive force of a weapon as awesome as an atomic bomb would long be held in fear and would meet public resis-

13

tance in its development for peaceful uses. But it is also unforgivable that when those peaceful uses are so vital to man's needs and particularly to those needs which when fulfilled could help eliminate so much human conflict, we have failed to create sufficient understanding to overcome that fear and resistance.

In recent years we have delivered many speeches and written many papers in trying to tell the story of the atom's progress and promise. At best these could only be fragments of the complete story that we felt had to be told. The opportunity, therefore, to write a book that would assemble these fragments into a larger and more comprehensive picture of man's relationship to the atom was indeed welcomed. And as it was being written, the events of the day—the growing environmental and energy crises, the social and economic unrest and uncertainties based largely on man's apprehensions and concerns over his physical conditions—led us to feel our book might bear an increased significance in terms of these events.

These apprehensions and concerns cover many topics related to nuclear developments; and, while we have touched on as many of these as we could in one book, we have not attempted to go into great detail. For example, we have not tried to offer detailed rebuttals to specific arguments on the health and safety aspects of nuclear power. To do so would involve excessively technical explanations with which we did not want to burden our reader. Instead we hope we have provided the reader with the necessary background and stimulus to investigate such topics further. And in our bibliography we have provided a list of books, articles, and papers which are available for such a purpose.

We felt similarly about the complex subjects of weapons testing, arms limitation, arms control, and disarmament as they relate to the atom. If no detailed discussion of these subjects is present, it is not due to a lack of importance we attach to them, but, on the contrary, to the fact that they are so important that they warrant separate treatment and more thorough discussion than we could include here. Again, to assist the reader who wishes to pursue these topics further we have included in our bibliography a list of references to which he may turn.

Another decision we had to make was to select the technical level of our writing. To reduce the complex subject of nuclear energy to the primer level would have been unfair to the intelligent though nontechnical reader; we wrote at a higher level of sophistication. We have added a glossary of terms at the end of the book to aid those unfamiliar with nuclear jargon.

The reader will find that our book is not devoted exclusively to things nuclear. Since we are believers in the future—a future we believe will be highly technological—we have presented a few examples of how inventions in other sophisticated scientific fields will relate to the advances being made in atomic energy. For example, unusual words such as "teleoperator," "aquaculture," "NAWAPA," and "cyborg" will appear as we discuss that future in which man and machine will work together symbiotically and constructively.

In our discussions of the myriad peaceful applications of the atom we make a number of projections into the future. These should be regarded not so much as actual predictions but rather as illustrations of what might occur if all health and safety considerations and the problems of public understanding can be completely resolved. We do not believe that nuclear technology will achieve its true potential in the absence of public acceptance. All aspects must be, and we believe will be, debated pro and con, including considerations of the problems and hazards connected with alternate approaches. As the result of this process the atom will find its proper and deserved place in helping to meet and solve problems in our society.

Glenn T. Seaborg
William R. Corliss

April 1971

PART I
ATOMIC
TOOLS

Chapter 1
Tools
to Build
a
New World

The Technological Revolution

To say we live in revolutionary times has become a major cliché of the day. And yet there is hardly any other way to describe the rapid change and reaction to that change that thrust our society forward. Evolution connotes a process of orderly development over a long period of time. The developments of today seem to be a series of explosions. And as a result we appear to be constantly in a state of chaos and confusion, examining the fragments from those explosions for clues to a more meaningful picture of life and the course of our civilization.

A total picture is almost impossible to assemble, but one thing runs through the background of all we learn: the principal force behind all our rapid change—the one revolution behind all others—is the technological revolution. It is this enormous force, resulting from man's awesome and far-reaching ability to manipulate his physical environment, that is creating the progress, the problems, and the paradoxes of our times. It is the technological revolution that is behind the rewards of our affluence and the backlash of its mismanagement—the production of wealth and the problems of waste. It is the technological revolution that is behind the hope of rising expectation and the frustration and despair of failing fulfillment. It is the technological revolution that is behind the conditions that bring us all the exhilaration and despair, the diversity and conformity, the involvement and alienation, the excitement and boredom, and the freedom and imprisonment we seem to experience all at once.

Nothing is more important in clarifying this confusing picture than attempting to put our technological revolution in perspective; to learn "where it's at" and where it could be going. We cannot run away from it. We cannot turn it off. We can only understand it and then, on the basis of this understanding, try to control and direct it better in terms of our major values and goals. It was toward the fulfillment of these economic and social drives that we originally created and enlarged our technology.

Some will argue that the growth of technology has become an end instead of a means to other ends, that we have created a monster with a life of its own that now controls us. This argument currently has great appeal. Many who espouse this feeling wish nothing better than to reverse the technological revolution. Yet, if pressed, few will admit that they would wish to revert to the state of the Noble Savage. In their calmer moments, most realize that only modern technology —further developed and more wisely applied—can support a world population of many billions, especially if we wish to see those billions living with some degree of dignity, without the suffering and deprivation still dominant in much of the world today.

We wish to build on this basic fact of life and show that advanced technology, particularly nuclear technology, if applied with great care and astuteness, can help man to realize his fuller potential. We wish to show that there is far more to be gained by understanding and working with the atom than by turning from it in either ignorance or alarm. We believe this is an essential part of understanding the greater technological revolution and that such understanding will assure that the technological revolution remains a human and humane revolution. In short, this is an optimistic book.

The Place of the Atom in the Technological Revolution

One should not measure civilization solely by the number of books published per capita, the length of the work week, or the number of calories each human being consumes. None of these symptoms of civilization could exist at all without a foundation of power generators.

Classical Greece—with its Aristotle, Plato, and Homer—was built by a population of some 34,000 free men supported by a foundation of about 300,000 40-watt machines; that is, 300,000 human slaves. The "glory" of Ancient Rome depended upon 15 to 20 million citizens who commanded about 130 million slaves—still generating 40 watts of mechanical power apiece on the average. In contrast, almost every American from ghetto dweller to mansion owner commands many nonhuman energy slaves. We could surely survive with less power, as great-grandfather did with his whale oil lamps and subsistence farm, but the trend runs strongly in the other direction, and is brought on largely by population increases, more sophisticated industrial process-ing, and a legitimate demand for a higher standard of living. Cer-tainly we can get by without electrical frivolities, and it is hardly necessary for the utilities to advertise inducing people to consume more electrical energy, when it is already in short supply. However, we should keep in mind that the kilowatt, that is, abundant power, has perhaps been as great a liberator of mankind as any political sys-tem.

We can go back to the heady days early in the century to find tech-nology set forth in equally hopeful and idealistic terms. H. G. Wells, a great champion of technology and no mean prophet, once described the ultimate release of nuclear power as:

> . . . this tremendous dawn of power and freedom, under a sky ablaze with promise, in the very presence of science standing like some bountiful goddess over all the squat darknesses of human life, holding patiently in her strong arms, until men chose to take them, security, plenty, the solution of riddles, the key to the brav-est adventures. (*The World Set Free*, 1914)

When we read these words today, we smile with a superiority born of experience with several more decades of technology. Yet, has not the fundamental dream of Wells come true in the industrially devel-oped countries? Our children no longer work from dawn to dusk. The sweatshops are gone in the advanced countries. Machines, instead of stooping men, now till, harrow, and harvest the endless fields of grain and cotton. We live longer, eat better, and learn more. Wells would say that we are indeed set free, that the whole universe stretches out

before us waiting for exploration, and that we can now apply ourselves to the other pressing problems of humanity.

The pressing problems are only too obvious: poverty and hunger in
much of the underdeveloped world, incessant war, and dangerous
pollution of our planet. Words and slogans alone cannot do much,
but political action combined with properly used technology make a
potent combination. We believe that the atom will be in the forefront
in these great battles.

The Many-Talented Atom

The whirring, mutedly clicking computer is the common symbol of
today's technology. Jet transports are also impressive though not as
fastidious as computers. Both computers and jets seem indispensable
if the wheels of the modern world are to keep turning efficiently.
They are single-minded and direct; they augment our heads and our
feet. The atom in contrast works silently and largely unseen in almost
every city in every country of the world. It is the most versatile of all
the technological revolutionaries. In a homely sense, the atom is like
one of those old many-bladed jackknives that can do almost anything
a kid would want to do—whittling, screwdriving, bottle-opening, and
so on.

By way of preview we list below some of the things the atom can
do. In Chapters 2 and 3, we will explain how some of these atom-
based "tools" work and the precautions we must take in using them,
for most are double-edged. In Chapters 4 through 8 we will discuss
present and prospective uses of these atomic tools, while in the last
four chapters we will discuss some of the broader aspects of the
atom's role in society.

When we speak of the "atomic" repertoire, we really mean the "nuclear" repertoire. The term "atomic energy" in popular usage has
come to be synonymous with "nuclear energy." In this spirit we use
the title *Man and Atom,* even though we are concerned mainly with
the *nucleus* of the atom and not its extranuclear or chemical and electronic aspects. The nucleus is the heart of the atom; it is also the
source of the energy, the radioactive labels, the therapeutic radiation,
and all the other "atomic" benefits described in the following chapters.

The Atomic Repertoire

Facets of Nuclear Technology	*Present and Possible Future Applications*
Cheap, abundant, reliable power from large fission and fusion nuclear power plants	Desalted water, more food, new industrial processes, electric highways, economic waste reprocessing, "closed-cycle" cities, marine transportation
Clean nuclear explosives	Canal, harbor, and watershed excavation. New mining techniques, gas and oil recovery
Small atomic batteries from disintegrating atoms	Power for artificial hearts and other organs. Power for instrument packages in outer space and under the sea
Nuclear-powered rockets	Exploration of the moon and planets
Atomic timekeeping using natural radioisotopes	Isotopic dating in geology and archeology
Applications employing natural and artificial radioisotopes	Many contributions in medicine, industry, pollution control, agriculture, chemistry, criminology
Nuclear radiation from radioisotopes and nuclear reactors	Cancer therapy, industrial radiography, surgical instrument sterilization, stimulation of industrial chemical processes
Extremely high temperature "plasma" from thermonuclear research	A universal solvent for reprocessing wastes and other discarded resources
Master-slave manipulators and other "teleoperators"	Industrial, urban, and household dexterous machines to "do the dirty work"

Chapter 2

Power
and
More Power

How Atomic Power Evolved

For years radioactivity was perplexing, mysterious, even a shade on the supernatural side. Its discovery by Henri Becquerel in 1896 had revealed that the atom was not lifeless and inert. Within a few years of Becquerel's chance scientific bonanza, the atom was transformed by optimists into a potential wellspring of energy. No one was sure where this energy came from, but it was manifestly there. Some scientists surmised that the radium atom was a little self-contained engine that extracted energy from changes in temperature of the atmosphere. Lord Ernest Rutherford in a classic paper published in 1904 mused about the practical potential of the radioactive elements:

> If it were ever possible to control at will the rate of disintegration of the radio elements, an enormous amount of energy could be obtained from a small quantity of matter.

No one had the faintest idea how to unlock the energy store of the atom. Nevertheless, speculation about applications ran rampant. In 1911 the *Scientific American* wrote:

> Why not develop the radium engine and conserve our coal supplies, and manipulate ounces of radium instead of tons of coal?

At one fell blow all our elaborate coal-conveying machinery disappears, and with it roaring furnaces, the blackened faces of the stokers, and all the sooty paraphernalia that the word "steam engine" stands for.

The key to unlocking the atom's store of energy remained stubbornly elusive. By 1932, Rutherford had soured on the idea of atomic power, calling it "moonshine." Six years later, Otto Hahn and Fritz Strassmann, working at the Kaiser Wilhelm Institute in Berlin, uncovered the first clue to the fission process. They had found the element barium to be an unexpected by-product of their bombardment of uranium with neutrons. It was almost incomprehensible that the medium-weight barium atom should be derived from uranium, the heaviest atom then known. Lise Meitner, who had fled Hitler Germany to work with Niels Bohr in Copenhagen, and Otto R. Frisch correctly deduced that the uranium atoms had been split or fissioned by the neutrons, with the consequent release of energy. The long-searched-for key to atomic power had been discovered—accidentally. It was a scientific surprise of almost the same order as the discovery of radioactivity.

The next chapters in the story are well known. Neutron fission of the uranium atom was confirmed experimentally. Then, each fissioning uranium nucleus was found to emit *more than one* neutron by the nearly simultaneous experiments of L. Szilard, W. H. Zinn, H. L. Anderson, E. Fermi, and H. B. Hanstein, all of Columbia University, and H. von Halban, Jr., F. Joliot, and L. Kowarski working in Paris. This meant that each fission could stimulate *more than one* additional fission. In theory at least, a self-sustaining chain reaction was possible. In his famous letter of August 2, 1939, Albert Einstein informed President Roosevelt that "extremely powerful bombs" might be constructed from uranium. The Manhattan Engineer District, under Major General Leslie R. Groves, was created during the summer of 1942. Events moved quickly, for the Allied scientists believed they were competing with the Germans. The first self-sustaining chain reaction was attained on December 2, 1942, in a squash court at the University of Chicago. The predawn test of the first atomic bomb announced its success with a searing flash and cataclysmic roar over the New Mexico desert, on July 16, 1945. No longer just a dream of energy unlimited, the atom had arrived—violently.

The atom was on the one hand so terrible a force that the press and the deliberations of statesmen were full of forebodings. On the other hand, the optimists forecast automobiles powered by pea-sized lumps of uranium. A piece of nuclear fuel the size of a sugar cube would heat a house for the life of the structure. The polar regions would be made habitable. More than twenty-five years later, neither the hopes of the optimist nor the warnings of the pessimist had proven accurate.

After the conclusion of World War II, the United States moved quickly to consolidate its monopoly in nuclear weapons. Several low-temperature nuclear reactors had been built during the war at Hanford, Washington, to produce the artificial element, plutonium, for weapons. More reactors were built after the war. These plutonium production reactors could be characterized as lukewarm. Steam and electric power production were neither intended nor practical; still, a technological foundation for commercial power reactors was laid.

The five-man civilian Atomic Energy Commission (AEC) assumed leadership of the U.S. nuclear power efforts in 1947, first under the leadership of David E. Lilienthal, during the Truman administration. Lilienthal was followed by Chairmen Gordon E. Dean, Lewis L. Strauss, John A. McCone, and the senior author, under the administrations of Presidents Eisenhower, Kennedy, Johnson, and Nixon. The powerful Joint Committee on Atomic Energy has played a critical role from the very beginning. It was created along with the AEC by the Atomic Energy Act of 1946. Beginning with the first organization of the Committee on August 2, 1946, the Chairmen have been Senator Brien McMahon, Senator Bourke B. Hickenlooper, Representative W. Sterling Cole (who also served as the first Director General of the International Atomic Energy Agency), Senator Clinton P. Anderson, Representative Carl T. Durham, Senator John O. Pastore, and Representative Chet Holifield.

From 1948 to 1953, the AEC sponsored a diversified research and development program in fuel technology, heat transfer engineering, coolant properties, and all other aspects of the brand new discipline of reactor engineering. Three important events occurred in 1953 and 1954: (1) the AEC brought private industry into the nuclear power picture; (2) a program to build several prototype reactors and small "demonstration" nuclear power plants was formulated; and (3) President Eisenhower announced the United States Atoms for Peace program. To make the nuclear power picture even more rosy, uranium, which had originally been considered a rare metal—so rare that the atom would be but a flicker on the world power picture—was discovered in abundance in the western United States, Canada, and other parts of the world.

When the first Geneva Conference on the Peaceful Uses of Atomic Energy convened in 1955, all was optimism. The atom would indeed set the world free as that eternal optimist, H. G. Wells, had foretold almost a half century earlier. The optimism was premature.

The second Geneva Conference in 1958 met amid a growing disillusionment. True, prototype nuclear power plants had been built successfully in the United States, Great Britain, France, and the Soviet Union; but the atom had been oversold. On an economic basis it did not seem that nuclear power could come close to conventional fossil fuel power. The atom was not at all the Philosopher's Stone to the less developed nations; they did not possess the sophisticated technological infrastructure even to think about building nuclear reactors. The gloom thickened as huge new petroleum reserves were discovered, particularly in North Africa. Who needed the atom?

However, the third Geneva Conference in 1964, and the scheduling of a fourth for 1971, marked a return to optimism.

Between 1958 and 1963, a number of bright spots appeared on the nuclear horizon. America's first three pioneer commercial nuclear power plants at Shippingport, Pennsylvania, Dresden, Illinois, and Rowe, Massachusetts, proved dependable and overconservative in design. While the Shippingport plant generated power at about five cents per kilowatt-hour, the later Yankee plant at Rowe, Massachusetts, was able to bring the generating cost below one cent per kilowatt-hour—not much higher than fossil fuel power in New England. The Shippingport plant was a real landmark. It typified the transition from military to civilian interests, a shift of emphasis that was heavily occupying the AEC at the time. But, even with this new direction, the AEC called upon the "father of the nuclear Navy," Admiral Hyman G. Rickover, to assume technical direction of the Shippingport project.

The years 1963–1967 constituted a crucial period of change for nuclear power. Costs came down further as nuclear power proponents drew up plans for 500-megawatt, even 1000-megawatt, power plants. The first sign of a real economic breakthrough came in 1964 with the selection of nuclear power for the Oyster Creek plant by the Jersey Central Power & Light Company. More of the new large competitive nuclear plants were selected as other utilities climbed on the nuclear bandwagon. Roughly half of the new large-size commercial power plants ordered in 1966 and 1967 were nuclear. Some analysts saw a turning point; coal had found a real competitor; by A.D. 2000, the projections predicted that half of the electric power in the United States would be nuclear.

It is a competitive world, however, and the coal industry reacted by finding cheaper ways to mine and deliver its product, such as the "unit train" shuttling between mine and power plant. It also became apparent that the nuclear industry had used plants such as Oyster Creek as loss leaders. Nuclear prices subsequently rose; there were long lead times and manufacturing problems from an overcommitted industry. Nuclear sales dropped sharply in the 1967–1969 period as utility men sat back to watch their handful of recently purchased plants perform on a dollars-and-cents basis against new coal plants. In 1970, rising coal costs—particularly that of low-sulfur-content coal—caused another upsurge in nuclear power plant orders. By 1970, more than one hundred plants had been ordered in the United States, with a total capacity of approximately 100,000,000 kw. This figure will increase to 150,000,000 kw by 1980, according to the most recent estimates. The United States is not alone in this switch to nuclear power; fully twenty-five foreign countries are also engaged in building nuclear power plants.

The competition among the alternate methods of power generation is not entirely economic today. Environmental considerations are playing an ever-increasing role. Efficient fly-ash control equipment has enabled operators of modern coal-burning power plants to reduce particulate emissions considerably, but the gaseous emissions of sulfur, nitrogen, and hydrocarbon compounds continue to plague air clean-up endeavors. On the nuclear side, public hearings on new nuclear power plants reveal growing concern about radioactive effluents, reactor safety and the problems of storing radioactive wastes for long periods. Whether they are fossil or nuclear, power plants face the problem of handling the waste heat from their operations while minimizing the effects of its rejection into the environment. Because they operate at higher temperatures than the current nuclear plants, modern coal-fired power plants turn proportionately more of their thermal energy into electrical energy and conversely have need to release less unspent thermal energy to the environment. This unspent or waste thermal energy is commonly carried off by the diversion of cooling water from lakes, rivers, ponds, or the ocean. Because it rejects less waste heat, the modern fossil plant requires less cooling water than present-day nuclear plants. Both types of plants, however, must be designed to meet new water quality standards that reflect the growing concern about thermal modifications of the aquatic environment.

Inside a Modern Atomic Power Plant

Scientists have always deplored the fact that the energy of nuclear fission is first turned into "low-grade" heat before it is converted into "high-grade" electricity. When the nucleus of a uranium atom is hit by a neutron and subsequently fissions, the pieces resulting from the collision rush off in all directions at high speeds. There are the big fragments from the original uranium nucleus plus neutrons, electrons, and other fission "debris." The energy of motion (kinetic energy) of these particles is considered high-grade energy; and it seems a shame to have to degrade it to lowly thermal energy and then reconvert it into the motion of electrons in a wire (electricity). However, no really good alternatives to this wasteful process have been found for power generation on a commercial scale. Nuclear reactors, for all their sophisticated technology, can do no more than boil water or heat gas as coal-fired power plants began doing centuries ago.

The nuclear furnace is called a reactor. In it, a fluid (usually water) is pumped past uranium-containing fuel, which is fabricated in rods, plates, or other convenient geometries. The heat created by the fissioning uranium nuclei flows into the lower-temperature coolant which then carries it out of the reactor, sometimes through an intermediate heat exchanger, to turbines that convert the heat into mechanical energy and then into electricity by means of a generator. The main distinction between coal-fired and nuclear power plants at the "hot end" is the obvious absence of combustion in the nuclear plants as well as of the lumbering coal trains, ashes, and smoke associated with coal-fired plants. A single charge of nuclear fuel can last a year or longer and even so only from 1 to 2 percent of its potentially fissionable material will be consumed. Only the uranium-235 component of the uranium fuel fissions to any substantial degree. However, some of the neutrons emitted in the chain reaction are captured by the more abundant uranium-238 in the fuel, transmuting some of it to plutonium-239 (a very small amount of the uranium-238 itself fissions). Later in this chapter, in discussing fast breeders, we shall look at this in more detail. Plutonium-239, like uranium-235, fissions readily. By the time reactor fuel has reached the end of its useful life, there is more plutonium-239 present in the fuel than uranium-235. Thus, near the end of fuel life, fissioning of plutonium-239 produces most of the reactor heat.

The reactor is always housed securely in a heavy metal pressure vessel. Motor-driven control rods can be driven in and out of the reactor to control the population of neutrons moving through the reactor in all directions, and thus the power level of the reactor. For additional safety, the entire pressure vessel containing the reactor is placed in a heavy-walled structure deep within the power plant building. The building itself is a massive, high-integrity structure designed to withstand internal pressures and to contain any fluids inadvertently released. Outside the reactor structure, the basic components of nuclear power plants are identical to those of coal-fired plants. Both generate the same succession of products—heat, then steam, then, by means of turbine rotation, electricity from a rotating generator.

Almost all modern nuclear power plants in the United States utilize fresh water as the primary coolant; it is abundant, possesses good thermodynamic properties, and is well understood after centuries of engineering familiarity. Two basic types of water reactors have emerged:

1. The Boiling Water Reactor (BWR) which, as its name implies, heats the water coolant stream to boiling directly in the reactor. After separation of excess moisture the BWR steam moves directly to the turbines.

2. The Pressurized Water Reactor (PWR) in which water is kept under such a high pressure that it doesn't boil within the reactor. Outside of the reactor the heat in this primary pressurized water steam is transferred to a secondary water stream in a heat exchanger. The secondary stream is at a lower pressure and consequently boils, creating the steam to drive the turbines.

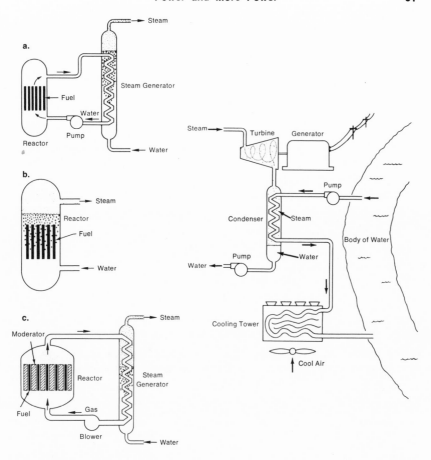

The three drawings on the left represent three important types of power reactors built in the United States: (a) the PWR (Pressurized Water Reactor); (b) the BWR (Boiling Water Reactor); and (c) the HTGR (High Temperature Gas-Cooled Reactor). They all boil water to generate steam. Following the generation of steam, the nuclear power plants are essentially the same, as indicated by the turbogenerator, condenser, and cooling tower on the right. Holding ponds are used sometimes to cool the water before releasing it to the environment. An important point: the cooling water from the environment does not enter the reactor.

The AEC and the governments of other countries have also sponsored development work on reactors cooled by such diverse fluids as liquid metals (usually sodium or a sodium-potassium mixture), heavy water, gas (carbon dioxide or helium, for example), and molten salts.

As with the PWR, a secondary coolant stream of ordinary water is usually the ultimate recipient of this heat. This secondary coolant loop supplies the steam to the turbine that drives the electric generator.

A particularly promising concept among these various reactor types has been developed in the United States: one cooled by a circulating gas stream at high temperature and pressure. Known as the High Temperature Gas-cooled Reactor, or HTGR, this kind of reactor uses helium as its coolant. It is at an earlier stage of development than the water-cooled reactors and has not been built yet in large numbers. One of its attractions is that it can operate efficiently on another fissionable isotope, uranium-233. As we shall show a little later on, uranium-233 is made from natural thorium and thus the HTGR, in effect, adds this abundant element to our nuclear fuel resources.

A distinctive feature of all American and most foreign water-cooled reactors is that they are fueled with "enriched" uranium—uranium in which the concentration of fissionable uranium-235 has been increased from 0.7 percent (the concentration in nature) to near 2 or 3 percent, on the average. Such enriched uranium is for power plants only and is not suitable for weapons. In the United States, uranium is enriched by a process called *gaseous diffusion*. Immense gaseous diffusion plants are located in Tennessee, Ohio, and Kentucky.

Except for the Soviet Union, whose nuclear power program is based on water-cooled, enriched uranium reactors, other countries do not yet have sufficient uranium enrichment capacity to support a power reactor economy. Consequently, some of these countries, notably England, France and Canada, have developed power reactors which use natural (unenriched) uranium for fuel. The coolants used are usually either heavy water or gas. However, the world trend is away from natural-uranium reactors. France is now including water-cooled, enriched uranium reactors in its nuclear power program and England has a program of advanced high temperature gas-cooled reactors that utilize slightly enriched uranium. Enriched uranium for future foreign power reactors will either be purchased from the United States or the U.S.S.R., or be enriched abroad by plants yet to be built, which possibly may employ new enrichment techniques, such as centrifuging. (See Chapters 9 and 10.)

The distinctive external features of a typical nuclear power plant are: (1) the large containment structure that houses the leak-tight reactor vessel and coolant loops; (2) the blocklike turbine building; and, in some cases, (3) the stack, which discharges plant ventilation air, but no smoke. Some highly diluted, carefully controlled radioactive wastes are released with the ventilating air; these wastes are diluted even further in the atmosphere and, as we shall see, constitute a negligible health hazard according to generally accepted standards. A walk through a nuclear plant impresses one with the cleanliness and quietness of nuclear operations. The whir of the turbines and generators is inescapable, but the muted roar and offensive odors of the coal furnace are gone. The "new look" in power plants is the nuclear look. The aesthetic competition has stimulated coal-fired plants to adopt a cleaner, less sooty facade.

All power plants that convert heat into electricity, as we have seen, must be cooled by air or, preferably, water from some river, pond, or ocean. No power plant can convert all the heat it generates (the dictum of the Second Law of Thermodynamics). This unusable or

The Robert Emmett Ginna nuclear power plant, Unit 1, owned by the Rochester Gas and Electric Company, is rated at 420 megawatts. Commercial operations began in July 1970. The reactor is a PWR (Pressurized Water Reactor) built by Westinghouse. (*Westinghouse*)

"waste" heat must be carried away. Per kilowatt of electricity generated, the PWR and BWR nuclear plants require about 50 percent
more cooling than modern coal-fired plants. This fact has led to
charges that nuclear plants thermally pollute the waters used to cool
them. We shall return to the thermal pollution question later in this
chapter, but it should be noted that all power plants that convert
heat to electricity must get rid of considerable waste heat. It is a
problem that is solvable in many places through the use of cooling
towers and holding ponds, or, more in keeping with our overall thesis, possibly through beneficial uses of the heated water.

Why the Breeder?

Today's reactors burn uranium-235, an isotope that constitutes only
0.7 percent of the uranium found in nature. The overwhelming bulk
of natural uranium is made up of the isotope uranium-238, which is
not readily fissioned. The long-term promise of abundant nuclear
power depends upon the development of fissionable fuels other than
the rare uranium-235. The purpose of the breeder reactor is to make
available just such fuel.

Other major advantages of the breeder reactor are almost as important as the conservation of the rare natural resource, uranium-235.
The breeder operates at higher temperatures than the water reactors.
To the engineer this means that a greater proportion of the heat generated can be converted to electricity and consequently less heat per
unit of electricity is rejected to the environment. Therefore, the
breeder reactor can be expected to be as efficient in this respect as
the most modern fossil fuel power plants and will minimize the thermal effects or pollution that we have described elsewhere in this
chapter.

Furthermore, the breeder reactor differs from water reactors in that
it must be hermetically sealed for technical reasons. A safety payoff
results from this feature. The breeder will discharge essentially no radioactivity to the environment—not even the very low levels of radioactivity that are emitted by the water reactors.

Now let us examine briefly how these advantages are brought
about.

In the nuclear business, a "breeder" is a reactor that creates *more than one* new atom of fissionable fuel for each fuel atom consumed. In a sense, breeding seems like perpetual motion—getting something for nothing—and thus prohibited by the laws of the universe. Actually, no laws are being violated. Breeding is in principle analogous to the rendering of crude oil (a potential fuel) into eminently combustible gasoline by the cracking process.

Two heavy isotopes, thorium-232 and uranium-238, are termed "fertile." Fertile isotopes can be made fissionable by bombarding them with neutrons, which change or transmute them into entirely new isotopes. Nuclear transmutation is analogous to petroleum cracking because the stored potential energy is also "unlocked" and nature's raw material made usable in conventional "burners"; that is, fission reactors and internal combustion engines.

In any nuclear reactor, each fission produces not only energy but about 2.5 neutrons, on the average (or twenty-five neutrons per ten fissions if the decimal is a bother). One of these 2.5 neutrons must go on to cause another fission and thereby continue the chain reaction. The remaining 1.5 neutrons, however, may either escape the reactor completely or be absorbed by nonfissionable nuclei within the reactor. In a breeder reactor, one or more of the 1.5 "extra" neutrons is absorbed by a fertile nucleus, which is thus transmuted into a fissionable nucleus. More specifically, thorium-232 is transmuted to fissionable uranium-233, and uranium-238 to fissionable plutonium-239. These transmuted materials can then be used to fuel the same or other reactors. The chain of nuclear events is shown in the diagram.

Petroleum refineries have converted the vast underground seas of sticky, rather disagreeable crude oil (once a mere curiosity of nature) into fuels and by-products that turn the wheels in much of the modern world. In the context of the world's supply of fuel for future generations, the breeder reactor is more important than the petroleum refinery. Petroleum can power the world for only a few more decades. According to current forecasts, naturally occurring fissionable fuels can easily sustain mankind for a few decades, but fertile nuclear materials will last *many* centuries, even millennia. The leverage of one naturally occurring fissionable atom of uranium-235 is tremendous. For each atom of natural uranium-235, there are about 140 fertile uranium-238 atoms that can be bred into plutonium. Further, there are vast deposits of natural thorium that can also be bred. Breeders are therefore a critical step in man's ability to "burn the rocks and the

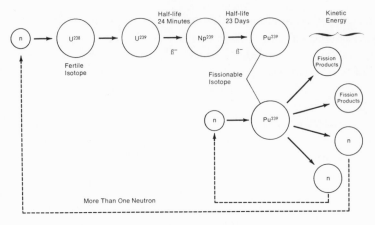

Uranium cycle for breeding in a fast-breeder reactor relies on fast, or highly energetic, neutrons. In the cycle, an atom of fertile uranium-238 absorbs a neutron and emits a beta particle (electron) to become neptunium, which then undergoes beta decay to become fissionable plutonium-239. When an atom of plutonium-239 absorbs a neutron, it can fission, releasing energy, fission products, and at least two neutrons. One of the neutrons is needed to continue the chain reaction, but others are available to transform a fertile isotope into a fissionable one, thereby "breeding" fuel. Within a few years a breeder doubles its original fuel inventory.

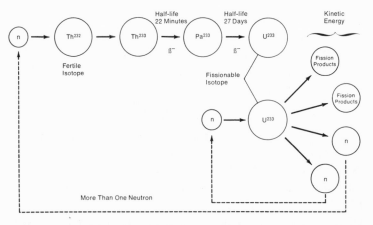

The thorium cycle of breeding is similar to the uranium cycle except that it works best in a thermal breeder reactor, where it relies on thermal, or relatively slow, neutrons. Thorium-232 is the fertile isotope that becomes, first, protactinium, and, then, fissionable uranium-233. (*Diagrams adapted from G. T. Seaborg and J. L. Bloom, "Fast Breeder Reactors,"* Scientific American *223 (Nov. 1970):13.*)

seas." To put the matter in prophetic language, if man can learn to "burn" rocks and water, that is, extract energy via fission and fusion processes, human civilization will not be just a brief fossil-fueled flicker in the long cosmic night.

None of the commercial nuclear power plants now built in the United States is a true breeder in which the bred fuel equals or exceeds that consumed. Because all reactor fuel contains a high fraction of fertile uranium-238, *some* fissionable plutonium-239 is always created. This plutonium has economic value, but there is not enough of it for us to call these plants breeders. To build breeder reactors the excess fission neutrons must be employed more effectively. Fertile materials must be placed strategically and absorption of neutrons in nonfertile and nonfissionable nuclei must be minimized.

Several ways to build breeders are open to reactor engineers. The most promising approach is called the fast-neutron breeder reactor, or, more simply, the fast breeder. The word "fast" refers to the speed of the neutrons causing the fission reactions. In the nonbreeding reactors being built today, the neutrons liberated by fission are intentionally slowed down (moderated) by collisions with nearby atoms in the reactor to slow ("thermal") speeds at which they are more effective in causing new fissions. Although breeding can be made to occur with thermal neutrons, the principal advantage of the fast breeder is that proportionally more neutrons are absorbed by fertile nuclei than would be the case in a thermal breeder. Water cannot be used to cool fast breeders because it readily slows down, or moderates, the valuable neutrons to speeds where they would be absorbed wastefully. Therefore, coolants such as liquid sodium or gaseous helium, which do not slow neutrons down as easily as water, must be used in fast breeders.

The potential of this type of breeder was recognized in the 1940s. Interestingly enough, the first nuclear reactor to be connected to a turbine and electric generator was a fast breeder, the EBR-I (Experimental Breeder Reactor I). Completed in 1951 at Arco, Idaho, the EBR-I plant generated a miniscule 200 kilowatts of electricity—hardly enough for a block in suburbia—but for the first time it turned nuclear energy into electrical energy in usable quantities. The EBR-II, like its predecessor, uses liquid metal coolant (sodium), as do the Fermi plant near Detroit and several other American and foreign breeder reactors.

Ordinary-water and molten-salt coolants are also being tested in the United States AEC thermal breeder reactor program. The thermal breeder using ordinary (or "light") water cooling is called, appropriately, the Light Water Breeder Reactor (LWBR). It is being designed on the premise that if the concept is successful, any of the more conventional PWRs now in operation or being built could be converted ultimately to a breeder by changing only its nuclear core. The development of this concept is being carried out under the direction of Admiral H. G. Rickover at the AEC's Bettis Laboratory. He contemplates making the first installation in the pioneer Shippingport reactor near Pittsburgh, Pennsylvania.

The Molten Salt Breeder Reactor (MSBR) is a very interesting technical innovation under development at Oak Ridge National Laboratory. Its most provocative feature is that it has *no* fuel elements as such. The fissionable material is dissolved in a circulating loop of molten salts which also carries the heat of fission to a heat exchanger. What is more, the fuel is processed (that is, the fission products formed are removed continuously from the molten salt) as part of the normal operation of the reactor.

Both the LWBR and the MSBR can operate on uranium-233 fuel, using thorium as the fertile material for breeding more fuel. It is difficult to predict their economic potentials until further development work has been done and a better gauge of their performance is available. However, it seems likely that a thermal breeder, which augments nuclear fuel resources by utilizing thorium, will supplement the uranium-consuming fast breeder reactor as a future source of electric power.

The sodium-cooled fast-breeder reactor (also called the Liquid-Metal-cooled Fast Breeder Reactor or LMFBR) is planned for the heart of the typical central power station during the last decade of this century. Therefore, a more complete description of this new type of power plant is desirable.

The "core" of the reactor consists of plutonium in ceramic form sealed into long, thin metal tubes. Surrounding the core are tubes containing uranium-238, the abundant (fertile) uranium isotope, in what is called the "blanket." It is in the blanket that the major portion of the breeding process occurs. Liquid sodium flows through the core and blanket to cool them, and the heat removed is transferred to another stream of flowing sodium. The purpose of the second "loop" of sodium is to isolate the first loop of sodium from the rest of the power plant. This is necessary because this sodium becomes artificially radioactive as it circulates through the reactor. The second stream of sodium then passes through a heat exchanger and gives up its heat to water, and the water is turned to high-pressure steam for running the turbogenerator.

Two different reactor configurations are being studied, as shown in the next two diagrams. In the pot type, many of the major heat transfer components are immersed in an enormous pool (upward of 1000 tons) of sodium, while in the loop type only the reactor core and

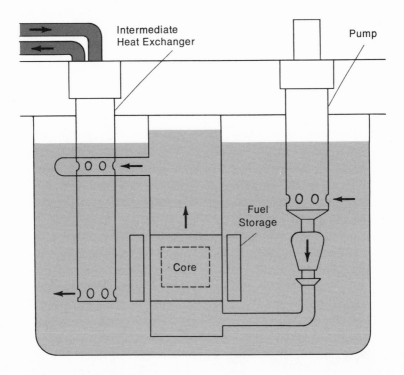

The pot system is one of two designs for containing the core-and-blanket assembly of the fast-breeder reactor and the primary heat-transfer system. The pot is a tank that is filled with sodium and also contains the reactor, the pumps that take sodium from the pool and move it through the reactor, and the intermediate heat exchanger where heat is transferred to nonradioactive sodium. (*Adapted from G. T. Seaborg and J. L. Bloom, "Fast Breeder Reactors,"* Scientific American 223 (*Nov. 1970*):13. *Reprinted with the permission of* Scientific American.)

blanket are kept in a smaller pool of sodium. Both concepts have technical advantages and disadvantages, and it is probable that both types will be developed into full-scale plants. A loop type LMFBR, designed to generate 500 megawatts, is shown in the diagram.

The fast breeder reactor cooled by gas (helium) is at an earlier stage of development, but it has some inherent advantages in performance that may spur its acceptance in later years.

To give an idea of the magnitude of the international effort being devoted to bring the LMFBR to reality, Table 1 lists the various major projects of this type throughout the world and their status as of 1970. Although these projects differ in their technical approach and

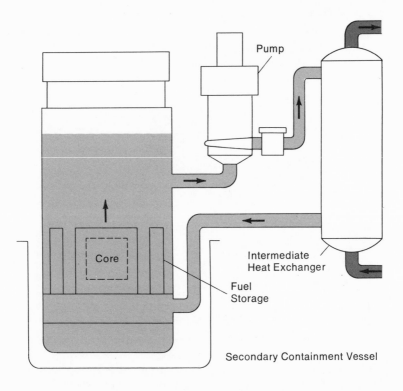

The loop system has most of its heat-exchange apparatus outside the fast-breeder reactor containment vessel. Only the reactor vessel is filled with sodium, which is circulated by pumps through the heat-exchange loops mounted outside the reactor vessel. (*Adapted from G. T. Seaborg and J. L. Bloom, "Fast Breeder Reactors," Scientific American 223 (Nov. 1970):13. Reprinted with the permission of Scientific American.*)

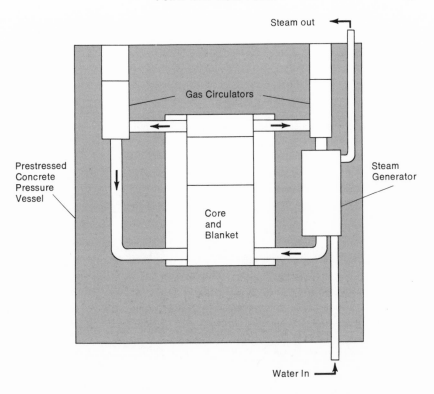

The gas-cooled fast-breeder reactor is cooled by circulation of gas in a closed loop at high pressure. Most of the equipment is contained within a prestressed concrete pressure vessel, which also acts as a radiation shield. Heat removed from the reactor core by flowing helium gas is transferred to one or more steam generators which convert water to high-pressure steam. No intermediate heat exchanger is required because the helium gas does not become radioactive in the reactor.

details, they have one thing in common: safety of operation will be paramount over all other factors. A full-scale fast breeder will contain tons of plutonium, and this element is one of the most toxic materials known to man. The challenge to, and the obligation of, scientists and engineers will be to use plutonium in a way that benefits man and does not harm him. It can be done.

It is logical to ask why breeder reactors were not perfected and inserted in the nation's power grid before the PWRs and BWRs described earlier—particularly when they are so critical to the future of the world's power supply. The PWRs and BWRs were easier to develop; that was the basic reason. Fast breeders are more difficult to build and control; there were also many metallurgical problems to solve before breeders could compete with coal-fired plants in the power marketplace. In addition, the new fissionable materials, uranium-233 and plutonium-239, had not been used as reactor fuels, although they are now being tested. The United States AEC breeder

Table 1

Liquid-Metal-Cooled Fast-Breeder Reactor Projects

	Name	Country	Power Megawatts (Thermal)	Power Megawatts (Electrical)	Initial Operation	Type (Pot or Loop)
Operating	BR-5	U.S.S.R.	5	—	1959	Loop
	DFR	U.K.	60	15	1959	Loop
	EBR-II	U.S.	62.5	20	1964	Pot
	Fermi	U.S.	200	66	1963	Loop
	Rapsodie	France	40	—	1967	Loop
	SEFOR	U.S.	20	—	1969	Loop
	BOR-60	U.S.S.R.	60	12	1970	Loop
Under Construction	BN-350	U.S.S.R.	1000	150	1971	Loop
	PFR	U.K.	600	250	1972	Pot
	Phenix	France	600	250	1973	Pot
	BN-600	U.S.S.R.	1500	600	1973/75	Pot
	FFTF	U.S.	400	—	1974	Loop
Planned	KNK-II	West Germany	58	20	1972	Loop
	JEFR	Japan	100	—	1973	Loop
	PEC	Italy	140	—	1975	Modified pot
	SNR	West Germany	730	300	1975	Loop
	Demo #1	U.S.	750–1250	300–500	1976	Not decided
	JPFR	Japan	750	300	1976	Loop
Decommissioned	Clementine	U.S.	.025	—	1946	Loop
	EBR-I	U.S.	1	.2	1951	Loop
	BR-2	U.S.S.R.	.1	—	1956	Loop
	LAMPRE-I	U.S.	1	—	1961	Loop

program has progressed through three generations of breeders and has solved many of the problems. Fast breeder reactors are now highest on the AEC priority list, as they also are in several other countries, such as the Soviet Union, the United Kingdom, France, Germany, and Japan. The AEC-industry target is to have large, economically competitive breeder reactors developed by the mid-1980s.

The 1980s will be none too soon, for world power demand is rising exponentially, and the Earth inherited only a limited supply of fissionable uranium-235 as a "seed stock" to use in breeding new fuel.

The impact of the fast breeder will be powerful. If the target date for the introduction of the fast breeder into American power grids is met, then, in the 35-year period that follows, there should be a saving in the consumption of uranium of over one million tons (equivalent in energy content to 400 billion tons of coal). The dollar saving to the consumer in lower electricity bills should amount to over $300 billion. Numbers like these give impetus to the development of the fast breeder. Delays in the introduction of the breeder reactor will be very costly to the electricity-consuming public.

A Captured Sun

Breeder reactors will help men realize the full potential of the energy nature has stored in the form of heavy, fissionable and fertile nuclei. It is the fusion process, however, that will "burn the seas." Heavy nuclei yield energy when they are split; in contrast, the lighter elements produce energy when they are united (fused) into heavier elements. Nuclear fusion generates most of the energy in the stars, including, of course, the sun, as shown by Hans Bethe as far back as 1939. Nuclei of hydrogen fuse in the high stellar temperatures, yield energy, and keep the stars burning. The sun and most stars are composed predominantly of light elements and can burn for billions of years. The Earth, however, is made mostly of heavier stuff: oxygen, silicon, aluminum, and other elements. Nevertheless, our planet's seas contain enough light, fusible nuclei to keep man's cities and machines running for many millions of years—*if these light elements can be fused on an economical basis.* For example, most of the deuterium (the heavy isotope of hydrogen) in a gallon of water can be extracted at a cost of about four cents. This small amount of extracted deuterium (much less than a teaspoonful) has the energy content of 300 gallons of gasoline. The ultimate energy problem is thus one of building miniature suns all over the world to serve as energy centers for the generations to come.

The fact of fission was a surprise to nuclear scientists in the late 1930s. Once the phenomenon was understood it was quickly turned to military ends. The fact of fusion, in contrast, was known several years before the discovery of fission. Deuterium was first concentrated from natural water by Harold C. Urey in 1931. Small electrostatic accelerators quickly showed that the so-called D-D reaction (fusion of two deuterium nuclei) was exothermic; that is, it evolved energy. All that was needed was some device to cause deuterium nuclei to collide at velocities high enough to overcome the mutual electrostatic repulsion of the positively charged nuclei. (Note: like charges repel one another.) Whereas the key to unlocking the energy of fission had been the electrically neutral neutron, which could slip through the electrostatic barrier of the uranium nucleus, the trick of fusion was providing enough deuteron energy to overcome the mutual electrostatic repulsion. In the early laboratory experiments, small accelerators sufficed for scientific studies of fusion; but these experiments were notoriously inefficient from the power plant standpoint. Far too many deuterium bullets were lost in fruitless collisions. The thought of extracting useful power from these exothermic nuclear reactions was never pursued beyond these early feasibility calculations. A few scientists realized that energy beyond man's dreams lay locked up in the

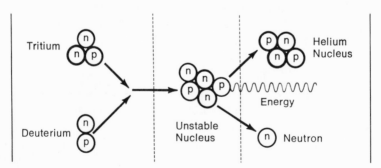

When deuterium and tritium, two heavy isotopes of hydrogen, fuse, an unstable nucleus is formed. This nucleus releases a neutron plus energy to become a nucleus of stable helium. Several other potentially useful fusion reactions involving light nuclei exist.

oceans, but no one could figure out how to "burn" the oceans efficiently.

In actuality, these same scientists saw a huge practical fusion reactor everyday—the sun. Nature obviously employed a high temperature key rather than the particle accelerator key to fuse light nuclei. In the 1930s, no match hot enough to ignite a supply of deuterium fuel existed on earth. A suitable match appeared in 1945 at Alamogordo: the fission bomb, possessing temperatures of hundreds of millions of degrees just after detonation.

Some bomb scientists realized immediately that deuterium and other light nuclei *might* be ignited by an atom bomb, thus greatly increasing its explosive power. Other scientists believed either that a "hydrogen bomb" *could not* be constructed or that if it could it *should not* be made on moral grounds. At the urging of some politicians and bomb scientists, President Truman decided to go ahead with the development of the hydrogen bomb. The United States tested its first full-scale hydrogen device in 1952.

Controlling fission for the purposes of generating useful power proved far easier than controlling fusion. The hydrogen bomb was thermonuclear like the sun, but it was an instantaneous flash of energy, not a continuous flow of power. How could the fusion reaction be ignited and confined within the walls of a terrestrial power plant, sufficiently long to extract useful energy? Beginning in 1954, the AEC increased its effort to find answers to these questions with the creation of Project Sherwood. Throughout the world, experimenters began to explore ways to build a small-sized sun.

In fission reactors, the essential condition for success was the self-sustaining neutron chain reaction. In fusion, the major requirements are set by a combination of natural laws, technology, and economics. Some requirements for the manufacture of miniature suns are:

1. The presence of very pure fuel consisting of light nuclei. Potential fuels are deuterium and tritium, which are hydrogen isotopes of weight 2 and 3, respectively.

2. Fuel densities of about 10^{15} nuclei per cubic centimeter.

3. A temperature of 100 million to 1 billion degrees Centigrade.

4. Confinement of the hot fuel at the required density for periods of time on the order of tenths of a second.

Useful thermonuclear power will result if the above conditions can all be achieved in a single experiment. Other combinations of density and confinement time can also lead to the production of useful power. By way of analogy, when these conditions are finally reached, the event will be similar to the first attainment of criticality in a fission reactor on December 2, 1942. In the case of fission power plants, some twenty-five years elapsed after this crucial date before economically attractive plants were built. Since the technology of controlled fusion is at least as difficult as that of fission, we can expect—assuming a similar level of effort—that many years will also separate the first laboratory attainment of the above-mentioned conditions and the first viable commercial fusion power plants.

The sun's strong gravitational field keeps most of its hot hydrogen fuel from expanding into space. On Earth, we have no localized gravitational centers to confine the contents of our fusion reactor. Certainly walls of solid materials are out of the question. A practical confining force being considered is the magnetic field. Atoms at room temperature are mostly neutral and cannot be manipulated by magnetic fields, but at 100 million degrees some of their electrons have been stripped off as they collide with their neighbors. Such a hot mixture of negative electrons and positively charged atoms (positive ions) is termed a "plasma." Plasmas can be maneuvered by magnetic fields because the electrical charges on the particles provide, in effect, magnetic handles. The branch of physics that deals with the behavior of a plasma in magnetic fields is called magnetoplasmadynamics (MPD), a part of the general field of plasma physics.

Fusion research has had its ups and downs like those in fission power. Project Sherwood and allied efforts began with great enthusiasm. By the early 1960s, however, all research efforts had apparently reached a plateau of achievement. Nature had thrown up an unforeseen roadblock: plasma instabilities. Magnetic fields, which were fabricated into various "bottles," "mirrors," "pistons," and figure eights, did indeed retain hot plasma within their nonmaterial walls, but only for microseconds (millionths of a second). All of the fusion research machines, bearing such strange names as Stellarator, Alice, Astron, and Scyllac, were plagued by errant plasma that somehow always seemed to writhe out of the machines' electromagnetic embrace too quickly.

During the 1960s, the bulk of the world's fusion research converged on the containment problem, as the other three problems just listed succumbed one by one. With roughly $100 million being invested per year throughout the world, the final containment roadblock had to give here and there. The United States, the Soviet Union, and the United Kingdom (the countries with the largest programs) have made big strides in achieving plasma confinement. The Americans and the British have demonstrated that their machines could hold plasma for periods of time longer than required for sustaining the fusion reaction, but the other conditions mentioned have not been met at the same time—yet. Reporting for the Russians, who have mounted the largest fusion program, Academician Lev A. Artsimovich stated in April, 1969, that his group at the Kurchatov Institute had confined a 5,000,000° C plasma at a density of 7×10^{13} nuclei per cubic centimeter for 0.02 second in their Tokamak machine. This is tantalizingly close to shattering the decade-old roadblock in the path to fusion power. Taken as a whole, these results heighten the belief of many

The Model ST Tokamak fusion experiment at Princeton. This experiment was stimulated by recent achievements at the Soviet Union's Kurchatov Institute. (*Princeton Plasma Physics Laboratory*)

scientists that there is no undiscovered law of nature that might make the roadblock truly impregnable.

If fusion power becomes a practicality—and it seems an inevitability—it will boast some enviable advantages:

1. The fuel would be cheap and almost inexhaustible.

2. Nuclear accidents of the runaway or fuel meltdown varieties would be intrinsically impossible.

3. The quantity of radioactive wastes would be drastically reduced.

4. The energy in the charged particles of the hot plasma *may* be turned directly to power production without going through the clumsy and archaic heat-engine routine. Higher efficiencies (perhaps as high as 90 percent) might result, reducing the thermal pollution problem to a more manageable size.

Fusion power will require some environmental precautions. The first fusion power plants may use tritium as a fuel, and tritium is radioactive. Large quantities of this hydrogen isotope would have to be processed, shipped, and "burned" for power. The most stringent precautions would have to be taken to meet this potential environmental hazard. The technology of handling hydrogen gas is well-known from experience in the chemical, space, and nuclear industries, but improvements will still be needed. Fortunately, tritium does not emit penetrating radiation, and precautions need to be taken only against the ingestion or inhalation of this gas. In addition, large numbers of neutrons will be present during the operation of a fusion power plant. These will interact with the materials of construction and surrounding matter and make them artificially radioactive. This also happens in present-day fission reactors. Handling these activated materials should be considerably simpler in the case of fusion power plants.

Engineers are already beginning to think about how to design fusion power plants, for practical fusion power will certainly be a reality someday. The basic problem here is the efficient, economical conversion of the energy of the hot plasma into electricity. There are many possibilities. (See diagram.) As the fusion roadblock seems to crumble, the anticipation of success is stimulating many novel suggestions. One of the more interesting of the new ideas is the use of

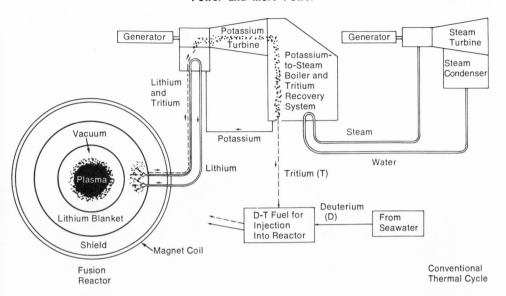

Fusion
Reactor

Conventional
Thermal Cycle

This illustration shows one concept for generating electric power from a controlled fusion reaction between deuterium and tritium. Much of the energy liberated in the magnetically confined plasma is in the form of high-speed neutrons, which are intercepted by a blanket of molten lithium surrounding the plasma. The neutrons heat the lithium and also convert some of the lithium to tritium. The molten lithium carries the heat to an external heat exchanger where part of the heat is transferred to molten potassium. The potassium boils and its vapor drives a high-temperature turbogenerator. Potassium vapor leaving the turbine is still hot enough to boil water and thus drive a conventional steam turbogenerator. The tritium in the lithium is separated, mixed with deuterium, and injected back into the reactor.

high-powered lasers to heat "pellets" of a deuterium-tritium mixture to thermonuclear temperatures, causing the fusion reaction to occur in bursts as the pellets are dropped one by one into the laser beam. Many scientists believe that this radical approach may be practical for generating commercial electrical power—perhaps even practical for providing propulsive power for spacecraft.

In summary, the practical fusion reactor seems almost within our grasp—perhaps only two or three decades away. Until something even better comes along, say, some unguessed-at process for converting matter directly and entirely into energy, fusion power should serve man's objectives well beyond the next millennium. For all its promise, however, fusion power still remains a goal, not a reality. The conventional fission power plant and its eventual replacement, the breeder, are further along in the technological development cycle. We see in this succession—fission plant, then breeder plant, then fusion plant—machines that first burn the rocks and then the oceans.

Conventional and Unconventional United States and World Power Scenarios

We visualize and plan for the future according to what we shall call conventional prognoses. We cannot expect the unexpected although it always seems to come to pass. To complicate matters further, predictions of population growth rates, consumption of manufactured goods, and all the parameters that go into constructing a world power picture, have almost always been too conservative, much too low. Yet anything beyond the conventional world power scenario is inevitably derided as "overselling" or "optimistic."

Past estimates of the impact of nuclear power illustrate the common tendency of technology to outrace predictions. Right after World War II, uranium was thought scarce and because of this nuclear power was considered inconsequential. By 1956, optimists went out

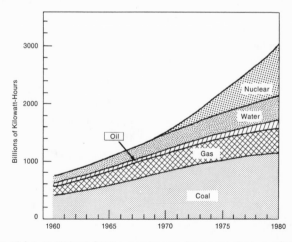

Projections of future United States power requirements show nuclear power assuming a larger and larger share of the burden of generating electric power.

on a limb to suggest that by 1967 possibly 10 percent of all planned new electric power capacity in the United States would be nuclear. Actually, the figure was nearer 50 percent in 1966 and 1967, although it has fallen some since then and is now rising again.

With these cautions, let us look at what the seers predict for the United States and world power demand, particularly nuclear contributions toward meeting this demand. The basis for estimates of a region's power demands must be a product of population and technological sophistication. Population growth is relatively easy to extrapolate, but technological sophistication is not. The populations of the highly developed countries such as the United States will probably rise rather slowly but their power requirements will probably increase much faster than their populations (Table 2). This dispar-

Table 2
U.S. Electric Power Consumption and Population Statistics

	1950	Estimated 1970	Predicted 1980	Predicted 2000
Population (millions)	152	205	230	300
Total power-generating capacity (millions of kilowatts)	83	360	690	2200
Kilowatt capacity per person	0.5	1.8	3.0	7.2
Total electric consumption (billions of kilowatt-hours per year)	390	1640	3300	10,000
Electric consumption per person (kilowatt-hours per year)	2600	8000	14,000	33,000
Nuclear generating capacity (millions of kilowatts)	0	5	150	1100
Nuclear fraction of total capacity (percent)	0	1	22	50
Nuclear electric consumption (billions of kilowatt-hours per year)	0	23	940	6800
Nuclear fraction of total electric consumption (percent)	0	1	30	70

ity reflects increasing sophistication in the use of energy; that is, more home appliances, more industrial processes that consume more energy, and so on. This same rise in sophistication also implies a relative shift from low-grade heat to more versatile electricity. Table 3 gives comparable data for worldwide electric power requirements. (The predictions for the year 2000 can be best described as presumptuous.) The greatest population increases of all will be in the less-developed portions of the world. As more and more local technologies shift from wood or camel-dung fires to electric stoves, world electrical power demands will skyrocket.

Table 3
World Electric Power Consumption and Population Statistics *

		Estimated	*Predicted*	
	1950	**1970**	**1980**	**2000**
Population (millions)	1970	2870	3600	5300
Total power-generating capacity (millions of kilowatts)	223	1070	2200	7200
Kilowatt capacity per person	0.1	0.4	0.6	1.4
Total electric consumption (billions of kilowatt-hours per year)	950	4760	10,000	33,000
Electric consumption per person (kilowatt-hours per year)	500	1700	2900	6200
Nuclear generating capacity (millions of kilowatts)	0	17	330	3600
Nuclear fraction of total capacity (percent)	0	2	15	50
Nuclear electric consumption (billions of kilowatt-hours per year)	0	100	2000	22,000
Nuclear fraction of total electric consumption (percent)	0	2	20	65

* Includes all countries except mainland China. Sources of information: *UN Statistical Yearbook* for 1969, U.S. Atomic Energy Commission internal reports, *Edison Electric Institute Statistical Yearbook* for 1960, Federal Power Commission and Bureau of Census reports.

The atom's share of the power market depends upon a region's requirements, taking into account its native fuel resources as well as its technological sophistication. Coal and oil are easy to burn, but few of the less developed countries can support a nuclear power industry. Therefore, we would expect—using the conventional scenario—that nuclear power would be slow to take hold in the less-developed countries.

But the whole conventional world power scenario is probably wrong. Here, we look into our personal crystal ball. Barring nuclear war and some general collapse of civilization, world power demands will probably be far higher than conventional extrapolations allow. We base this prophecy on three observations:

1. Much of the world is hungry.

2. Much of the world is poor.

3. Much of the world is polluted.

Oversimplified though it may sound, energy can help solve all these problems, and perhaps with them, the social stresses they create. Our unconventional world power scenario is based on the thesis that nuclear power will soon be so cheap and so abundant that it will greatly accelerate the development of the hungry, poor parts of the world. If energy is cheap and abundant, so will be food, water, clean air, and all the amenities of what we call civilization. This stimulator role of energy is not a hope or a figment of some futurist's imagination, but rather an observable fact all over the globe. (See Chapter 10.)

As we see our natural resources and scenic beauty ebb as industrial civilization encroaches, more people are asking, "Can we afford a continued exponential rise in power-generating capacity?" In reality this question is completely equivalent to a second question: "Can we afford an exploding population and a rising standard of living?" For many millions who live in poverty in this country and abroad, the answer to the second part of the question is, "We must." Neither will other millions stand by and see their standard of living reduced because of inadequate supplies of electricity. Although we have no final answers to these questions, it is time to explore their ramifications.

The Case for Nuclear Power

Although our purpose here is to demonstrate the broad sweep of the advantages of atomic energy in its many ramifications—and we are numbered among the strong advocates of atomic energy—we also believe that our views are not based on blind faith or dogma, scientific or otherwise. We would like to explain the rationale that brings us to our position while not attempting to debate all issues or to refute all charges. The debate is proceeding in other arenas, where the opportunity for rebuttal and counterrebuttal is more accessible to those who wish to join the cause of nuclear power or to campaign against it.

Our basic premise, already stated earlier in this chapter, is that the demand for electric power is going to increase so rapidly and to such a great magnitude that the resources of the United States and the world will be sorely taxed in meeting it. This in itself is not a maxim, but a belief that in the drive to improve the standard of living, to remove blight, and indeed to improve the environment, the key is adequate electric power. Although some naturists decry this approach, hoping that man will voluntarily return to the more simplistic ways of his forebears, their views do not appear to be either logical or desirable (to the huge majority) as a pattern for future life on this planet. People want to be cool in summer and warm in winter. The conveniences and even the luxuries of life are fast becoming the necessities. Again, electric power is the key.

We have treated earlier in this chapter, and will discuss further in Chapters 9 and 10, the economics of nuclear power and the advantages that accrue from it relative to other sources of energy, most of which are in dwindling supply. The arguments against nuclear power do not seem to lie in these areas, although perhaps the stimulus for rejection arises there because of resistance to changing economic patterns. The word today is "environment" and nuclear power may well stand or fall on how effective it is in reducing undesired environmental effects *relative to alternative sources of energy*. Nuclear plants do not "burn" fuel in the same sense that a fossil-fueled plant does, and there are no combustion-products—no sulfur dioxide, oxides of nitrogen, carbon monoxide, carbon dioxide—a very important factor in our fight against air pollution. Each new, large fossil-fueled power plant requires additional millions of tons of fuel each year, presenting an increasingly difficult source problem. The same size nuclear plant burns only about a ton of fuel, readily available and easily transportable. The movement of so many carloads of coal and tankers full of oil produces a serious environmental and aesthetic impact.

Finale

It can be maintained that there are unconventional and as yet unexploited forms of energy which are cleaner, ecologically speaking, than any form in use today. Solar energy is one example of relatively clean energy. The rays of the sun fall uncontrolled on every part of the Earth's surface and are completely indispensable to life. An enormously large fraction of this energy is re-radiated to space without having performed any direct useful function, such as entering into the photosynthesis process. Shouldn't it be possible to collect this energy and convert it into electricity, completely avoiding any environmental impact? Of course it is possible. Solar cells do this every day on Earth satellites, converting 5 to 10 percent of the incident radiation into electricity, but at power levels of up to only one or two kilowatts and at very low direct current voltages. To provide electrical energy at night or when cloud cover obscures the sun, the energy must be stored in batteries. The estimated cost of electrical power under these conditions is in the range of $5 to $10 per kilowatt-hour—this is *more than 500 times* the average cost of electrical energy to the private consumer in the United States. The cost could be brought down tremendously if there were widespread terrestrial use of solar cells, but not to the point where electricity would be the omnipresent servant it is today. Peter E. Glaser of Arthur D. Little, Inc., who is a staunch advocate of solar electrical power, claims that these drawbacks can be overcome in time. For example, he has proposed that enormous arrays of solar cells covering several square miles could be put into synchronous (stationary with respect to the Earth) orbit, converting solar energy into electrical energy and then transmitting the electrical energy to the earth in the form of microwaves similar to radar. Several drastic improvements in technology would be required, however, to make such a system practical, and even then it would apply to only a fraction of the needs of a heavily populated country. Just one of the improvements required would be to reduce the cost of putting equipment into orbit from $1000 per pound to $50 per pound. Still, the future will doubtless see more effective use of solar heating (as opposed to solar electricity) than we do now, and ultimately even solar electricity may become attractive. Solar heat should be useful for evaporating seawater and for local heating and cooling systems, as in the well-known solar homes.

In similar fashion we could examine other natural sources of energy, such as the tides, geothermal energy, or the temperature differences in the oceans, but to no avail. They are too dispersed, too expensive, or too variable to be of practical use on the large scale we need. For the near-term future, then, we must consider ourselves limited to the energy derived from falling water (hydropower), fossil fuels (oil, gas, and coal), and nuclear fuels. Of the three, hydropower is the cleanest, but it is not without unfavorable ecological effects. Its most significant drawback, however, is that there simply are not enough places in the world where sufficient water flows the year round down a great enough gradient to take advantage of the energy available. Even those parts of the world which historically have depended almost completely on hydropower are now turning to other sources. The Pacific Northwest in the United States, for example, long justly proud of its Grand Coulee and other large dams, now must turn to other sources of supply. The Scandinavian countries find themselves in the same situation.

At least hydropower is regenerative. Water evaporates. Rain or snow falls. Rivers run. More power is produced. This is not true of the mineral resources used for energy. Nature required millions of years to convert carbon dioxide to living things and then to organic debris and last to coal, oil, and gas. Man can simulate this process, but only by adding energy in place of the sun. When these natural materials are exhausted, our descendants would then have to replace them by the difficult process of chemical synthesis, starting from carbon dioxide and water—a very expensive and circuitous return to the conditions under which nature originally synthesized our fossil fuels. For all practical purposes, the organic fuels stored in the Earth's bowels are therefore ours to use only once. The time scale of consumption then becomes important. All significant hydroelectric power stations have been installed since the beginning of the twentieth century and almost their entire capability is now being utilized. More coal, oil, and gas have been extracted from the Earth in the last thirty years than in the previous 5000 years, and there is only a period of grace left for them of just a few hundred years if we continue to use them in a profligate and wasteful manner; that is, crudely burning them merely to use the heat of their fires. Only 10 percent of the world's coal and oil is in the United States, and this nation is now consuming about 35 percent of the world's annual consumption of these resources. We need these vital products of nature's long synthesis for their precious organic chemical bonds in the manufacture of plastics, drugs, dyes, and a host of other chemical compounds. And we need them to power our air, sea, and ground transportation vehicles until a better portable energy system comes along. What, then, is left? Where shall we turn?

Those who believe that nuclear power is the answer to these questions state their position as follows:

1. Uranium and thorium ores can last for decades as fuels for the existing types of nuclear reactors, and for millennia as fuel for breeder reactors. (The ultimate arrival of economic fusion power will make all other energy sources obsolete.)

2. Nuclear power is now economically competitive. It is needed now to supplement other sources of energy in order to meet the rapidly increasing demands for electric power. Furthermore, uranium and thorium are not needed for other purposes as are the fossil fuels.

3. With proper controls and precautions, nuclear power is safe.

4. The environmental effects of nuclear power are manageable and have been reduced to the point where they are less harmful than competitive energy sources; that is, nuclear power is clean.

But these points sound like the pitch of the sideshow man—one-sided, soothing, "commercial." Don't accept them on faith. Examine both sides of the many questions evolving from them. Reach your own conclusions.

But Don't Build It Here (or Anywhere)

In the context of the preceding section, we have a fascinating case history to present.

During the 1880s, when Thomas Edison's direct current (D.C.) scheme for distributing electrical power was embroiled in the great confrontation with the alternating current (A.C.) system championed by George Westinghouse, Nikola Tesla, and C. P. Steinmetz, the D.C. forces mounted an intense, nationwide campaign against A.C. "Alter-

The Monticello nuclear power plant is located near Minneapolis and is owned by the Northern States Power Company. This plant is rated at 545 megawatts. The power source is a BWR (Boiling Water Reactor) built by General Electric. Note the two long cooling towers in the foreground, which cool the warm discharge water to environmentally safe temperatures before it enters the Mississippi River. (*General Electric Company*)

nating current is too dangerous to use," they said. "Look, they use it for electrocution in prisons; A.C. cannot be permitted in the American home." Westinghouse replied to this gross misrepresentation with articles, pamphlets, every avenue of publicity he could find. He even seriously contemplated suing the Edison forces. Westinghouse finally won the battle in 1893 when his company was awarded the Niagara Falls power plant contract. Westinghouse had the better system, but it was touch and go for several years. For too long a time the country had to live with two noncompatible electricity generating systems, with all of the attendant inefficiencies and complexities, until reason won out.

The A.C./D.C. episode of almost a hundred years ago is now nearly forgotten, but the lesson it taught is still valid. Introduction of a new technology is not without its problems, no matter how valid it may seem to its proponents. The advocates must be prepared for attacks from any and all quarters and their defense must be thoughtful and reasoned. Today, nuclear power is going through an inquisition —justified in many respects but not in all—heightened by better communications, a public much more aware of the deteriorating environment surrounding it, and a growing popular suspicion of the advantages offered as new technology.

Below, we list the major questions asked about nuclear power. We shall answer them in some detail within the framework of what has been probably the most carefully conceived and the most conservatively approached technological development in history.

1. Since the regulation of atomic energy in the United States is in the hands of the same people who promote and develop atomic energy, does not an unacceptable conflict of interest exist?

2. Are not the radioactive releases from nuclear power plants causing cancer and mutations, and will the situation not get worse?

3. Are not nuclear power plants simmering atomic bombs, which can explode or otherwise result in unimaginable catastrophes?

4. Do not nuclear power plants discharge hot water, killing fish, damaging the ecology, causing irreversible harm to the environment?

5. Is not the beauty of the countryside being ruined by the burgeoning number of nuclear power plants?

Let us consider these questions in order.

The Dichotomy of the AEC. When the Atomic Energy Act was revised by Congress in 1954, the most significant change was to permit private industry to engage in major nuclear activities, such as the operation of nuclear reactors to produce electricity. The revised act gave to the AEC the heavy responsibility of regulating civilian use of atomic energy to protect the health and safety of the public. At first, the AEC organized this regulatory function within the existing managerial framework, but by 1959 it had become apparent that an infant nuclear industry would soon burgeon into a dynamic, full-grown one. Upon suitable further revision of the Atomic Energy Act by Congress, all civilian regulatory activities were removed from the province of the AEC's developmental programs under the general manager and were assigned to a new director of regulation, who reports independently and directly to the five commissioners. The general manager remains responsible to the commission for all operational matters.

This type of organization evolved for one simple reason: so that the right hand would know what the left hand was doing but yet be independent of it. The dichotomy between the AEC regulatory staff and the AEC development and operational staff has proved to be a symbiotic relationship in practice. The commissioners, directly responsible to the President for their actions, are able to gauge what developmental effort is required by the government in the advancement of nuclear safety and to see that it is carried out effectively by the general manager. Conversely, because of their familiarity with the technology of nuclear energy, they are equipped to deal more effectively with the highly complex regulatory program. It has been apparent for the past few years that further evolution of the system, in consonance with the continuing growth of the nuclear industry, will lead ultimately to a complete separation of the two functions, with one of them being transferred to another governmental entity. The first step in this direction occurred in 1970, when President Nixon removed the responsibility for developing radiation protection standards for the public from the Federal Radiation Council and the AEC's regulatory arm and transferred it to the new Environmental Protection Agency (EPA).

How does the AEC carry out its regulatory mission? From the tone of letters reaching the commission, it is apparent that some members of the public believe that the AEC is a law unto itself, able to operate without the supervision of the President and Congress, and—even more importantly—without public scrutiny. This belief is far from correct. An examination of the procedures followed for the most important of nuclear facilities—power reactors—will demonstrate that the AEC's procedures are unique in the government's control of industrial undertakings. In fact, some members of the nuclear industry grumble that the AEC is too zealous and painstaking in carrying out its regulatory function.

The Regulatory Process. The evaluation of an application for a construction permit for a nuclear power reactor is a painstaking process designed to see that adequate measures are provided to assure the protection of the public health and safety and the quality of the environment, and to permit state and local authorities and the public to have a voice in the proceeding before action is taken to grant or deny the permit. Great care is taken to see that the entire process is carried out openly and publicly in "goldfish bowl" fashion.

The first step toward licensing a power reactor is a searching analysis of the safety of the proposed plant by the AEC regulatory staff to determine whether a reactor of the design and power proposed can be operated safely at the selected site. A broad spectrum of a dozen or more technical and engineering disciplines is involved, ranging from physics and metallurgy to instrumentation and civil engineering. Even then, the regulatory staff does not consider that it has a monopoly on all the knowledge required, for it also calls upon the expertise of other government agencies and special consultants in such fields as hydrology, seismology, geology, meteorology, and fish and wildlife resources. An additional independent review is made of each application by a special group of experts set up by the Congress to advise the commission on nuclear safety matters. This Advisory Committee on Reactor Safeguards (ACRS) is composed of fifteen recognized scientists, engineers, and others from research and development laboratories, universities, industry, and other areas of American life. These technical reviews may take as long as eighteen months or more before these two bodies have resolved radiation health and safety questions with the applicant.

The applicant must also satisfy federal laws covering the whole range of environmental effects of his proposed plant. He must prepare a detailed report on environmental matters. This is circulated by the AEC to other federal and state agencies for comment, along with a draft of the AEC's evaluation of the plant's potential environmental effects. The AEC then prepares a detailed environmental statement based on the applicant's report and the comments received. This is submitted to the Council on Environmental Quality in the White House; it is also made available to the public. All construction permits demand strict observance of environmental protection requirements which are validly imposed under federal and state law, and which are determined by the AEC to be applicable. In addition, if the plant will discharge effluents into navigable waters, the applicant must obtain a certification that there is reasonable assurance it will not violate applicable water quality standards. This certification, which generally must be provided to the AEC before a permit or li-

cense can be issued, comes from the state or interstate water pollution control agency, or the Secretary of the Interior, whichever is
appropriate.

After the reviews covering radiological safety and environmental
matters have been completed, still another independent review is
conducted at a public hearing held at a location convenient to people
living near the proposed site. The hearing is held before a three-man
Atomic Safety and Licensing Board, which is completely independent
of the AEC regulatory staff and the ACRS. The purpose of the hearing is the examination of the adequacy of the previously conducted
safety evaluations and the sufficiency of the application itself. Dis-
puted safety and environmental issues are discussed and decided,
based upon all the evidence and the testimony presented at the hearing. Full public participation is afforded at these hearings. Persons
affected have the opportunity to intervene as parties in the proceedings, and those wishing only to express their views may do so.

Even after the Atomic Safety and Licensing Board renders its initial decision on the issuance of a construction permit, the matter is
not ended. Before becoming final, this decision is subject to appeal
by the parties to the proceeding, and, in any case, to review by the
Atomic Safety and Licensing Appeal Board and/or by the commission itself. Of course, judicial review of the commission's final action
in proceedings is available through the courts.

After a construction permit is issued, the plant's construction is
checked periodically for conformance with AEC regulations and the
permit. When an application for an operating license is ultimately
filed, the regulatory staff and the ACRS again conduct a comprehensive safety review. A hearing is not mandatory at this stage unless requested by persons affected. The commission also may schedule a
hearing on its own initiative. The AEC has a continuing interest in
these reactors. Once licensed for operation, each facility remains
under AEC regulatory surveillance, with periodic inspections made
throughout its operating life and final retirement. In addition, the
AEC examines and individually licenses the people who manipulate
reactor controls. In cooperation with the individual states, the AEC
administers a less elaborate system of licensing and regulating the
possession and use of nuclear materials, including radioisotopes.

The procedures for regulating the use of atomic energy described
in this section impose a serious financial and time-consuming burden

on the licensees of the AEC. There is hope that the regulations can be streamlined and that the time required to obtain permission to operate large nuclear facilities can be reduced without sacrificing the public's health and safety. This possibility is being studied aggressively. Early hearings to determine the suitability of sites for reactors provide substantial improvements.

As the nation mounts an expanding campaign to improve the environment, many changes are being made in the regulatory framework set up by the Federal Government to control potential or actual offenders. The creation of the Environmental Protection Agency (EPA), for example, was a major step in this direction.

The full impact of the National Environmental Policy Act of 1969 has not been felt yet. Under it, the AEC must (and will) meet its obligation to ensure that the plants it licenses do not perturb the environment beyond a minimum level which can be demonstrated as safe to a consensus of all concerned. Even aesthetic factors will be taken into account, under the terms of the act.

The Fort St. Vrain nuclear power plant in Colorado employs an HTGR (High Temperature Gas-Cooled Reactor) built by Gulf General Atomic. The plant is owned by the Public Service Company of Colorado and is rated at 330 megawatts. (*Gulf General Atomic*)

It is possible that what has been somewhat of a *laissez-faire* attitude toward important environmental decisions such as selection of power plant sites will no longer be permitted, and strict federal and state controls will be imposed as the nation gears itself to managing rather than reacting to changes in the environment.

Radiation—Boon or Bane? Much of this book is devoted to the countless benefits that nuclear energy—and hence radiation—can bring to man. What about the other side of the coin? To introduce this part of our discussion, we state flatly that radiation is dangerous. Radiation can cause illness and death when people are exposed to it in improper ways or in excessive amounts. These facts were recognized not long after the discovery of radioactivity and no responsible person denies them. We can also state with equal assurance that more is known about the effects of radiation on plants, animals, and humans than about synthetic chemicals or any other factor in our ecology. The question of whether radiation can be handled *safely* is the one that is paramount in the entire atomic energy program.

How does one define the rather abstract concept of safety? *No human activity* is safe in an absolute sense, as a little cogitation will demonstrate. What we do every day is to weigh each contemplated action, either consciously or unconsciously; we then decide whether the benefit, the enjoyment, or the advantage of the action is worth the risk to be taken. The risks can vary from negligible (but never zero) to real and substantial. The examples of each extreme are legion. In some cases the individual is not in a good position to weigh and determine each of the innumerable benefit/risk decisions that might affect him personally, because he lives in a society made up of thousands or millions of other individuals—each relating to the others in extraordinarily complex ways. To handle this complexity, the concept of public health or public safety has been devised for the common good to replace the "every man for himself" or "dog eat dog" attitudes toward safety held by our prehistoric ancestors. To examine the public health aspects of atomic energy, some familiarity with the sources of radiation, their amounts, and their effects is required.

Nuclear power plants contain radioactive solids, liquids and gases. In the routine operations of these plants, small concentrations of radioactivity are released in the effluents from the plants. These concentrations are too low to be measured by chemical analysis, and it is only the exquisite sensitivity of nuclear methods that permits their identification at all. Here is the first environmental impact of a nuclear power plant, and we will return to it.

Most of the radioactivity that is formed—practically all of it—is retained within the fuel elements. This is where most of the hazard resides and this is where most of the safety precautions are taken. The design and construction of a nuclear power plant are begun with the philosophy that every practical step should be taken to avoid an accident condition that would result in the rupturing of the fuel elements. Then, since perfection cannot be guaranteed in reactors any more than it can in any other human endeavor, additional safety barriers are inserted to confine the effects of potential failures and thus reduce the hazard to the public to negligible levels even if an accident should occur. This same philosophy of defense-in-depth has been followed in all aspects of the atomic energy program, and the twenty-five year experience record is excellent. There have been only seven radiation-caused deaths in all American industrial and government atomic programs—six of them in experimental operations, none at all in commercial nuclear power activities.

By the beginning of 1971, licensed nuclear power plants in the United States had accumulated more than 100 reactor-years of safe operation without an injury to the public, and furthermore, another 780 reactor-years of operating experience without a reactor accident have been provided by our nuclear Navy.

To reiterate the conservative approach to reactor safety, five ground rules have been employed:

1. Locate the plant in general away from areas of high population.
2. Take into account the frequency and severity of natural catastrophes such as earthquakes, floods, tsunamis, tornadoes, etc., in selecting plant sites and design the plants to withstand such events.
3. Design the plant to be safe; that is, use high quality fuel, reliable control equipment, good pumps and piping, and so on.
4. Employ strong, multiple barriers against the accidental release of radioactivity. Accidental release can be forestalled by this hierarchy of barriers: fuel element cladding, reactor pressure vessel, reactor containment structure, engineered safety features.
5. Apply and enforce strict design specifications, governmental licensing and regulation, equipment standards, constant surveillance,

etc., to insure that the above conditions prevail. Man is acknowl-
edged to be fallible.

Coal power plants have their smoke and ashes; their nuclear counterparts have radioactive wastes, but, of course, in much, much
smaller volumes. The radioactive fission products locked within the
fuel elements after the reactor has run awhile are the "ashes" of the
fission process. The used fuel elements also contain large amounts of
plutonium. Although plutonium gives off hardly any penetrating radiation, it is so highly toxic if taken up by the body that special pains
must be taken to ensure that it is not scattered as an aerosol (suspension of minute particles in air) and that it is not dispersed in liquids
which could become mixed with drinking water. (Inhalation is more
serious than swallowing.)

When nuclear fuel elements are "spent," they are shipped with
great care in heavy, sealed containers to one of several fuel reprocessing plants located as a precaution well away from population centers.
Here, the unfissioned fuel (plutonium and uranium) is chemically separated from the fission products for fabrication into new fuel. At the
time the spent fuel is dissolved, gaseous fission products which have
accumulated while the fuel was in the reactor are released. Some of
these gases can be trapped chemically but others, like krypton-85, are
inert and are often allowed to escape to the atmosphere under controlled conditions. In future plants, the inert gases will probably be
collected and stored. The chemical separation process is not quite 100
percent efficient, and extremely low concentrations of dissolved fission products are also permitted to leave the plants in the water effluent, also under carefully monitored conditions. The fission products
left behind are called "high-level wastes" because of their intense
radioactivity. Since these concentrated ashes will remain radioactive
for centuries, they must be packaged and stored safely underground
in such a way that they can never enter the biosphere, say through
seepage into the water table. Such high-level wastes are *never* stored
at commercial nuclear power plants. As we shall see, these fission
products may be dangerous, but some can also be useful as radiation
or heat sources for a great variety of applications.

The high-level wastes remaining after the useful radioisotopes have
been extracted must be interred somewhere forever. At present, the
AEC has stored about 75 million gallons of high-level wastes (mostly

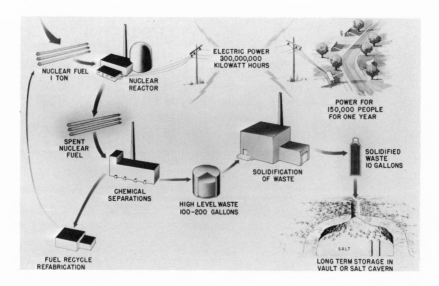

The nuclear fuel "cycle" is shown at the left. Waste products from the fuel cycle, the "ashes" of the nuclear power plant, will be solidified and buried permanently far underground in geologically stable formations. (*Battelle Pacific Northwest Laboratory*)

from the weapons programs) in some two hundred large, underground steel tanks at a few isolated spots, such as Hanford, Washington. The safety record with these tanks has been good; but no one pretends that tank storage is the ultimate answer. Tanks cannot last forever and several have leaked (though they have so far constituted no hazard). A more permanent method must be found. The best answer seems to be solidification of the liquid, high-level wastes—not only new wastes but much of the old wastes now in the AEC tanks. With current technology, 100 gallons of high-level wastes can be converted into about one cubic foot of solids, mainly through heating the wastes and driving off the fluids. The solid residues are then sealed in metallic containers for safety in transport. To make doubly sure that the public will be protected, these solids will be stored in vaults within salt mines far beneath the earth's surface. Massive salt formations are impervious to groundwater, and since they are rather plastic, cracks or fissures which might occur in them in the event of earthquake action are sealed naturally and quickly. Fortunately, suitable salt deposits are extensive in the United States and the volume of high-level solid wastes is very small in comparison with the storage volume available. By the year 2000, the solidified high-level wastes produced annually from all nuclear power plants would occupy a cube about 40 feet on a side; this is less than 2 percent of the space created by normal salt mining each year. The philosophy behind the handling of high-level wastes is termed "concentrate and confine."

The future transportation of spent fuel from reactor to reprocessing plant and the subsequent transportation of solidified high-level waste from reprocessing plant to salt mine will introduce a large increase in the opportunity for radioactive material to become involved in ordinary accidents that occur with trucks and trains. The shipping containers and vehicles are designed with the greatest care and are tested under abnormally severe conditions. Yet we favor reducing the possibility of accident by locating reactors, reprocessing plants and waste disposal sites as close together as is consistent with safety and environmental restrictions. In the years ahead, the concept of a "nuclear park" may materialize, where all the facilities needed are centralized to increase efficiency and reduce or eliminate transportation of radioactive materials on public thoroughfares.

Controlled Releases of Low Levels of Radioactivity and the Radiation Standards Controversy. We return now to a more extensive discussion of the policy of permitting nuclear power plants and fuel reprocessing plants to release low concentrations of radioactivity in the water and air that routinely leave these plants. Much that has been said about the safety of nuclear power in the United States has centered on this practice.

The volumes of the air and water needed to operate the plants are too large to isolate them permanently from the biosphere or to treat them so that they contain "zero radioactivity." ° The major sources of these radioactive contaminants in the effluents from nuclear power plants are:

LIQUIDS: Tritium (radioactive hydrogen) formed by neutron irradiation of the reactor cooling water, and as a fission product which diffuses out of the fuel elements into the water; metallic

° The ideal of "zero radioactivity" is not attainable; the Earth and its waters and air envelope contain naturally occurring radioactive materials that produce a radiation background everywhere.

corrosion products from the structure of the reactor made radioactive by "neutron activation"; minute concentrations of impurities in the otherwise highly purified cooling water, made
radioactive by neutron activation; occasional leakage of fission
products through the walls of defective fuel elements into the
cooling water.

GASES: Tritium in the form of water vapor and as hydrogen gas;
other gaseous fission products, both chemically reactive and
inert, that leak through the walls of the fuel elements.

Much of the radioactivity in both the air and water is removed by
chemical treatment or by retaining the effluents temporarily in tanks
or ponds so that short-lived radioactive species can decay. What is
left apparently must be discharged to the environment if nuclear
power plants are going to be allowed to operate at all. Of course, as
our technology advances, the quantity of radioactivity discharged
will diminish further.

The most abundant radioactive contaminant of the water released
from nuclear power plants is tritium. Chemically, it is almost exactly
the same as the ordinary hydrogen which comprises 11 percent of all
water. Although tritium has a half-life of about twelve years, plans
are being made in some zero-release plants to store tritiated water at
least for the plant lifetime.

The most important radioactive contaminants in the released air
are tritium (again) and the long-lived (half-life = 10 years), chemically
inert gas, krypton-85. In the future, the krypton-85 may be collected
and stored.

The health protection principle applied in managing air and water
discharges is: (1) remove as much radioactive material as is practical
by chemical means and, then (2) dilute the remaining radioactivity
with fresh air or water before it leaves the plant boundary, so that the
concentrations of radioactivity are too low to be a significant biological hazard to the general population. So-called "maximum permissible
concentrations" or MPCs have been established by federal regulation for each and every radioisotope that could be present in air or
water. These concentrations are based on the cumulative knowledge
of the world's radiological scientists and are set so that a large population could breathe air and drink water at these radioactive levels
for a lifetime without experiencing any discernible adverse effect. In

fact, the MPCs are thought to have a margin of safety built into them, but it must be borne in mind that they represent an upper limit that is not to be deliberately exceeded. To do so is a violation of federal regulations and stringent penalties can be imposed.

Some concern has been expressed that radioactivity in routine releases, though acceptable in drinking water, may be reconcentrated to dangerous levels by aquatic organisms. Reconcentration processes are known to occur, and a sizable fraction of AEC's budget for aquatic research has been spent examining routes and rates of reconcentration of radionuclides.

Contrary to popular opinion, most radionuclides do not increase in concentration in passing up the food chain, but rather reach their peak in the lower biological levels of the aquatic system (for example, phytoplankton and zooplankton) that are not normally eaten by man. And, of the few radionuclides that do accumulate in the fishes consumed in human diets, the highest concentrations are usually in the bones and internal organs, portions normally discarded before eating. Nevertheless, monitoring of food organisms in the vicinity of a nuclear power plant is required, and regulations provide for corrective action if reconcentration should appear to be a problem.

Much of the concern expressed about reconcentration has been hypothetical. Actual experience at nuclear power plants, and even at the giant nuclear facilities like the Hanford Works and Savannah River Plant, shows that aquatic organisms living in waters open to the public are acceptable for human consumption. The mouth of the Columbia River and environs provided large amounts of oysters, razor clams, crabs and fishes to commercial markets even in 1964, when nine large reactors upstream at Hanford were in operation. Monitoring by the state health agencies in Oregon and Washington provided assurance that these products were, and still are, acceptable for human consumption.

Part of the rationale behind permitting the release of small quantities of radioactivity to the environment is the knowledge that the environment has been radioactive from natural causes since the beginning of time. All natural solids, liquids, and gases contain radioactivity in varying amounts. Further, radiation due to cosmic rays continuously bombards us. The human senses do not react to these forms of radiation, but they can be easily measured with suitable instruments. To compare how much radiation is received in various ways, a yardstick is needed, and the common unit is called the "rem"—standing for "roentgen equivalent—man." This unit is too large for the levels of radiation that occur in the environment and so scientists find it convenient to use a unit called the "millirem," simply one thousandth of a rem.

Along with the yardstick, some kind of starting point is needed and there are several possibilities: exposure due to natural radiation, exposure due to medical treatments, the radiation "dose" due to nuclear weapons testing, etc. For convenience, the average individual exposure of the United States population to these and other sources of radiation is shown in Table 4. Note that the exposure figures are expressed on an annual basis; that is, in millirems per year. This permits comparing them directly with the now-controversial federal radiation standards, which are also listed in the table. *These standards apply to man-made radiation from all sources lumped together,* except for radiation exposures received in medical and dental treatments. The radiation standard of 170 millirems per year forms the basis for computing the maximum permissible concentrations of radioactivity in air and water mentioned earlier.

So far we have talked only about exposure *rates*—so much radiation in so much time. To fill out the picture, it is necessary to know something about *total* exposures and their effects. Quite a bit is known about the exposure of humans to acute doses of radiation, "acute" meaning "relatively large amounts delivered in a very short period of time." This knowledge has come from the necessities of medical treatment, the few nuclear accidents that have occurred, and the unfortunate casualties of nuclear warfare. Much less is known about "chronic" or continuous and prolonged exposure to low dose rates of radiation. Although it is known that a given dose of radiation has less short-term effect on a man if delivered over a relatively long time rather than instantaneously, this effect is not taken into account when the radiation standards are set up.

It is quite apparent from the data in Table 4 that the people of the United States are receiving radiation exposures from nuclear power plants that are only small fractions of what they receive from other sources of radiation or of what is believed to be an acceptable level by competent authorities. Proof of safety, or conversely, proof of risk just has not been possible, since experimentation with large numbers of humans is neither ethical nor desirable and because extrapolation from experiments with lower animals to humans is not completely reliable. What is left is to try to infer, from the case histories of humans

Table 4
Comparison of Radiation Exposures

Annual Exposures Received During Routine Activities

From cosmic rays and natural radioactivity in the human body, rocks, soil, air, for the average U.S. citizen	70–200 millirems/year
Same for people living in the volcanic areas of Brazil	1600 millirems/year
Same for individuals living in some of the coastal regions of India	1300 millirems/year
Additional average exposure inside a masonry house (due to natural radioactive materials)	50–150 millirems/year
From nuclear weapons testing, for the average U.S. citizen	2 millirems/year
From X-ray diagnoses, for the average U.S. citizen	75–100 millirems/year (estimated)
Additional exposure from miscellaneous sources (cosmic rays during jet travel, luminous watches, color TV, etc.) for the average U.S. citizen	2 millirems/year
From the radioactive effluents from nuclear power plants, for the average U.S. citizen in 1970	less than 0.001 millirem/year
From radioactive effluents from nuclear power plants, for those living within four miles of plant boundary, average in 1970	less than 1 millirem/year
From radioactive effluents from nuclear power plants, for those living near the plant boundary, average in 1970	5 millirems/year

Annual Exposure Limits from all Sources Set by the Federal Radiation Council (excepting exposure from natural sources or from medical treatment)

For occupational exposure (radiation workers)	5000 millirems/year
For an individual in the population (nonoccupational)	500 millirems/year
For a suitable sample population group (nonoccupational)	170 millirems/year

Typical Medical Exposures of Portions of Body

Average chest X ray	200 millirems
Average GI tract examination	22,000 millirems

who have been exposed occupationally or through medical practice to low doses of radiation, whether any biological effects, observed perhaps thirty years later, can be attributed to the increase in radiation exposure and not to any other cause. This forms the heart of the controversy. One vocal scientific camp argues that *any* radioactivity added to the environment increases the incidence of disease, particularly cancer. Another scientific group counters that the human body inherently has a repair capability and therefore some practical radiation threshold must exist below which the repair processes balance the body-damaging ones. To be on the safe side, this latter consideration is not employed in setting radiation standards.

The effects of radiation, when observed in the individual exposed, are called *somatic* (from the Greek word for body), as distinguished from *genetic* effects which are not apparent as physical changes in the person exposed but which may appear in subsequent generations of his offspring.

The somatic effects of radiation are quite evident if a sufficiently large exposure is involved. Typically, an exposure of at least 25,000 millirems to the entire adult body, delivered over a short period of time (an *acute* exposure), is required to produce an observable short-term clinical effect. Protracted, or *chronic,* exposure to lower levels of radiation is recognized to increase the statistical likelihood of long-term clinical effects such as cancer and other diseases, but unless the total dose of radiation is relatively large, it is extremely difficult, if not impossible, to establish a causal relationship between radiation and disease for individual persons, especially since the disease may appear perhaps twenty years after the radiation exposure occurred. Therefore, in order to develop some feeling for the effect, estimates are often made on a statistical basis involving large numbers of people. The estimates are computed by extrapolating from experiments with mice and other animals and are roughly consistent with data from retrospective studies of medical radiologists who had been subjected to radiation exposure over long periods of time. One method of summarizing the information concisely, which we believe useful in placing the effect in suitable perspective, is through the concept of life shortening. This approximate approach, when applied to large numbers of people exposed chronically to radiation—as all of us are —suggests a life shortening of the whole population through nonspecific mechanisms amounting to roughly one day for each 1000 millirems. Again consulting Table 4, we can easily calculate that a year's exposure of the general populace to the radiation released from nuclear power plants (in 1970) *would shorten our lives by less than one-tenth of a second.* As discussed later in this chapter, scientific studies indicate that human life is shortened much, much more by air pollution—and the generation of electricity from fossil fuels without effective emission controls is one of the prime contributors to air pollution.

Most scientists believe that no threshold exists with respect to genetic effects. That is, the more radiation a population receives (starting from zero radiation), the more genetic mutations are induced in that population. One must always talk about large numbers of people in this kind of discussion and not single individuals.

Some idea of the difficulty in defining the problem of the genetic effects of radiation can be gained from the results of studies of the only large population of humans who have been exposed to abnormally great amounts of radiation: the survivors of the Hiroshima and Nagasaki atomic bomb explosions. Starting soon after the end of World War II, the Atomic Bomb Casualty Commission (ABCC), staffed and supported jointly by the governments of the United States and Japan, has been performing the most careful scientific studies on the survivors, and has covered essentially all aspects of the biological effects of radiation on humans. The results of this research are available to anyone who wishes to pursue these matters in detail. In the ABCC's 1969 edition of its *Fact Book,* a chapter summarizes the genetic findings up to that time—twenty-four years after the explosions. The first genetic investigation was conducted on 71,280 newborn children during the 1948–1954 period. A second study examined 47,624 newborns delivered between 1956 and 1962. Evidence was sought for genetic damage in these children in relationship to the amounts of radiation their fathers or mothers had received. The parental radiation exposures varied from zero to more than 50,000 millirems, which is considered to be the amount of radiation that doubles the frequency of genetic mutations that occur spontaneously in the absence of man-made radiation. Biological indicators, such as sex ratios, congenital malformations, and body weights at birth, were recorded. The numbers of stillbirths and neonatal deaths were also determined.

In the first survey, no statistically significant effects were observed with the exception of sex ratio. In the second study, no support was found for the earlier indication that the radiation exposures had affected the sex ratio. Quoting from the ABCC *Fact Book:*

> These studies have not demonstrated a genetic effect of the atomic bombs. On the other hand, they do not exclude the possibility that some mutations were produced and transmitted to the offspring of survivors. Otherwise stated, while the studies clearly show that some of the dire predictions made just after the war were inaccurate, they do not rule out the possibility of some genetic damage, comparable in amount to that produced in mice by similar amounts of radiation.

The studies are continuing. New techniques are being used which will hopefully generate more precise information about gene damage from atomic weapons radiation than is presently available. There is little doubt that radiation does produce genetic effects, but we do not know exactly what these effects are in humans or how many generations must pass before the effects can be observed physically.

Although there are no actual long-range statistics on the detrimental genetic effects of radiation in humans, geneticists are able to make estimates of what these economic effects might be in terms of the increasing health impairment in future generations. Nobel Prize-winning geneticist Joshua Lederberg has performed such an analysis and concludes that, if every person in the United States were to receive an increase in radiation exposure of 100 millirems per year, the economic cost to the nation (consummated during the next century or later) would amount to $50 per person per year. Lederberg proposes that the population which begins receiving this increase in radiation should begin paying for the genetic impact (which will be spread out over a period of five to ten generations) at a discounted rate, which turns out to be $10 per person per year at present values. (The analysis assumes, of course, that dollar values can be assigned to biological damage. This is done every day in courts of law.)

Numbers like these, although conservatively high and based on many unproven assumptions, do give a basis for assessing the risks associated with radiation-based technology. Let us apply them to the power reactor situation. In 1970 the per capita cost that could have been charged for the additional genetic burden due to reactor effluents was *less than one one-hundredth of a cent per person per year. By the year 2000 this could rise to one cent per person per year.*

Lederberg concludes his provocative analysis by pointing out that the same kind of cost-analysis approach should be applied to other forms of pollution. Regrettably, no data are available on which to base estimates of genetic damage from other pollutants. When our knowledge about gene mutations caused by chemical pollutants, food additives, infection, drugs, narcotics, pest infestations, and other genetic forces catches up to our understanding of radiation effects, it will be interesting to learn the relative effect of nuclear radiation. As the air-pollution study to be discussed later infers, nuclear radiation may be relatively unimportant.

Of course, the human race has survived, evolved, and prospered for millions of years in a radiation environment far more intense than that created by nuclear power plant effluents. In fact, the average human body undergoes several hundred thousand radioactive disintegrations per minute as the natural radioisotopes found in the body (mostly potassium-40) decay. The changes caused by natural radiation have been acceptable—indeed unavoidable. But the question remains, should we add to our radiation burden, even minutely? Again we arrive at that old conundrum which asks whether the benefits outweigh the risks. In assessing nuclear power, we must also evaluate the health hazards and economic penalties involved in generating electricity from pollution-creating fossil fuels.

Years ago, those who pondered these questions while setting the radiation standards took the conservative approach dictated by the uncertainties. They recommended that while a maximum exposure level was needed as a limit for all kinds of man-made radiation, regulatory authorities should assume that there is no threshold for either genetic or somatic effects of radiation and that consequently the deleterious effects should be considered proportional to exposure, even at very low levels. Implicit in this approach was the concept that the public should not be exposed to radiation for frivolous purposes and that any approved exposure should show a benefit/risk relationship heavily weighted toward the benefit side.

Regulatory agencies have carried out this mandate in practice, but the concept of having a maximum permissible radiation exposure rate (the 170 millirems per year) has been misinterpreted by many as implying that the regulators believe that a radiation threshold exists.

To clarify this point, the AEC changed its regulations in 1970 to require that its licensees keep their releases of radioactivity to the environment "as low as practicable," a term used by the scientific groups that recommended the standards. While this does not have the quantitative significance that a lowering of the numerical standards would have, it gives the AEC the legal authority to require the installation of additional purification facilities if the effluent levels begin to approach the standards too closely, and if it can be demonstrated that equipment of proven performance is available at a cost that can be justified in terms of the beneficial effect obtained. At the same time, the "as low as practicable" concept retains a much-needed flexibility for both the regulators and the plant operators. Just so that no one will misunderstand this qualitative concept and try to creep up to the levels set by the standards, the AEC has stipulated that radiation levels must be kept to a small percentage of the standards.

A few words about how the "as low as practicable" concept works in the case of nuclear power plants may help to explain this rather complicated subject, using the numbers taken from Table 4.

Since the plant operator is required to limit radiation exposure of the public to a level that is as low as practicable and since he has no control over other exposures that members of the public are receiving from other nuclear power plants, color TV, luminous-dial wristwatches, travel in jet aircraft, and so forth, he controls the release of radioactive effluents to a small fraction of the standards. This practice results in *actual* exposures to persons living at the plant boundary of about 5 millirems per year on the average for all nuclear power plants now in operation, or about 1 percent of the individual standard and 3 percent of the group standard. Since the radioactive effluents become more and more diluted and dispersed as they move away from the plant boundary, they cause even less exposure to other persons. Within a four-mile radius from the plant, the exposure currently averages less than 1 millirem per year. When dispersed over the entire geography of the United States, the radioactivity causes an average individual exposure of less than 0.001 millirem per year.

Numerical criteria on design objectives for nuclear power reactors to keep levels of radioactivity as low as practicable have been put into effect. These criteria will keep actual exposures to persons living at the plant boundary below 5 percent of exposures from background radiation.

As new plants are built, the radiation level may rise somewhat, but it will never be allowed to become a public health hazard. Our best estimates indicate that the average individual whole-body exposure due to reactor effluents will rise from the current value of less than 0.001 millirem per year to about 0.1 millirem per year by the year 2000, assuming that none of the likely improvements are made in controlling the radioactive effluents from nuclear power plants. This future exposure level is still only a minute fraction of the conservative standards recommended for the United States population. And technological advances will certainly reduce the exposure even further. Studies indicate that the *exposure of the public to 0.1 millirem per year would shorten the life of the average person about ten seconds for each year of exposure.*

The Atomic Bomb Syndrome. The misconception that nuclear power plants can explode like atomic bombs is no longer widely prevalent, but there are still some who fear this possibility—a reaction that is not surprising when the history of atomic energy is considered. Nuclear fission bombs utilize parts made of nearly pure uranium-235 or plutonium-239 which are driven together in just the right way (in just the right configuration) at high velocities. The fissionable material in nuclear power reactors is dispersed throughout a large volume of essentially nonfissionable uranium and/or inert structural material. This dilution absolutely precludes the massive energy releases that characterize nuclear weapons.

It is still physically conceivable that the nuclear chain reaction in a power plant could, in an accident situation, become uncontrolled, in which case the power could rise rapidly beyond that for which the reactor was designed. Such events are highly unlikely because of the inherent stability of reactor systems and the multiple protective mechanisms and devices engineered and built into such systems. Nevertheless, such unlikely events are considered and analyzed during plant design to ensure that the consequences of such an event would be safely contained within the plant structure. Considerations of the worst accidents of this type that one can conceive within applicable physical laws lead to the conclusion that, because of the effect of the diluent materials, only a very small amount of destructive energy could be generated. The "explosion" involved would be characteristic of that associated with chemical reactions, not of that with nuclear weapons. In spite of the extreme conservatism inherent in these considerations, all United States reactors are designed to accommodate this type of internal energy release without causing significant hazards to the public.

In water-cooled reactors, virtually any rearrangement of the fuel will terminate the nuclear reaction. However, in fast breeder reactors, the situation is different. The fissionable fuel in these reactors is somewhat more concentrated, and it is necessary to analyze in greater detail whether any accidental rearrangement of a fast breeder core could possibly lead to the release of explosive energy. Calculations have shown that some rather improbable rearrangements could generate a small explosion, but nothing like an atomic bomb. The possibility of such an explosion is taken into account in the design of the reactor structure and containment housing, which are made strong enough to withstand the effect. There are positive indications that the calculations on the core rearrangements have been very conservative in the past and that more refined computations will demonstrate that the probability of an explosive release of nuclear energy is negligible. These considerations, when taken together, suggest that there is no credible rearrangement of a fast breeder core which could lead to the release of explosive energy with a force sufficient to breach the containment.

Is a "Catastrophic" Accident Possible? A more serious concern is that nuclear power plants might suffer some other form of "catastrophic" accident which could result in grave damage to the public. This concern is based on the fear that there is a possibility that the billions of curies of radioactivity inside the reactor could be entirely or partially released to the environment, causing widespread devastation. There is no question about the enormous amount of radioactivity formed in a reactor by fission. It is for this reason that every possible precaution is taken during the design, construction, and operation of a nuclear power plant to reduce to a vanishingly small probability the possibility of a major release of radioactivity.

When a reactor is being designed, exhaustive safety analyses and experiments are performed which consider the types of malfunctions that might occur in the reactor system, the consequences of each malfunction, and the precautions needed to prevent their occurrence or to mitigate their consequences. A firm, logical cause for a malfunction or failure is not a prerequisite for a particular accident or even a series of accidents to be considered. The question is often asked, "What would happen if this system or piece of equipment failed?"—even though no rationale for this failure has been identified and even though a reactor can be shut down should symptoms of difficulty appear.

The potentially most serious of hypothetical events that can occur in a nuclear power plant is the complete loss of cooling of the reactor core. Consequently, a significant number of engineered safety features and an emergency cooling system are installed in each reactor. These are relied upon to prevent or counteract any adverse condition which could lead ultimately to a complete loss of cooling. In the unlikely event that none of the several protective systems is effective, the complete loss of cooling could lead to excessive temperature rises and possible melting of the nuclear fuel and the release of radioactivity. The release of radioactivity in this accident sequence could be greater than that from any other potential or hypothetical accidents that have been studied, and thus the complete loss of cooling accident is generally considered to be as close to a "catastrophic" event as it has been possible to conceive.

Although complete loss of cooling, followed by successive failures of the safety equipment and the emergency cooling system, is highly improbable (engineers and scientists never say "impossible"), prudence dictates that each reactor should be provided with a massive, steel-lined, concrete containment structure, the purpose of which is to confine any radioactivity that might escape from the reactor following an accident. This barrier provides additional assurance that potential reactor accidents, should they ever occur, will not endanger public safety.

But, ask the skeptics, suppose *all* lines of defense fail and the fission products escape? What then? Such an event can only be conceded to be a serious catastrophe, perhaps causing billions of dollars in property damage and injuring or killing thousands of people. Alvin Weinberg, Director of Oak Ridge National Laboratory, suggests that the consequences of a hypothetical reactor meltdown and subsequent release of its radioactivity could be similar to those related to the sudden collapse of the largest of the hydroelectric dams. He thinks it more likely that the consequences of a real type of accident would be comparable with the crash of a large transport aircraft.

It is only fair to then ask in turn, "What are the chances that such an accident could occur?" The chances are so small that it is not possible to make a sensible estimate. Nevertheless many attempts have been made to provide answers to this question that would place the problem in perspective. Two examples may suffice here.

Herbert Kouts of Brookhaven National Laboratory has suggested that the chances for a catastrophic reactor failure are about the same as the probability that all of the airplanes circling in the vicinity of New York's airports might collide over Shea Stadium during a Sunday afternoon baseball double-header—with the wreckage falling into the stadium.

Ralph Lapp, the prominent physicist and writer, has calculated that if he assumes that there is one chance in ten thousand that a catastrophe will occur within one year's time in a nuclear power plant and that if—on the average—there are 500 power plants in operation over the next thirty years, then there is a likelihood that there will be one reactor catastrophe before the end of the century. Others believe the chance should be less than "one in a million," and probably even much less than this. The one-in-a-million probability reduces the likelihood of a major accident to one in 3000 years for 500 reactors. This probability should not become larger as the total number of reactors increases because the safety measures will continually be made more effective.

The Hot Water Problem. Present-day nuclear power plants are cooled with water. In contrast to the breeder reactors of the future, they are less efficient in converting heat to electricity than are modern coal-fired power plants. They therefore discharge more waste heat to the environment per unit of electricity produced than do the coal-fired plants. Both types of plants are contributors to *thermal pollution,* another point of environmental controversy. Although some nuclear advocates prefer the more euphemistic term "thermal effects," pollution by heating is a clear and present hazard to the biosphere and must be taken into account. Civilization thermally pollutes the environment because it adds fire and fission heat to the sun's heat and that welling up from the earth below. In global terms, the heating effect of the sun is 30,000 times greater than the small flames of civilization. But our cities are small hot spots on the map, and waste heat from power plants can locally outdo the sun. Basically, the issue here is not that of direct hazard to human life but rather local hazard to the biosphere, in particular life in the rivers and lakes into which power plants and industry expel their cooling water.

It is incontrovertible that fish and other aquatic forms of life are affected by temperature increases. Once more, the coin is two-sided: some fish like warmer water, others don't. They thrive at different temperatures. For example, as the water temperature rises above 60° F, the speed of the brook trout declines and so does its predatory efficiency. On the other hand, some fish are attracted to the points where power plants release warm water, thus improving the fishing. The conservationist, however, justifiably wishes to keep cold rivers cold, the way man found them.

The balance of nature is delicate and history is full of man-made changes, more of which, perhaps, are written in red ink rather than black. The guidelines that the federal government and many state agencies in the United States are drawing up recognize this precarious balance. Power plant designers will have to guarantee that their cooling water will not raise river, lake, or bay temperatures more than a stipulated few degrees above normal within a certain distance of the discharge point. Coal-fired and nuclear power plants, along with other heat-producing enterprises, will have to follow these guidelines in the future. Cooling schemes, such as cooling towers and holding ponds, must be applied if necessary, to meet these requirements.

It is well that conservationists and other concerned people have forced the thermal pollution issue into public debate. Marine biologist John R. Clark estimates that by the year 2000, if present practice is continued, almost one third of all the fresh water runoff in the United States could be needed to cool power plants of all kinds.

The most common engineering solution to a thermal problem is the installation of cooling towers. Most of the heat is then transferred to air rather than to the cooling water. This transfers the thermal burden from one environmental reservoir to another, but most scientists seem to agree that the biosphere would not suffer so severely if a plant's waste heat is dumped into the air rather than the water.

A positive approach is that of "thermal enrichment," wherein waste heat is applied first to practical purposes rather than being dumped unused into the environment. Aquaculture, ice-free waterways, and frost-free agriculture are but a few possible beneficial uses of waste heat that should be explored. Waste heat may prove to be a valuable power plant by-product. We shall look at this more closely in later chapters.

Aesthetics and Energy. A large and influential segment of mankind turns to untrammeled nature as a last refuge from encroaching technology—places where one can still breathe clean air, drink pure, clean water, and be civilizationless for a few hours or days. Such spots are becoming rarer and rarer; those near large supplies of good water are the scarcest of all. Both conservationists and industry compete for these spots. The words exchanged are often intemperate. Some conservationists accuse industry, particularly the builders of coal and nuclear power plants, of despoiling the virgin landscape with ugly buildings, reactor domes, and smokestacks for profit—never for the good of the community. The accusers point out that the despoliation can be extensive, such as where the strip mining of coal has permanently scarred the Earth. The utility man retorts by asking why the millions in the cities should be deprived of air conditioning on torrid summer days so that nature lovers can admire a fine view. Such are the extreme views about the aesthetics of generating power. The more imaginative planner, recognizing the need for power plants, tries to create a setting that is an aesthetic benefit rather than a detriment. For example, the internationally renowned urban planner, Constantine Doxiadis, visualizes nuclear power parks replete with lagoons, bird sanctuaries, recreational areas, and visitor centers, all taking advantage of the facts that nuclear power plants are located near sources of water and that the water discharged from the plants may be cooled in large ponds before being returned to estuaries or lakes. The use of simple, unobtrusive architecture can contribute to the integration of the plant into its surroundings. Furthermore, nuclear power plants discharge no smoke, display no coal stockpiles, and can be designed to be aesthetically pleasing.

On a rational basis, then, conservationists and utilities can work together to select power plant sites and power plant architecture that complement rather than detract from the natural beauty of an area. In fact, the future will force them to work together. Nonnegotiable demands for great tracts of untouched, pristine wilderness will eventually collide with the nonnegotiable demands of an expanding population and its desire for that higher standard of living that depends upon abundant electrical power.

An Environmental Reprise

People want both children and a high standard of living, including increased longevity, freedom from recurring and debilitating illness, assured sources of food, and continuous means of earning a livelihood. More power and technology make these possible. But industry, including nuclear plants, involves risks and this planet's resources *are* limited. As the population rises, so do the risks. If we do not hit the proper balance, Nature will establish a balance of its own which may not be to our liking. Helpful in achieving an equitable balance should be the knowledge that the more advanced a country is, technologically speaking, the more able it is to withstand natural catastrophes that continue to plague the world—epidemic diseases, tidal waves, typhoons, earthquakes, and the many other evidences of the not-so-gentle hand of Nature. This point is often overlooked when we read of thousands of human beings wiped out by Nature in some distant part of the world. Our philosophical attitude toward these tragedies has a touch of hypocrisy in it. This hypocrisy should not be allowed to pervade our attitudes toward our man-influenced local surroundings as well, or we may see environmental tragedies equal in magnitude to those occasionally wrought by Dame Nature.

A balance between alternatives is needed in particular in the choice of the means for generating electricity. By good planning and concerted, unselfish action, we can undo some of the damage that is already our responsibility. According to a thoughtful and perhaps first-of-a-kind study by Lester B. Lave and Eugene P. Seskin,° *if air pollution were reduced 50 percent in our major cities, a newborn baby would have an additional three to five years of life expectancy;* the same reduction in air pollution would cut death from lung diseases by 25 percent and death and disease from heart and circulatory disorders by 10 to 25 percent; and all death and disease would be reduced by about 4.5 percent, with a saving to the United States of at least $4 billion a year (if 1970 figures are used) in medical costs and in time

° *Science,* August 21, 1970, contains a technical report by these economists.

lost from work. To put this another way, Professor Lave tells us his data show that the economic cost of cancer to the United States is about 5.7 percent of this nation's costs for medical care and time lost from work due to disease. A 50 percent reduction in air pollution would reduce these total costs by nearly the same amount, and therefore such a reduction would be almost equivalent economically to eradicating cancer. With incentives like these, we should be able to plot a rational course.

We all dream of that vast, unpeopled land the Pilgrims found. It has fallen before the turnpikes and high-tension lines. Thoreau's loon doesn't stop at Walden Pond anymore. Yet we can try to regain some of that lost Eden and sustain increasing billions of people as well, but we must move carefully.

We conclude this chapter by saluting the newly awakened interest in environmental matters, confident that much of the concern that has highlighted this awakening will contribute to an enlightened, reasoned approach to the solution of environmental problems.

Chapter 3

Labels, Bond Breakers, and Explosives

The big, new nuclear power plants steal all the headlines. Beneath the debates about their potential dangers, the atom works away quietly, as it has for half a century, in medicine, industry, agriculture, and science. Radioisotopes and atomic radiation were embraced by the medical profession as soon as their properties became known. The papers and magazines of the early 1900s were full of the miraculous curative powers of radium. The first phase of the atomic revolution is already over; the atom plays key roles in thousands of universities, laboratories, hospitals and industries all over the world.

The "silent" atomic tools are varied; most depend not upon fission and fusion but upon more subtle properties of the atom, such as its precise clockwork, the high-speed projectiles it emits, and the vivid, distinctive label it provides. However, one tool which will be described in this chapter is far from quiet, physically or politically. This is the nuclear explosive, cleaned of most of its fallout and tailored for constructive use.

Power from Radioisotopes

Some applications require only a tiny bit of power. An implanted heart pacemaker, a scientific satellite, or a small instrument package on the floor of the ocean can operate with just a few watts of power.

Radioisotopes (radioactive isotopes) are ideal fuels for many small power generators which must operate for months and years in remote, hostile environments.

Per pound, some radioisotopes store more than a thousand times as much energy as the best chemical fuels. It is this marvelously concentrated energy in the nucleus of the radioisotope that confounded physicists fifty years ago. They believed that the unlocking of this power would revolutionize the world—just as fission and fusion reactors now bid to do. But here we deal with miniature power generators rated in watts and milliwatts (thousandths of a watt) rather than billions of watts.

During their experiments with radium in 1900, Marie and Pierre Curie noted that a voltage difference was created as charged particles escaped from their radium-bearing samples. The English physicist H. G. J. Moseley employed this effect in 1913 when he built the first "nuclear battery." Moseley simply silvered the inside of a glass sphere and mounted a speck of radium on a wire at the center. As the charged particles from the radium sped from the radium to the sphere, they constituted a flow of electricity. Moseley's nuclear battery delivered only millionths of an ampere of electrical current, but the voltage built up to thousands of volts—a consequence of the high energies of the charged particles. The output power (the product of current and voltage) was only milliwatts.

Most of the so-called nuclear batteries convert the energies of motion of the charged particles emitted by radioisotopes into electricity *without* first changing their energies to heat. In addition to the Moseley type of battery, a nuclear battery can be constructed from the solar cells which ordinarily convert light to electricity on man-made satellites. The solar cells also convert the kinetic energy of charged particles—such as alpha or beta particles—directly into electricity. All nuclear batteries generate low powers (thousandths of a watt). Still, they find applications in watches, radiation dosimeters, and other small devices where ordinary dry cells are impractical.

Radioisotopes can also be applied as small heat sources, as they were in the instrument package left behind on the moon by the Apollo-11 astronauts. Radioisotope heat can also be turned into electricity in the same way heat is converted into electricity in the fission power plant. That is, the high velocity particles given off by the radioisotopes can be slowed down and converted into heat within the fuel and in the walls of a thick container; then, the heat can be turned into steam that drives a turbine and generator. A small radioisotope-powered steam engine was, in fact, constructed by Mound Laboratory for the AEC in the 1950s as a demonstration project. Using polonium-210 as fuel, the generator produced only 1.8 milliwatts with an overall efficiency of 0.1 percent. The performance was not impressive, but it showed that radioisotopes were potentially useful heat sources.

Studies in the late 1940s evaluated radioisotopes as heat sources for jet airplane engines and satellite power plants. The great leap from milliwatts to megawatts needed for aircraft seemed justified at the time in view of the huge quantities of radioisotope fission products that were accumulating in the underground waste tanks from the nuclear weapons program. Radioisotope-powered aircraft, however, proved too ambitious and attention focussed on small power sources generating only a few watts.

In 1956 the AEC inaugurated its well-known SNAP (Systems for Nuclear Auxiliary Power) program. The requirements of military surveillance satellites were the main stimulus for SNAP. The satellite problem was: How could one generate tens or hundreds of watts of electricity in outer space for six months or a year? Batteries were too heavy and solar cells were too new, having just been discovered at Bell Laboratories in 1954. Two approaches were taken in SNAP: (1) boil a fluid and run a tiny turbogenerator, and (2) heat thermoelectric materials which would convert the heat *directly* into electricity via the thermoelectric effect. Summarizing many years of development, the second approach proved the more successful in space, under the sea, and on the ground. These small but long-lived SNAP radioisotope power sources are commonly called RTGs, for Radioisotope Thermoelectric Generators.

The typical RTG contains a central fuel mass, usually cylindrical in shape. Surrounding the fuel is a thick metal capsule. Heat flows radially out through the fuel mass and through the capsule wall. Thermoelectric elements, usually small cylinders of a material such as lead telluride, are arranged around the capsule wall like spokes. As the heat passes through the thermoelectric elements, anywhere from 2 percent to 8 percent of it will be converted into electricity. Thus, over 90 percent of the heat originating within the capsule is waste heat which must be "dumped" to the environment. Almost all radioisotope-fueled SNAP units follow the above scheme. Exceptions are SNAP-13, which employs thermionic converters rather than thermoelectric elements, and some larger power units under study in which a gas is heated for the purpose of driving turbogenerators.

Nuclear reactors produce several fission products that make suitable fuels. Other fuels are made artificially in reactors when neutrons are absorbed by various chemical elements. Some of these fuels are:

Potential fuel	Half-life	Source
Cobalt-60	5.3 years	Neutron absorption
Strontium-90	28 years	Fission product
Cesium-137	30 years	Fission product
Cerium-144	285 days	Fission product
Polonium-210	138 days	Neutron absorption
Plutonium-238	89.6 years	Neutron absorption
Curium-242	162 days	Neutron absorption
Curium-244	18 years	Neutron absorption

The half-life of a radioisotope is the time taken for half of any given amount to decay or disintegrate. A gram of strontium-90, for example, will decay to 0.5 gram of strontium-90 in 28 years; after 56 years, only 0.25 gram will remain, plus, of course, approximately 0.75 gram of the decay product. Some natural radioisotopes, such as uranium-238, have half-lives of over a billion years—they have to or we wouldn't find any of them remaining on a planet 4.6 billion years old. Roughly 2000 different radioisotopes exist, but only a small handful have the right half-lives and are abundant enough to fuel RTGs.

RTGs have been launched on satellites, buried near the North and South poles to power automatic weather stations, and installed beneath the sea. Generally speaking, they are useful where the power required is between a few milliwatts and 100 watts and where the lifetime must be several months to several years. In outer space, solar cells are superior (lighter and cheaper) wherever there is plenty of sunlight. On the surface of the moon, underneath the thick atmosphere of Venus, and on voyages to the outer planets, where sunlight wanes, RTGs have the advantage.

2000 Unseen Labels

A second tool in the atomic repertoire depends upon a different set of properties of the radioisotope, namely the type and energy of the particles it emits from its nucleus when it decays. Radioactive atoms can be distinguished easily from their nonradioactive fellows by nuclear particle detectors, such as Geiger and scintillation counters, even though they are identical chemically. The radioisotope iodine-131, to illustrate the point, behaves chemically just like stable iodine-127 when it is used in human thyroid studies. But with a radiation counter, a biologist can trace the history of a dose of "tagged" or "labeled" iodine as it begins to accumulate in the thyroid gland. The basis of any tracer experiment is the ready detection, identification, and measurement of the concentration of specific radioisotopes. Radioactive tracers help answer the questions: Where? When? How many? These are important questions to experimenters in many fields.

With almost 2000 different radioisotopes known—more than a dozen for each chemical element, on the average—the problem of tracer identification would seem almost insurmountable, particularly when natural radioisotopes abound in nature, including those in the human body. However, modern instrumentation overcomes this problem. An experimenter carefully selects his tracer radioisotopes so that they can be easily found and distinguished from any other radioisotopes likely to be in the neighborhood. Each radioisotope emits specific particles with specific distributions of energies. In more homely terms the particles and their energies constitute the "fingerprints" or positive identifications of the radioisotopes.

The most common types of radioisotope decay are:

> *alpha decay,* in which a nucleus (usually the nucleus of one of the heavier elements) spontaneously emits an alpha particle (a doubly ionized helium atom). The energy of the alpha particle is fixed and highly specific for each alpha-emitter. In most cases there is more than one alpha-particle group, each with a fixed energy, accompanied by one or more gamma rays (similar to X rays). The gamma-ray energies are also fixed and highly specific for the radioisotope.
>
> *beta decay,* in which the nucleus emits an electron or positron (beta particle) plus, in most cases, one or more gamma rays

(similar to X rays). The beta particle energy varies from a specific maximum down to zero. The gamma-ray energies again are fixed and highly specific for the radioisotope.

One of the great pioneers in the art of radioisotope tracing was the Hungarian, Georg von Hevesy. During 1912, when Hevesy was working at Ernest Rutherford's laboratory, he was given the task of chemically separating radium-D from a large supply of lead that Rutherford had received from the famous radium mine at Joachimsthal, Bohemia. Hevesy was unable to accomplish this separation and concluded that radium-D and lead were almost identical substances so far as chemistry was concerned. In those days, radium-D was called by that name only because it was a product of radium decay. Little else was known about it. Hevesy's experimental conclusion was understood when radium-D was finally identified as a radioisotope of lead—different from stable lead on a nuclear basis but essentially the same chemically. This is the very property of radioactive tracers that makes them so useful.

Hevesy went on to use radioactive lead (née radium-D) in measuring the solubility of lead sulfate and lead chromate in water. In 1923, the same lead radioisotope helped him trace the movement of lead in bean seedlings. Here was another case where an accidental by-product of science led to many unexpected practical applications.

If a suitable natural radioisotope cannot be found for a specific purpose, an artificial one can often be made to order in a cyclotron or nuclear reactor. Oak Ridge National Laboratory, in Tennessee, is the AEC's mail-order house for radioisotopes. In addition, a number of industrial suppliers also offer various radioisotopes for sale. From their catalogs a scientist or engineer can find radioisotopes of practically every chemical element, with a wide range of half-lives. Over 1500 tagged chemical compounds, especially those employed in biochemical research, are also on the regular market. An additional 1000 compounds can be prepared to order. Oak Ridge makes many thousands of shipments of radioisotopes each year to investigators all over the world.

Most of the thousands of tracer applications follow Hevesy's technique. Radioisotopes of element X are added to natural element X. Both types follow the same chemical paths, and these paths can be charted by monitoring the travels of radioisotope X with particle detectors. The sensitivity of this type of tracer experiment is remarkable; sometimes only one radioactive atom amid 100 billion nonradioactive atoms is sufficient; this is equivalent to one grain of corn in 850 full boxcars.

Radioisotopes also make good alarms in counting experiments. For example, the migration habits of cockroaches in New Orleans sewers have been studied by daubing a bit of radioisotope on cockroaches which were then released with their unmarked relatives. As the labeled cockroaches passed strategically located radiation detectors, they were automatically counted. In such application, the chemical properties of the radioisotope have little importance.

Tracers have infiltrated almost every laboratory and university in the world. They are quiet revolutionaries, working without fanfare or controversy. They have added new dimensions to medicine, agriculture, industry, hydrology, pollution control, and many other technologies that help make the world a better home for man.

Activation Analysis

The technique called activation analysis has much in common with radioisotope tracing. Thus, it is not surprising to discover that the first activation analysis experiment was conducted in 1936 by Georg von Hevesy, one of the founders of radioisotope tracing. Hevesy conducted this experiment in Copenhagen with Hilde Levi. To measure the small quantity of the element dysprosium in impure yttrium, they bombarded the sample with neutrons. The dysprosium was "activated"; that is, made artificially radioactive as the neutrons transmuted dysprosium nuclei into new radioactive nuclei. The radiation from the activated dysprosium gave Hevesy and Levi the quantitative data they needed. Activation analysis is an analytical tool that tells its user "What" and "How much." It is an extremely sensitive tool— better than chemical or spectroscopic analysis for many elements.

One of the most exciting applications of activation analysis in recent years took place 239,000 miles away on the moon when several unmanned Surveyor spacecraft analyzed for the first time the composition of the moon's surface by remote control. The principal experimenter was A. L. Turkevich from the University of Chicago. Turkevich knew roughly what elements to expect on the moon from the compositions of the Earth and fallen meteorites and from science's reconstruction of the history of the solar system. For example, one would certainly "expect" aluminum, magnesium, and oxygen to be rather common elements. Sending a remotely operated chemical-assay experiment to the moon was out of the question, but a small radioactive source of alpha particles and a few particle detectors would not weigh too much. Turkevich and his associates fabricated an "alpha-scattering experiment" that would, in effect, shoot alpha particles into the moon's surface and measure two things:

1. The numbers and energies of the alpha particles that bounced back (i.e., scattered back) from the outermost layer of atoms on the lunar surface. The energies lost in these collisions are indicative of the masses of the atoms. This part of the experiment was not activation analysis, as one usually understands the term, because no atoms were activated.

2. The numbers and energies of protons created when the alpha particles activated some of the lighter atoms in the moon's crust. This information is even more specific about the identities of the atoms. This *was* activation analysis.

Turkevich's experiments were successfully landed on the lunar surface on three Surveyor flights in 1966 and 1967. They are now being superseded as astronauts bring back actual samples from the moon. These samples have essentially confirmed and extended Turkevich's remote analyses. The inclusion of an activation analysis experiment on an unmanned Mars lander would give us rough composition data at least a decade before man sets foot on that planet.

The applications of activation analysis are usually much closer to home than the lunar surface. The technique languished from Hevesy's days until after World War II, when research reactors and other good neutron sources became generally available. During the 1960s, the extreme sensitivity of this analytical tool attracted many scientists working in medicine, geology, criminology, and even archeology. For example, scientists analyzing a hair from Napoleon's head found it tainted with arsenic, indicating that perhaps he was slowly poisoned to death on Elba. The tracing of insecticides and minute quantities of pollutants is another application of activation analysis that is becoming vital in our effort to reclaim our environment.

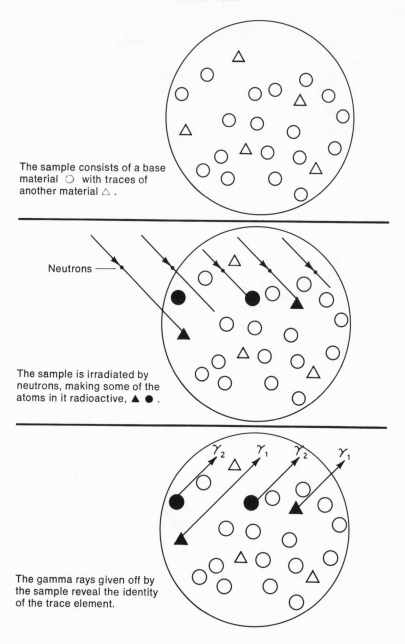

The sample consists of a base material ◯ with traces of another material △.

Neutrons ———

The sample is irradiated by neutrons, making some of the atoms in it radioactive, ▲ ●.

The gamma rays given off by the sample reveal the identity of the trace element.

In neutron activation analysis, traces of various elements can be identified and measured by analyzing the gamma rays they emit after being irradiated by neutrons. (*U.S. Atomic Energy Commission*)

Nuclear Clocks

The radioisotope carbon-14 is a nuclear clock—one of the most important clocks ever devised by man. If one could build a suitable microscope and watch a single, isolated atom of carbon-14, it might emit a beta particle and decay into nitrogen-14 in the next second or you might have to wait a hundred thousand years. On the average, however, when thousands of carbon-14 atoms are scrutinized with a particle counter, they have a dependable half-life of about 5800 years. In a sample, some atoms will disintegrate sooner than 5800 years, others later; but when 5800 years have passed, only half the original number will remain in the sample. The mechanics of the carbon-14 clock are now apparent. If we know how many carbon-14 atoms are in a sample originally and we measure how many remain, we can calculate elapsed time.

This procedure sounds good, but how does the carbon-14 clock get started, say, in a stick of preglacial wood tentatively pegged at 20,000 years old? The answer is that the clock started when the organism being dated died. The logic goes like this: The radioisotope carbon-14 was created in the upper atmosphere when cosmic ray neutrons hit nuclei of nitrogen-14, which constitute the bulk of air atoms. This carbon-14 quickly mixed throughout the lower atmosphere which is used by the Earth's animals and plants. As ordinary nonradioactive carbon-12 was assimilated by terrestrial organisms, a little carbon-14 was also added to the organism. Now the crucial point: carbon-14 has a short half-life (5800 years) compared with geological time. It has reached what is called "secular equilibrium" in the atmosphere. In simple words, cosmic rays create one new carbon-14 atom for each one that decays, and the ratio of carbon-14 to the normal, nonradioactive carbon-12 atoms remains constant in the atmosphere. The ratio stays approximately fixed in a living organism, too, because it is made partially from atmospheric carbon. Once the organism dies, however, it can no longer add carbon-12 and carbon-14. The carbon-12 it assimilated during its life is stable, but half the carbon-14 will be gone in 5800 years. In this way, the changing carbon-14/carbon-12 ratio will tell fairly accurate time during the last 50,000 years.

Carbon-14 is a rather new time-telling tool. The American scientist W. F. Libby first used it in 1947. Since then, organic materials ranging from Egyptian mummy hair to Stone Age sandals have been dated. Carbon-14 has given the archeologists an absolute physical clock which, while it may not have revolutionized this branch of science, certainly solved one of its major problems.

The early workers in radioactivity knew nothing of carbon-14. They were absorbed in the intricacies of the complex decay chains of radium and uranium. The real age of the Earth, however, was one of the burning questions of the time. Scientists had realized, of course, that our planet had to be a great deal older than suggested in the Bible if Darwin's theory of evolution was to have had enough time to work—and just about all scientists believed in the theory in the early 1900s. On the other hand, the astronomers believed that the Earth was a lot younger (from a cooling standpoint) than did the geologists (from a sedimentary standpoint). In the end, radioactive dating and other techniques showed the Earth was even older than the geologists had estimated.

The nuclear clocks suitable for such geological dating must have half-lives in the billion-year category. These are mainly heavy elements, such as uranium-238 (4.5 billion years) and thorium-232 (14.1 billion years), just the elements that intrigued the early workers in radioactivity. The Curies and others soon recognized that these radioisotopes had decayed with ponderous slowness across the geological eons, leaving in their temporal wakes two stable lead isotopes called "radiogenic" leads. Uranium-238, to illustrate, was always associated with its ultimate daughter product, lead-206, not with lead-204, the so-called primordial lead, which was considered one of the basic ingredients of the newborn universe. Physicists with geological leanings, such as the Irishman John Joly, began making estimates of the Earth's age using radioisotopes as early as 1910. The answers revolutionized geological and astronomical thinking, for, on this basis, the Earth seemed at least a billion years old. But the crude methods of those days left much to be desired. It was not until just before World War II that the instrument essential for geochronological accuracy came along. This was the mass spectrometer, perfected by A. O. Nier, a physicist then at Harvard.

Nier's mass spectrometer made it possible to measure accurately the quantities of the various isotopes of radiogenic lead present in radium-uranium ores. In 1946, Arthur Holmes in England and F. G. Houtermans in Germany, made estimates of the age of the Earth based upon the amount of uranium-238 and other radioisotopes that had decayed to radiogenic lead. Assuming that all of the radiogenic lead is due to the decay of the heavy radioactive elements, it is easy to compute the age of the ore from the decay formula. Holmes and Houtermans concluded that the Earth was between 2 and 3 billion years old. Many experimental and theoretical difficulties are encountered in these analyses and the method has been simplified in this exposition. As more accurate techniques have come along, the Earth's origin has been pushed back in time to about 4.6 billion years. Just as in the case of carbon-14 dating, assumptions sometimes have to be made before the method can be applied. For example, geochronologists usually assume that *no* radiogenic lead was created when the universe was born—no one really knows for sure. We gain confidence in the ancient nuclear clocks by looking at several different kinds and cross-checking them. So far, the evidence hangs together well.

A New Kind of Chemistry

A cardinal principle of nuclear physics states that nuclear events proceed undeterred by pressure, temperature, and other ordinary environmental forces. The high energies of nuclear events—thousands, often millions of times more energetic than chemical reactions—help isolate them from those trivial perturbations of the chemical universe. A hundred degrees of temperature are nothing to a thermonuclear reaction cooking at 100 million degrees; but they are life or death to those chemical engines called human beings. Turning the logic around, nuclear events, being as powerful as they are, ought to wreak havoc with chemical reactions. And they do.

The passage through matter of a charged particle or gamma ray propelled by a nuclear reaction leaves a long trail of broken chemical bonds. Molecules can be torn apart, as proven by the evolution of "radiolytic" hydrogen from the water in water-cooled fission reactors. Temporarily dissociated molecules and ionized atoms also litter the trail of an energetic charged particle. Many of the broken bonds heal spontaneously but some don't. This powerful capacity to break chemical bonds can be turned to useful ends.

The most obvious application of this destructive force of radiation is in biological sterilization. Gamma radiation from radioisotopes and reactors can be deadly—to advantage—in cancer therapy and in the sterilization of insects for biological control of agriculturally harmful species. Another common use of radiation today is in sterilizing medical supplies. The radiation processing of vegetables and meats is based on the same effects of radiation. These biological applications will be covered in more detail later. Note, though, that the biological applications are inherently destructive at the molecular level, although they may still be beneficial macroscopically.

Constructively speaking, chemical catalysis and the "cross-linking" of molecules can be induced by radiation. Chemical catalysis or the stimulation of reluctant reactions results when heat, light, pressure, radiation, or some specific chemical catalyst speeds up a reaction to a commercially useful rate. In radiation catalysis, gamma radiation or charged particles strike the constituents and dissociate or ionize them, permitting reactions to proceed that were otherwise blocked. For example, ethyl bromide will form when ethylene and hydrogen bromide are mixed and exposed to the radiation from cobalt-60. Biodegradable detergents are also made using radiation catalysis.

The phenomenon of cross-linking occurs when long, initially independent molecules are induced to form bonds one to the other, like the rungs holding a ladder together. The rungs or cross-links can be created when the passage of radiation disrupts some of the normal lengthwise chemical bonds holding the long molecules together. After the traumatic passage of gamma rays, the bonds never return exactly to their former positions. Some end up attached to adjacent molecules —these misguided bonds are the cross-links. They tie previously separate molecules together into a single entity. Plastics cross-linked by radiation acquire new and useful properties. Some of the polyethylene plastic wrappings shrunk around supermarket meat have been treated with radiation to strengthen them.

Chemical bond breaking seems a feeble game to play when cities must be rebuilt and deserts irrigated. Yet, today's civilization depends heavily on controlling the chemical bond in the manufacture of synthetic materials, antibiotics, fresh foods, cheap fuels, new plant species. We have just begun to explore the role nuclear forces can play in breaking and remaking these relatively fragile interconnections of matter.

Building with Nuclear Explosives

Out in the desolate reaches of southern Nevada, a hole 1200 feet across and 320 feet deep punctuates the arid geology set down by nature during recent eons. From the rim, men and machines working in the bottom of the crater seem toylike. The Sedan crater is one of the largest holes ever excavated by man at a single stroke—the stroke in this case was from a 100 kiloton nuclear charge. *163650*

Over the centuries man has greatly modified the planet through deforestation and pollution. The changes have generally been for the worse: the Dust Bowl of the thirties, the Great Lakes dead and dying, are all man's handiwork. We now possess the power necessary to reverse this deterioration. In fact, a new discipline called *planetary engineering* has arisen. Planetary engineering includes weather modification, watershed control, and all large-scale Earth-modifying activities. Because nuclear explosives can inject instantaneously more energy into the planet's air, earth, and water than any other man-made device, they are key tools in any effort to reshape the Earth or perhaps return it to its pristine condition. Exciting as this sounds, extreme caution is advised, because science does not yet fully understand the physical processes that shape our environment.

The constructive possibilities of nuclear explosives were apparent to Enrico Fermi and his group during the days of the Manhattan Dis-

The Project Schooner crater was excavated by a 35-kiloton nuclear explosion in Nevada on December 8, 1968. Placed 355 feet underground, the charge blasted a crater out of hard rock about 852 feet in diameter. (Note the football field grid superimposed.) (*Lawrence Radiation Laboratory*)

trict. During the later 1940s, however, proposals for nuclear excavation of canals and harbors were easy to postpone for three reasons:

1. Fissionable material was scarce, expensive, and needed for the country's Cold War posture.
2. Explosive yields were low.
3. Radioactive fission products would contaminate wide areas.

With the first thermonuclear explosion on Eniwetok Atoll in the Pacific Ocean in 1952, these objections were removed or weakened. Hydrogen bombs are more powerful, cheaper, and less radioactive because fusion rather than fission supplies most of the energy.

With the United States and Soviet monopoly on thermonuclear know-how, it came as a surprise to find a Frenchman, Camille Rougeron, publishing the first book on the subject: *Les Applications de l'Explosion Thermonucleaire* (Paris, 1956). The Suez crisis, also in 1956, stimulated American nuclear circles to consider blasting a sea-level canal across Israel with nuclear charges. The Suez crisis eased temporarily, but United States activities continued. In the summer of 1957, the AEC formally established the Plowshare Program to investigate the constructive uses of nuclear explosives.

Urged onward by Edward Teller, the embryonic Plowshare Program derived much of its early technical data from underground nuclear weapons testing. The first of these tests, code-named Ranier, occurred in September 1957. Many more followed. From the eerie unnatural cavities blasted out of solid rock thousands of feet beneath the surface, Plowshare scientists were able to formulate empirical laws concerning cavity size, earth shock waves, and fracturing, as functions of weapon size, type of rock, etc. Further tests of nuclear charges were not attempted from late 1958 until the self-imposed nuclear test moratorium was discontinued in 1961. Experiments with chemical explosives (with charges up to 1 million pounds) helped establish Plowshare engineering data during the moratorium. On December 10, 1961, the first nuclear test in the Plowshare Program blasted a cavity almost 1 million cubic feet in volume 1200 feet below the desert near Carlsbad, New Mexico. The first nuclear cratering experiment was the Sedan shot, on July 6, 1962, mentioned earlier. Additional tests have been conducted since, with specific applications in mind.

The handful of Plowshare tests (and to some extent the many more weapons tests) have provided enough data to sketch out the so-called phenomenology of these nuclear explosions that occur underground and at the surface.

A nuclear explosion vaporizes everything in its near vicinity. Temperatures of tens of millions of degrees and pressures of hundreds of thousands of atmospheres are created almost instantaneously. As this thermal and kinetic energy propagates outward from the center of the explosion, much of it is converted into a mechanical shock wave in less than a millionth of a second. The shock wave crushes, cracks, melts, and vaporizes the surrounding rock. Farther away from the center of the explosions, the shock weakens and, finally, is converted into elastic seismic waves.

In a deep underground nuclear explosion, all direct effects are contained, and the explosion's energy is roughly equal to the work done in melting and fracturing the rock, plus the heat deposited in the surrounding rock. The size of the cavity created depends upon the energy of the explosion, the depth, and the kind of rock. Immediately after the detonation, the spherical cavity is partially filled with solidifying molten rock. However, the rock in all directions has been heavily shattered. Soon, angular pieces of rock begin to fall into the cavity. The roof caves in and keeps caving in until the originally spherical chamber becomes a tall "chimney" loosely filled with broken rock. The chimney may be several hundred feet high with a small empty space or "void" at the top. The practical applications of a deep underground explosion must be derived from the rock fracturing.

When the explosion point is moved near the surface, not all of the energy is contained. The shock wave and pressure blast out a crater, throwing debris up and away. Thus, the explosives can be employed for large-scale excavation. The more water the rocks contain, the greater the cratering effect, because the heat turns the water to superheated steam which adds to the explosive effect. If the explosion is too deep, most of the rock and soil will arch up and fall back in the hoped-for crater. Too shallow an explosion will waste energy in atmospheric effects rather than moving earth.

The hemispherical cavity created by the Gnome explosion was about 75 feet high and 134 to 196 feet across. Note the man standing on the rubble at right center. (*U.S. Atomic Energy Commission*)

Nuclear craters resemble meteor craters. They are rimmed and covered at the bottoms with small amounts of broken rock. Trenches, mountain passes, and canals can be constructed using rows of charges.

Nuclear explosives are the cheapest sources of raw energy available. Ten thousand tons of TNT would cost almost fifteen times as much as an equivalent thermonuclear explosive; at the 2-million-ton level, nuclear energy is about 1700 times cheaper. So long as the excavation task is a large one and in a relatively remote, geologically stable area, nuclear explosives should be considered as an energy source.

Naturally, nuclear explosives, like any powerful tool, must be used

with care. The reasons for the limitations imposed on their use are
four in number:

1. Radiation (fallout) exposure.
2. Ground motion and earth tremor stimulation.
3. Air blast damage.
4. Legal and foreign treaty restrictions.

The great bulk of the radioactivity released in a deep, underground
explosion is contained permanently within a glassy slag that solidifies
at the bottom of the cavity. Nevertheless, the explosion site must be
selected and surveyed carefully to avoid introducing any of the radioactivity into ground water.

Local radioactive fallout occurs after cratering explosions. The 100
kiloton Sedan explosion in 1962 deposited a small quantity of radioactivity in an irregular fan downwind. Most of the radioactivity decayed rapidly. Persons living fifty miles downwind from the crater
during the blast and thereafter would have received only one-seventh
as much radiation dose, on the average, as they would have received
from natural radioactivity already in the area.

Since 1962, nuclear explosives for excavation purposes have been
made much cleaner from the standpoint of radioactive fallout. More
of an explosive's energy now comes from fusion rather than fission—
and fusion is a much cleaner energy source. Because nuclear fission is
needed only to trigger the fusion reaction, the bigger a nuclear explosive is, the cleaner it is in a relative sense. A Sedan-sized explosion
today would be almost 100 times cleaner than in 1962, allowing people to live nearby and use the crater area and still keep radiation exposures well within accepted levels. Still, public reaction to planned
nuclear excavation tests in the past clearly indicates that the fear of
radioactive fallout is uppermost in the minds of many people.

Although fallout has been a major concern in the Plowshare excavation programs, ground motion may be locally severe around the explosion point. The amount of damage produced by the initial blast
depends a great deal upon local geology and the types of structures
located near the blast point. This aspect of the problem is well understood from an engineering standpoint and the amount of damage can
be predicted readily.

The initial ground motion and stimulated aftershocks introduce a "human" problem and an element of the unknown. Any company that must blast to accomplish its mission (as in conventional quarrying) knows it will receive many damage claims for cracked plaster and other architectural damage. The AEC faces this situation every time it explodes a Plowshare charge near civilization.

The fear also exists that the aftershocks stimulated by a big explosion may not always be minor—as they have been from nuclear tests to date. Might not a major earthquake be stimulated by a nuclear blast, particularly if it occurs in the neighborhood of a known, active geological fault? Scientifically, the answer is not a simple "yes" or "no." However, most competent geologists and seismologists who have studied the cause-effect relationship carefully have come to the conclusion that the risks of a major earthquake are extremely small.

Base surges accompany any crater-forming explosion. After a large nuclear cratering explosion, the downward and outward flow of ejected material causes the formation of a base surge cloud. This cloud may be several hundred feet high and may roll outward several miles before dissipating. Most of the radioactivity is contained in the crater itself and its lip; the remainder becomes airborne or is carried in the base surge. This radioactivity decays rapidly and the area near the explosion is soon safe to enter.

Cratering explosions can also cause air blasts that result in broken windows and frayed nerves similar to those generated by the sonic booms from jet aircraft. Craters are blasted only when meteorologists predict conditions which will minimize air blast damage or even avoid it completely.

Even far from civilization, protests may arise. Conservationists were very concerned over the 1969 underground weapons test in Alaska. Wildlife in the area and the proximity of a major geologic fault were factors debated prior to this test. As predicted by AEC scientists, neither severe aftershocks nor harm to wildlife resulted from the test.

Farther Frontiers

Whatever the future brings, it will make today's atom-based tools seem tame. The word "tool" is hardly descriptive of truly revolutionary concepts. Already we have mixed the technical ingredients of the next fifty years. They will gestate for years until some genius or chance discovery molds them together into a new device or technique that may change the world forever.

Among the various pots cooking away on the back of the stove is that of fusion research. A fusion reactor would use the heavy hydrogen in the seas as fuel. Ostensibly, the goal is limitless electrical power. We will probably attain this goal, but even more likely is the discovery of some unexpected process, some unexpected technique. Several harbingers of things to come are:

1. The fusion-torch concept.
2. The plasma radiation source.
3. Electromagnetic radiation transmission and handling techniques.

Each of these could help shape the next century.

Consider what happens when ordinary material is bathed in plasma at 100,000 degrees Centigrade from a fusion reactor or some other source. The plasma is of much lower density than everyday matter. The energy content of the plasma is tremendous; its thermal conductivity is likewise high. Result: a fusion torch that disintegrates ordinary matter, breaking all molecular bonds and partially ionizing the material. The fusion torch is not a weapon like the disintegrator pistol of science fiction, but it can reduce (in concept) any material into its atomic constituents. To carry the science fiction analogy to the limit, we need next a matter "reconstituter," some machine to remake what the torch dissolved. Such matter tear down/build up devices could lead to the "matter transmitters" that have graced the pages of science fiction magazines for decades.

In a more serious vein (assuming the reader is skeptical of science fiction as a mirror of the technological future), the fusion torch could be employed as a universal solvent for all of the obnoxious wastes of civilization. Separation of the seething mixture of waste atoms into elements could perhaps be accomplished by electromagnetic means, much as the isotopes of uranium were separated by high temperatures during World War II in the calutrons or mass spectrometers. What a neat plan: throw garbage in and take away elements—ultrapure metals, nonmetals, gases, etc.! These are futuristic concepts, it is understood, and cheap, abundant power is an essential ingredient. Feasibility experiments, however, could be carried out today, and some are being designed already.

The second concept on the list envisions introducing certain materials into the fusion torch and converting the torch's energy into electromagnetic radiation in a narrow band of frequencies. This source would probably not generate radiation as pure as that of the laser, but much larger amounts of radiant energy could be generated. Narrow bands of intense radiation might be applied to various industrial processes, such as those requiring bulk heating or surface heat treatment. Exactly what could be done with a megawatt of narrow-band radiation per square meter is a matter of conjecture. It is a form of energy-in-transit that demands an entirely new way of thinking.

Electricity is now our best way to transport energy. But radiant energy is the driving force of the physical universe, whatever our temporary terrestrial expedients may be. In fusion research, we are at last learning how to create and manipulate intense fluxes of electromagnetic energy. Some day we may "pipe" radiant energy around our planet the same way we do electricity. If radiant energy is easy to make and use, in other words, it may become electricity's competitor.

Burning the seas? Transmitting energy without wires? Playing God with matter? These are the kinds of techniques that make some shudder. But, if the population scientists are correct and our cities ultimately sprawl from sea to sea, and even under the sea, only such great powers can sustain humanity on such a tiny planet.

PART II
APPLYING THE TOOLS

Chapter 4
More Food and Water

Technology's Role

The prophecy of Malthus is still unfulfilled. People continue to multiply largely unchecked by starvation; but every time Malthus' ghost rattles its chains, technology comes to the rescue with new insecticides, new fertilizers, and new, more productive strains of plants and animals. Millions of Asians are alive today because new varieties of wheat and rice have arrived in the nick of time—like the cavalry in a Western movie. Extremists would like to close completely the faucet marked *Technology*; this would be no less than genocide.

In the long run, Malthus must be right. The Earth can support only so many people; and the people keep coming, billion after billion. Even technology has its limits. In this chapter, we shall suggest several technological "fixes" or solutions which might permit the Earth to sustain several tens of billions as compared with its present population of three to four billion. A world brimming with 20 billion humans is physically possible in terms of water, energy, and arable land resources. (However, it might not be a politically or socially viable world.) Eventually, however, unless population growth is stopped, solar photosynthesis and even food synthesis using other forms of energy may be insufficient. Malthus would then come into his own. Fortunately, technology, if wisely applied, can forestall this final confrontation for some decades. In fact, the so-called Green Revolution, involving the introduction of new species of wheat and rice into underdeveloped countries, has already saved millions from starvation. Developed by the Rockefeller and Ford Foundations, these new plants represent still another facet of socially constructive technology.

The world we picture in the following pages could sustain a large world population in good health. It would, of course, be desirable to begin at the beginning and prevent somehow the Earth's population from ever rising to tens of billions. But if politics and other factors in the human equation prevent us from controlling population, the technical methods described herein will extend man's dominion in time, but only at the price of an increasingly artificial and (to us) uncomfortable world.

Controlling the Hydrosphere

The hydrosphere is that thin layer of water that rings the continents with seas and at the same time permeates their rocks with groundwater. Abundant fresh water is usually synonymous with abundant food. That pound of beef in the supermarket requires 4000 gallons of potable water in its production; a ton of alfalfa takes 200,000 gallons. In addition to food production there is a host of industrial and domestic uses for water. There is plenty of water in the world—fifty times as much fresh water as we now need—but most of it is either in the wrong place at the wrong time or it is brackish or polluted. Result: deserts, floods, brackish marshes, dead lakes, and sewage-laden streams; all of which do agriculture little good. And, of course, the great salt oceans are scarcely tapped at all by farmers.

Man has always tried to manipulate that small fraction of fresh water that falls from the skies. Roughly one-thousandth of one percent (0.00001) of the Earth's water exists as water vapor at any instant. This atmospheric reservoir discharges via rainstorms and fills up again through transpiration and evaporation about thirty times a year. Yet, it is this tiny fraction of our watery inheritance that sustains the great bulk of our agriculture.

Down the ages, men have used dances, sacrifices, cannons, and silver iodide crystals to try to control rainfall—mostly to make it rain in drought-plagued regions; rarely has man wanted to stop precipitation altogether. The results are controversial. The seeding of clouds with

silver iodide apparently has some positive effect under special conditions and over restricted geographical areas. Atomic technology has
never played an intentional role in rainmaking, although past above-
ground nuclear weapons tests have been blamed for heavy regional
rainfalls in the United States. Presumably, the large quantities of dust
injected into the stratosphere by the blasts could have seeded some of
this rainfall. The atom's main role will be in helping scientists understand the hydrological cycle through the use of radioactive tracers.

Tracing Water. Tritium, the superheavy radioactive isotope of hydrogen (atomic mass of 3), is one of the most important tracers in hydrology because its chemical properties are essentially identical to those
of the stable hydrogen in ordinary water molecules. The natural tritium found in rainwater is created by cosmic-ray interaction with the
Earth's atmosphere. But man has been a more prolific tritium manufacturer than nature since 1952, when thermonuclear weapons tests
began. By tracing the massive artificial injection of tritium into the
hydrosphere from weapons explosions, unique data have been obtained on groundwater recharge rates, aquifer ° flow velocities, and
aquifer storage capacities. By learning the details of how fresh water
flows through streams and rock strata, engineers can plan water conservation and distribution programs better.

Other radioisotopes, such as iodine-131 and bromine-82, have also
been useful in aquifer and groundwater studies. Tracers also measure
river flow velocity, dam leakage, pollution sources, and river recharging of groundwater stores. Deep groundwater is usually pristine pure,
even though many reservoirs of such "fossil" water, such as that under
the Sahara Desert, are tens of thousands of years old. These old water
strata are dated by carbon-14 like other relics of prehistoric time.
Groundwater reservoirs are immense—possibly thousands of times
the volume of surface water; it is essential to understand this largely
untapped source of fresh water, and the atom is an important tool in
this endeavor.

Grand Plans. The sixteenth-century English poet John Heywood once
wrote, "Much water goeth by the mill that the miller knoweth not of."
Mankind's mill is planet-sized, and planet-sized plans to control it are

° An aquifer is a water-permeable stratum of rock, usually bounded by non-
permeable strata, creating in effect a sheetlike pipe.

in order. Even though far less than one percent of the globe's water runs toward the seas, immense engineering works are needed to "tame" this flow. Already, hundreds of thousands of dams, large·and small, intercept the world's rivers. Great dams are a symbol—even a passion—of the American West. Despite the best efforts of the dam builders, though, some areas are still too dry (southwestern United States) and some have more than their share of fresh water (eastern United States).

When regional efforts prove inadequate, continental plans are born. We speak now not of the TVA (Tennessee Valley Authority) or the Mekong Delta, but of even larger projects such as NAWAPA (North American Water and Power Alliance). The engineering firm of J. M. Parsons proposed NAWAPA in 1964. It embraces Canada, the United States, and Mexico. The basic plan is to collect surplus water from high precipitation areas of the northwestern part of the continent and subsequently redistribute it to water-scarce areas in all three countries. Hydroelectric power (70,000 to 150,000 megawatts) could be generated as the water worked its way seaward.

The largest reservoir in the NAWAPA system would be the Rocky Mountain Trench, an artificial body of water sixteen times the size of Lake Mead. Six huge pumping stations (possibly nuclear-powered) would pump the water stored in the Trench over the mountains into the American Southwest and Colorado Basin, and into Mexico. Another major NAWAPA feature would be the Alberta–Great Lakes Canal, a waterway seventy-three feet wide and thirty feet deep linking the West (possibly even the Pacific Coast) with the Atlantic via a connection with the St. Lawrence Seaway.

NAWAPA is year-2000 size; a big investment of technology to guarantee a water supply for the continent, and the most effective distribution of runoff among Canada, the United States, and Mexico. NAWAPA displays the boldness and imagination also needed to solve our colossal environmental problems. Many western states back NAWAPA, although Canada is still reluctant. Being a $200 billion project, it is hard to sell when many other planetary ills call for money. There are also unresolved problems and undesirable side effects such as the inundation of valuable mining and recreational areas, to say nothing of flooding some towns and roads.

Another large-scale plan to control surface water is the Amazon Great Lakes plan proposed by Hudson Institute. In this concept, a se-

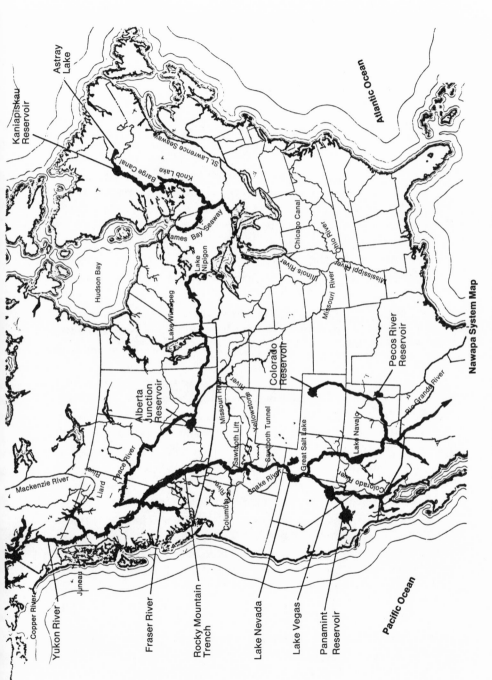

Nawapa System Map

Map of the NAWAPA system concept. *(Ralph M. Parsons Company)*

Labels appearing on the map:

Kaniapiskau Reservoir
Astray Lake
Barge Canal
Knob Lake
St. Lawrence Seaway
James Bay Seaway
Lake Nipigon
Lake Winnipeg
Hudson Bay
Chicago Canal
Illinois River
Ohio River
Mississippi River
Missouri River
Atlantic Ocean
Colorado Reservoir
Pecos River Reservoir
Alberta Junction Reservoir
Missouri River
Yellowstone River
Sawtooth Lift
Sawtooth Tunnel
Great Salt Lake
Lake Navajo
Rio Grande River
Colorado River
Mackenzie River
Liard River
Peace River
Columbia River
Snake River
Copper River
Yukon River
Juneau
Fraser River
Rocky Mountain Trench
Lake Nevada
Lake Vegas
Panamint Reservoir
Pacific Ocean

ries of low dams would create seven large lakes, opening up more completely the Amazon Basin (the largest in the world) to boat traffic. More electricity would again go hand in hand with improved transportation. Hundreds of thousands of square miles of fertile soil would be made arable below the dams.

How Nuclear Power Fits In

Nuclear power is cheapest when the power plants are biggest. Nuclear power should be considered whenever large amounts of water must be transported against the force of gravity. In the NAWAPA scheme, for example, it may be more economical to employ nuclear-powered pumps to lift water over the Rockies than to transmit hydroelectric power back to the Rocky Mountain Trench from dam sites far downstream. Another interesting possibility is the use of power plant waste heat—otherwise called a thermal pollutant—to keep NAWAPA waterways ice-free and open to navigation in the winter. For example, an ice-free Alberta–Great Lakes Canal would greatly reduce transportation costs in the northern portion of North America. However, the problem of disposing of waste heat in the summertime must be solved.

Why manipulate rivers when three thousand times as much fresh water lies within a half mile of the surface? Groundwater has been neglected because we know little about the deposits and the mining thereof. Although the arid American West and other regions have pumped groundwater for decades, we have not used this resource with any sophistication. Already the American Southwest has mined most of its fossil water with little thought given to recharging the aquifers that once provided a wealth of crystal-clear water.

Radioactive tracers, on one hand, will help us chart and understand this subterranean sea better. On the other hand, large nuclear power stations may be the cheapest source of pumping power for exploiting this underground resource that may ultimately surpass oil and other minerals in value.

A thin layer of Sahara sand now covers extensive tracts of fertile, once-cultivated soil. Farther beneath, a great reservoir of fresh water waits to be claimed. The Sahara's water seeped into underground aquifers during prehistoric times when all North Africa was verdant. Although the Near East's oil pools could supply the energy needed to bring this water to the surface, nuclear-powered pumping might be cheaper on a large scale and would save the oil for tasks the atom cannot do, such as supplying raw materials for the huge petrochemical industry. The Sahara's climate has made it one of the least populated regions in the world; it could again become a rich land. The renaissance, however, would be short-lived unless the subterranean reservoir was replenished.

A few thousand miles east of the Sahara, one of the most populous areas of the world, the Indo-Gangetic Plain, is often on the brink of starvation. Like the Sahara, its salvation lies just beneath the surface, an underground sea fed by the monsoons and Himalayan snows. Perry Stout, at the University of California at Davis, has suggested that nuclear-powered pumping of groundwater, combined with new strains of wheat, can help rescue this region that is usually hungry.

Modifying the Hydrosphere with Explosions

Conventional chemical explosives are commonly employed in the construction of canals and dams. If the task under consideration is large, and in a remote area, nuclear explosives may be quicker and cheaper. Because of the great power of Plowshare explosives, they can also blast out lakes and fracture huge rock chimneys for water storage and groundwater recharge.

In terms of sheer size, nuclear explosives fit in well with schemes as ambitious as NAWAPA. The NAWAPA waterway connecting of the Atlantic and Pacific oceans would obviously call for considerable blasting in relatively remote areas. Here, nuclear explosives would be relatively safe and effective, as would also be the case in the modification of northeast Australia's hydrosphere.

Australia is populated only on its fringes. The vast, arid interior of this continent is susceptible to grand concepts that would open it up to civilization. To illustrate, Australia possesses huge quantities of iron ore in its Hamersley Range, but there is no permanent water supply to support industry or people in the area. The Fortescue River, which flows through the area, is swollen with a half-million acre-feet of water ° a brief part of the year, but it is usually dry. An earthen dam could be created by caving in the sides of a gorge near the Hamersley Range with nuclear explosives. An iron industry could

° *I.e.*, enough to cover a half-million acres to a depth of one foot.

then be built around the water reservoir and hydroelectric power
source that results. Unfortunately, earthen dams eventually succumb
to seepage, leaks, and general deterioration. Athelstan F. Spilhaus has
suggested a neat circumvention: Inject water into the loose debris
comprising the dam and freeze it, forming a permafrost barrier to seepage. After the initial freezing only a little power from the new hydroelectric plant would suffice to keep the earth-insulated ice frozen.
The energy generated by a contained, underground nuclear explosion
could be tapped to power the initial freezing of the dam. By pumping
water into the hot cavity, sufficient steam could be generated to drive
turbogenerators that would run a temporary refrigerating plant.

It should be obvious by now that nuclear civil engineering is an
imaginative discipline.

Nuclear explosives can also excavate artificial lakes and reservoirs
almost instantaneously wherever safety and ecological factors permit.
Of course, the strata beneath the proposed body of water must not be
so permeable that sieves rather than lakes result. Driving across northern Nevada, you become aware of many marshes and shallow, brackish lakes that indicate that the region's scant rainfall is not well utilized. Nuclear-made craters in the mountains could hold this water at
high elevations instead of letting it flow further downhill, dissolving
the salts that make it nearly useless to man. Similar retention lakes
have been proposed for the Susquehanna Basin and other streams
whose water is not used effectively.

Below-ground nuclear explosions have negligible fallout problems
and are thus more flexible in application. The chimney of fractured
rock—hundreds of feet tall and almost as wide—created by a subsurface charge is, in effect, a huge short-circuit for groundwater. A chimney can recharge an aquifer by funneling surface water into it from
above; or it can connect two separated aquifers below ground. Aquifer recharging is common in arid lands all over the world. They are,
in essence, underground sponges which can be charged during the
wet seasons. A typical nuclear project proposed in this category
would divert the seasonal surplus water in Idaho's Snake River
through chimneys into unsaturated portions of a basalt aquifer for
later use.

Squeezing Water from Humid Air

Rainmaking remains an inexact art. But in some localities, the air literally drips with moisture although fresh water is a rarity. Many such

humid areas are coastal or oceanic and have ready access to deep, cold ocean water. Pumped to the surface from the proper depth, seawater will be much colder than the air and also full of life-building nutrients. Although too salty for drinking or irrigation, this seawater potentially represents both fresh water and food.

Robert D. Gerard and J. Lamar Worzel, scientists at Columbia's Lamont Geological Observatory, have studied the possibility of using cold ocean water to condense fresh water from moisture-laden sea air for domestic use. The slightly warmed seawater would then go back to the sea—not to the same layer from which it was extracted, but to a nearby lagoon or sea basin where it would be used to nourish edible marine life.

Gerard and Worzel centered their attention on the Caribbean Islands, but many other maritime localities could make use of the concept. Saint Thomas, in the Virgin Islands, already operates desalting plants (capacity: about a million gallons per day) to supplement meager natural supplies of fresh water. This water is expensive and the desalting plants also discharge hot brine, a potential pollutant. A cheaper source of fresh water may be in the 200 million gallons of water in the humid air that sweeps across each kilometer of windward shore each day on the average.

In the Virgin Island Basin one can find water with a temperature of 40–45° F at depths of 2700 feet as close as one mile from the shore. Calculations show that a mile-long pipe about three feet in diameter would be able to supply a condenser producing about a million gallons per day of fresh water. About 500 kilowatts of pumping power would be more than enough to deliver the required 30 million gallons per day of cold seawater to the condenser. Smaller fresh-water condensation plants would find wind power or diesel engines sufficient, but nuclear power plants would probably be more economical for plants big enough to supply water for industry and agriculture. This would be particularly true on islands where fossil fuel must be shipped in from the mainland.

Thermal Enrichment in Aquaculture and Agriculture

Aquaculture. The tie between condensing fresh water from the air and the application of the still-cool, nutrient-carrying seawater to food production steers us to the subjects of aquaculture and hydroponics. The biological productivity of the seawater pumped from depths of 2700 feet is ten to twenty times greater than that of the surface water. G. B. Pinchot has even proposed "coral corrals"—circular atoll lagoons—into which nutrient-rich water would be pumped to encourage the growth of zooplankton. Pinchot suggests that captive baleen whales would then convert the zooplankton into protein. While imaginative, this scheme is perhaps no more impossible than, say, landing a man on the moon.

The sea is a great grazing land—like American prairies before they succumbed to the plow. Optimists note such parallels and hope that one day man will harvest the sea as he harvests the endless fields of Dakota wheat. The annual fish harvest currently stands at about 60 million tons. Whether this can be multiplied by a factor of two or ten, no one knows for certain. Too little is known about the sea's biological cycles and their productivity to judge the real feasibility of such schemes as the coral corral. One thing is clear, however, and that is that fish make up a rather small fraction of the world's protein intake and that just doubling the fish catch will hardly stave off world starvation. To make the ocean produce as intensively as the land, it has to be confined and "worked" by man. The projected roles of the atom range all the way from blasting out ponds for intensive aquaculture to providing propulsive power and process heat to huge automated "whales," i.e., food processing ships, that engulf everything from plankton to real whales and turn them into palatable food of one sort or another—perhaps that controversial "fish flour" that seems to offend American palates or perhaps FPC (Fish Protein Concentrate), another unconventional seafood with high nutritive value.

Discounting the misty future for the moment, aquaculture is currently a very real and important industry in many parts of the world. In Asia, carp, milkfish, oysters, and various algae are grown with great success, usually in fresh or brackish water. Mussels are cultivated in Spain and rainbow trout in the United States. Roughly 2 million tons of fish—more than the whole United States catch—are harvested annually from ponds worked mainly by hand.

Heating Things Up. One of the most significant things the atom can do for aquaculture is to provide heat; waste heat from nuclear power plants is highly acceptable, thus solving two problems in one stroke. Unlike the nuclear-propelled "whale," thermally enriched aquaculture is not wishful thinking. The nuclear power plant at Hunterston, Scotland, delivers some of its waste heat to a complex of concrete troughs where sole and plaice are raised. Both species reach marketable size in six to eight months instead of the three to four years required under natural conditions. Warm water from a fossil-fueled plant is now discharged into a Long Island Sound lagoon to promote clam and oyster growth. Shrimp will be cultivated in the warmth of water released by the nuclear power plant at Turkey Point, Florida. The list of proven applications grows yearly.

Thermal enrichment may also benefit agricultural areas and hydroponic farms in the vicinities of large nuclear power plants.

In hydroponics, vegetables are grown directly in a liquid medium or in beds of inert materials. Hydroponic farming is usually year-round farming and little imagination is required to visualize nuclear power plants surrounded by glass- or plastic-enclosed fields where summer never ends.

The fields, in fact, need not be enclosed to gain some advantage from thermally enriched water. By adding warm water to irrigated land, the normal growing season can be extended and frost damage reduced in cold climes. New crops can also be introduced where the climate is locally tempered by a power plant's waste heat. It is even possible that crop growth can be accelerated so that two crops a year are feasible. If you are a visionary, you can see each nuclear power plant surrounded by its own little Eden—even when winter snows blanket adjacent land. While these gardens will not be miniature Floridas for midwinter vacationers, they will create food and jobs, as well as electrical power.

These oases from winter are not vague dreams of technologists. In Oregon, the Eugene Water and Electric Board has carried out a pilot thermal enrichment project on 170 acres located on the McKenzie River. Heated water from the nearby Weyerhaeuser pulp and paperboard plant spray-irrigates the walnuts, apples, corn, tomatoes, and other vegetables and fruits grown on these 170 acres. So far, the results have been very encouraging. Apparently, frost damage to fruits has been prevented and fruit quality improved. Fruit ripens several days before that in nearby untreated fields. Based on the success of this experiment, Representative A. Ullman, from Oregon, has urged the establishment of a multiuse nuclear power complex in Oregon's Umatilla River Basin for the purpose of experimenting with heated effluents over wide areas.

A 1000-megawatt nuclear plant could service 100,000 to 200,000 acres of irrigated land. By the year 2010, the Pacific Northwest expects to have 17 million acres of irrigated land. Projected nuclear power plants and other industries will undoubtedly divert some of their controversial heated water to these irrigated tracts. Farmers (who like hot water) and conservationists (who don't) will both gain.

Other Atomic Aids to Food Production

Gulliver Sails Again. Fish harvesting, being a hunter-prey type of activity, would be simplified greatly if fishermen knew where harvesting is best. An instrument called Sea Gulliver is being developed by the AEC to this purpose. The Gulliver concept was first developed by Gilbert V. Levin under a NASA contract to build an instrument that would signal the presence of extraterrestrial life through the detection of metabolic activity. The Sea Gulliver would repeatedly and automatically sample seawater and measure the overall metabolic rate of indigenous plankton by analyzing the amount of radioactive carbon-14 metabolized by the sample. By monitoring the radio signals from a network of strategically located Sea Gullivers, a central agency could tell fishermen where areas of high biological activity were located and how they changed with the seasons—a sort of subsurface "weather" station network.

The Story of the Returning Salmon. A futurist must always be a bit on the mystical side, like the oracles of ancient Greece. We are cautioned incessantly about the danger of radiation. Nevertheless, life has prospered despite it—or is it *because* of it? Natural radiation *may* have caused many of the mutations that, accumulating down the geological eons, have made us what we are today.

At the University of Washington, Lauren R. Donaldson has divided batches of salmon eggs into two groups, one of which he irradiated with gamma rays from the day of spawning until the salmon were fingerlings. The irradiated and control groups were then fed for ninety days, marked and released. Two, three, and four years later the marked salmon from both groups came back to spawn. Contrary to what we might expect, the irradiated salmon returned in much greater numbers. Why? No one knows. Perhaps they were more vigorous or more disease-resistant. No conclusions can be drawn until similar experiments are completed with other species.

Nuclear Desalting Plants

Captain Nemo, in Jules Verne's *Twenty Thousand Leagues Under the Sea*, satisfied all his needs from the oceans he sailed. The ocean provided his food, his fresh water, and the *Nautilus'* energy. Somehow, we have not yet been able to catch up with Captain Nemo's technology, although we recognize the great potentials of sea farming, fusion power using the sea's deuterium, and seawater desalting. Ultimately, we, too, may draw on the sea for our needs as the land becomes stripped of its primordial resources.

Fully 97 percent of the Earth's water resides in the great ocean basins. This fact has always frustrated practical people who see the sea washing arid coasts and of no great use to agriculture directly. No one knows who first distilled fresh water with heat—probably the ancient Greeks, who seem to have thought of everything—but the distillation process was never used for producing fresh water on an industrial or municipal scale until this century. Distillation and other desalting techniques require considerable energy, a commodity which historically has been scarcer than fresh water. In his *New Atlantis* (circa 1618), Francis Bacon conceived of a special filter to supply his Utopia with palatable water from the sea. Today, scientists and engineers are pursuing dozens of different desalting techniques, including some membrane techniques reminiscent of Bacon's suggestions. It takes energy to separate water molecules from the diverse salts and other substances dissolved in seawater—this is where nuclear power enters the picture.

Principal Desalting Processes. Desalting processes are of two basic types: (1) those that remove the fresh water and leave concentrated brine behind, and (2) those that remove salt and leave fresh water behind. This is not a play on words. Some processes extract water molecules from the seawater mixture; others aim at removing specific salts or impurities. The first category of processes includes distillation, freeze separation, solvent extraction, reverse osmosis, and the so-called gas hydrates approach. The foregoing techniques are most useful in desalting ordinary seawater. The second category of processes is most often applied to purifying brackish water. In this group are electrodialysis and ion exchange.

In most of the above processes, energy must be added in the form of electricity, at least initially. The atom competes here only as another source of electrical power, and it is a serious competitor if the power requirements are high. In distillation, however, energy is applied in the form of heat—a copious commodity in a nuclear power plant. When one speaks of a nuclear desalting plant, he almost always means a nuclear-heated distillation plant.

Distillation demands more energy per gallon of fresh water than most desalting processes. But, if energy is cheap and abundant, this is no drawback. The primary factor to use in judging the performance of a desalting plant is *cost per gallon*. However, an important consideration is omitted from this evaluation. A big desalting plant should not be an isolated installation; more likely, it will be surrounded by farms, industries, and people who need not only water, but also electrical power. The nuclear power plant serves, therefore, as an *energy center* that is integrated inextricably into the economic fabric of a community. At a minimum, nuclear desalting plants are conceived as *dual-purpose* plants, producing fresh water *and* electricity.

The distillation process is paramount in nuclear desalting and deserves further description. Three important variations of the basic approach exist:

1. Multistage flash distillation.
2. Vertical tube distillation.
3. Vapor-compression distillation.

In addition, combinations of these are possible. The basic idea, of course, is to extract as much fresh water as possible per dollar of operating cost and capital investment.

Multistage flash distillation begins with the heating of seawater to perhaps 250° F. The hot water then enters a low-pressure chamber where it boils quickly, part of it "flashing" into fresh-water steam. The flashing is repeated in a series of chambers at ever-lower pressures and temperatures. The steam in each stage is condensed to fresh water as it is cooled by the incoming seawater flowing through tubes in the steam chamber. The end products are fresh water and concentrated brine.

Multistage flash distillation is the most popular of all desalting processes. Nuclear dual-purpose plants are well matched to it because the low-temperature waste heat from the turbogenerator discharge is adequate.

A vertical tube distillation plant is somewhat similar to one employing a multistage flash distillation in the sense that evaporation occurs in a series of stages at lower and lower pressures. The seawater, though, does not flash into vapor; and vertical tubes are usually used instead of large vessels.

The forced-circulation vapor-compression method introduces energy at a different point in the cycle. As the seawater is forced up through a bundle of tubes in an evaporator, part of it is converted into vapor. At the tops of the tubes, the vapor is separated from the hot brine and compressed—a process that raises its temperature. When the compressed vapor returns to the evaporator it contains enough heat to boil some of the fresh seawater rising in the tubes. As it gives up its heat, it condenses into fresh water. This process requires mechanical energy input instead of heat.

Seawater contains many chemicals that make it a difficult medium to process. Above about 170° F, scale accumulates in the pipes and vessels, interfering with flow and heat transfer. The higher the temperatures, the more pronounced are these effects. It is customary to treat the input seawater with chemicals, such as sulfuric acid, to reduce scale formation. There is another technique that converts the seawater's scale-forming chemicals into high-grade fertilizer. W. R. Grace and Company has built a pilot plant using this technique for the Department of the Interior's Office of Saline Water, at OSW's Wrightsville Beach, North Carolina, testing center, but the economics are still unproven.

Development of United States Programs. Until the early 1950s, the seemingly distant threat of water shortages had brought little United States government action in the field of desalting. Instead, fresh water was obtained primarily by further control of the hydrosphere, especially through pumping deep wells and river diversion. But in 1952, the situation became serious enough to warrant federal attention. The Congress enacted legislation which enabled the Department of the Interior to establish the Office of Saline Water (OSW) and begin a five-year research program. OSW's activities were later broadened by additional laws but the primary mission of OSW was and still is research and development in the area of desalting processes.

The United States AEC did not enter the desalting picture until 1958, when Senator Clinton P. Anderson asked the Los Alamos Scien-

tific Laboratory to take a long look into the future of nuclear power.
R. Phillip Hammond, then at Los Alamos, suggested that very large
nuclear power plants might be effective and economical in desalting
seawater for arid localities. Moving to Oak Ridge National Laboratory, Hammond continued the exploration of nuclear desalting. A
1962 report by Hammond and his associates stimulated Jerome Weis-
ner, director of the President's Office of Science and Technology, to
appoint an interdisciplinary group in 1963 to evaluate this thesis. The
group concluded:

> Although we are less optimistic than Oak Ridge National Lab-
> oratory, we have confirmed the essential validity of the ORNL
> conclusion—that relatively low-cost fresh water can be obtained
> with very large-scale, dual-purpose operations where there is a
> sufficiently large market for electric power, and that nuclear en-
> ergy plants appear to have better economic potential in these
> very large sizes than fossil-fueled plants.

As a consequence, President Johnson requested the AEC and the Department of the Interior to prepare an aggressive and imaginative
joint nuclear desalting program.

Oak Ridge National Laboratory is the focal point of most AEC-
OSW cooperative desalting studies. Various industrial contractors
also participate in the work. Generally, AEC efforts have been confined to the study of reactors and their integration into an ever broad-
ening spectrum of "agro-industrial" applications. In other words, the
AEC has been trying to promote the energy center concept by showing its manifold advantages in a variety of societal applications
through calculations and studies on paper. On the other hand, the Office of Saline Water sponsors considerable research and development
on desalting processes, including full-scale pilot plants. The OSW
work is, of course, largely independent of the energy source. The arguments for nuclear energy sources depend upon the economy of very
large nuclear plants—plants in the thousand-megawatt-plus category.
Obtaining funds for a "pilot" plant of such size manifestly demands
considerable salesmanship.

The first nuclear desalting plants will probably be constructed at
places in the United States or in other countries where water and
power shortages are acute. Joint studies to this end have already been
carried out by the United States with Israel and Mexico.

Of the United States sites studied, the most ambitious conceived of a 40-acre artificial island—Bolsa Island—off the southern California coast to supply southern California with 150 million gallons of fresh water per day plus 1800 megawatts of electricity. The Bolsa Island plans finally collapsed in 1969 (amid some controversy and acrimony) partly because of the complexity of the administrative arrangement involving a half dozen private and governmental entities, and partly because of the escalated costs resulting from the consequent delays.

In the face of the demise of Bolsa Island, the increasing opposition to anything nuclear, and the large investment required, the case for nuclear desalting in the United States will have to be a most convincing one before plants will graduate from studies to actual development projects. One such study is being carried out by OSW and the State of California. It is considering sites along the California coast where fresh water is scarce and where large quantities of electrical power are also needed. One or more of these sites may be suitable for dual-purpose nuclear plants, which produce both electricity and fresh water. An important consideration in the final decisions will be the need for a plant to develop actual operating experience, so that the cost of desalted water can be compared with that of water brought in by aqueduct from the northern part of the state. Such a plant could then lead to other plants that could operate on an economical basis.

Thinking More Positively. Nuclear desalting has many champions. One of these, James T. Ramey, a United States AEC commissioner, has summarized the case for desalting in three succinct points, which we paraphrase here:

1. Nuclear energy represents, except for the favorable circumstances where fossil fuel supplies are nearby and abundantly available, the cheapest source of large blocks of energy. Further, the costs are essentially independent of plant location.

2. The cost of desalting can be reduced substantially by scaling up plant size, in the same way that nuclear power costs have been cut dramatically.

3. The combination of nuclear power and desalting in large dual-purpose installations is an excellent match, accentuating the favorable economic features of each.

Quoting Ramey's conclusion verbatim:

> Taken together, these factors mean that water costs from large
> —let us say, 50–100 million-gallon-per-day and 300 electrical
> megawatt—dual-purpose nuclear desalting plants on the order of
> 30–40 cents per thousand gallons appear to be achievable in
> plants built in the near future. While this represents some in-
> crease from estimates of a year or two ago, it is a figure which
> should give a green light to properly conceived projects in sev-
> eral selected locations around the world. These are areas where
> the dwindling availability of conventional water sources, the
> need for "drought-proof" supplies, and, perhaps, the premium
> value of high purity water . . .—combine to make the applica-
> tion of desalting especially favorable.°

The cost figure of 30 to 40 ¢ per thousand gallons is still considerably higher than the 15 ¢ urban consumers pay where water is relatively plentiful, but it is also lower than water costs—$1.00 or more
per thousand gallons—in some arid areas where agricultural and industrial development is stymied by lack of water.

We predict that man will soon be employing the ultimate energy
source and the ultimate water source together.

Portrait of a Nuclear Desalting Plant. The Israel (or Near East) and
Mexican nuclear desalting study projects are furthest along. We will
sketch some features of the latter, which represents a three-year study
effort of a team of experts from the United States, Mexico and the International Atomic Energy Agency (IAEA).

The target area of the study comprises parts of Southern California
and Arizona, and the Mexican states of Baja California and Sonora.
The Colorado River supplies some water to this area, but the fresh
water deficits are projected to be about 1.5 billion gallons per day by
1980 and 4.5 billion gallons by 1995, assuming *no new* agricultural
development. It is a land rich in resources, except for water.

The study team recommended a series of nuclear dual-purpose
plants, each providing a billion gallons of fresh water per day plus

° J. T. Ramey, "Practical Considerations in Desalting and Energy Development
and Utilization," Symposium on Nuclear Desalination, Madrid, Spain, November
18, 1968.

2000 megawatts of electricity—twice the electrical output of the largest nuclear power plants and dozens of times larger than the output of the desalting plants built to date. Each plant would cost about 1 billion dollars. The fresh water, when delivered to major distribution centers, would cost between sixteen cents and forty cents per thousand gallons. Electric power cost would be about 3 mills per kilowatt hour.

Plants could be brought into production as early as 1980 and would be based on desalting technologies well proven at the pilot-plant level today. The first reactors would be light-water types, like those being built in the United States today, with later plants incorporating fast breeders. Large multistage flash distillation units would utilize 260° F steam from the turbines as the basic heat source. Later, vertical tube evaporators would be incorporated into the project.

It is interesting that the United States–Mexico study involved the same water-starved area that would be served by the southern portion of the NAWAPA (North American Water and Power Alliance) concept—the scheme for redistributing surplus fresh water from the Northwestern United States and western Canada rather than desalting seawater. NAWAPA, though, would be a much larger undertaking, serving the whole continent. Both plans call for billions of dollars as well as bold decisions by the governments and peoples of North America.

Which approach will be selected: nuclear desalting or NAWAPA? Or will a combination of the two schemes be used? Or will some other solution to a worsening problem be discovered? Basically, we do not have enough solid, dependable data to make many of the decisions that need to be made. Cost arguments—these are the most critical arguments—rarely take place using consistent assumptions. Both NAWAPA and nuclear desalting cost estimates probably suffer from that common disease of advanced technology: underestimation of costs. For example, a recent critique of nuclear desalting in *Science* [*] claims that nuclear proponents have greatly underestimated the costs of nuclear desalting and at the same time greatly overestimated the value of the water the plants would produce for agriculture.

We know that many nuclear scientists and engineers are perennial optimists and that there is undoubtedly some substance to these criti-

---

[*] M. Clawson, H. H. Landsberg, and L. T. Alexander: "Desalted Water for Agriculture: Is It Economic?" *Science*, 164, June 6, 1969, p. 1141.

cisms. But, the situation is reminiscent of the early days of nuclear power, when the naysayers claimed that nuclear power would never be competitive in generating electricity. We anticipate that history will also redeem this new generation of optimists. Meanwhile, we proceed with the knowledge that nuclear desalting incontrovertibly represents a new dimension of technology that can build some unforeseeable number of oases for men in the Earth's arid lands.

Nuclear Desalting in Agriculture

Nuclear prophets see the *nuclear energy complex* or *Nuplex* concept as an important key to future world development. In this vision, all animals and machines will draw on the atom's energy as if it were a second sun. A more limited version of the Nuplex is the nuclear powered *agro-industrial* complex consuming water, electricity, and heat in industrial processes (Chapter 5). Still more restricted in its scope of applications is the dual-purpose, water-electricity nuclear power plant, such as that just described from the United States–Mexico studies. The simple *agrocomplex*—an immense investment itself—might come before its more ambitious brethren. By first adding fertilizer plants and then other chemical production facilities, subsequent agrocomplexes would gradually evolve into agro-industrial complexes and, finally, Nuplexes, in which the atom is the mainspring of a whole community.

Oak Ridge National Laboratory completed a comprehensive study of agro-industrial complexes in 1968; the following data are taken from the ORNL report.* The ORNL concept presented below should be taken as a stimulating idea—a sketch of what might be done with desalted water in an arid area that has not felt a farmer's plow in centuries.

Desalted water is expensive, even in the Oak Ridge projections— ORNL's twenty cents per thousand gallons is several times the price of irrigation water in most United States localities. (In many cases the price charged is less than the cost, the difference being borne by the

* Oak Ridge National Laboratory: *Nuclear Energy Centers, Industrial and Agro-Industrial Complexes, Summary Report*, ORNL 4291, 1968. Note: The ORNL work assumed desalted water, but nuclear power could also be applied to pumping groundwater.

Artist's concept of a large nuclear agro-industrial complex which could desalt up to a billion gallons of salt water a day while generating more than 2000 megawatts of electricity. The fertilizers and water produced could feed 6 million people from a scientifically managed, 300,000-acre "food factory." (*Oak Ridge National Laboratory*)

general public.) The key to financial success with desalted water is intensive farming and tight rotation of improved strains of high-value crops. In addition, a "match" has to be made between the continuously operating nuclear desalting plant and the staggered growing seasons of the crops. Water demand for irrigation can be spread out by astute selection of crops and planting times and through artificial extension of the growing season with plastic-covered greenhouses and warm-water irrigation using waste heat. Surplus winter water could also be stored in aquifers or Plowshare-blasted chimneys, and later pumped out during the growing season.

As a first estimate of the economic viability of a nuclear-sustained food factory, ORNL assumed three possible crop patterns of ten basic crops. The results in Table 5 show that intensive farming with desalted water (at an optimistic twenty cents per thousand gallons) would yield good returns to a hungry country that normally had to import staple foods.

Many unanswered questions arise after analyzing this "quick-look" study. If highly intensive farming is possible with expensive desalted water, will not other parts of the world, where water is relatively

Table 5
Results of ORNL Food Factory Study [a]

	Crop Pattern [b]		
	Ten Crops	High Value	High Calorie
Water input (billions of gallons / day)	0.9	0.9	0.9
Percentage of water temporarily stored	18	26	24
Production (millions of tons / year)	3.6	3.1	3.3
Calories (billions / year)	4080	4800	5680
Millions of persons fed [c]	4.5	5.3	6.2
Protein per person fed (grams / day)	91	107	79
Water used per person fed (gallons / day)	200	170	145
Operating costs per year (millions of dollars)	148	135	125
Operating cost per person fed (cents / day)	9.0	7.0	5.4
Gross receipts per year (millions of dollars)			
At export prices	159	150	123
At import prices (1.3 × export prices)	206	195	160
Investment (millions of dollars)	295	306	295
Investment (dollars / acre)	1055	957	979
Investment per person fed (dollars)	66	58	47
Internal rate of return (percent)			
At export prices		2	
At import prices	16	17	12

[a] From ORNL-4291, *loc. cit.* Assumes desalted water at twenty cents per thousand gallons.

[b] Wheat, sorghum, peanuts, beans, safflower, soybeans, potatoes, tomatoes, oranges, cotton.

[c] Daily requirement, 2500 kilogram-calories.

"free," adopt the same techniques and bring prices down? Will not labor costs—a major factor in food cost—in a highly technical nuclear food factory be much higher than those in less intensively farmed areas? Can the less-developed nations even begin to supply the trained manpower?

Such questions must be raised and answered as we endeavor to establish the first food factory beachheads. Competition from conventional farming should force prices down as well as spur further exertions by proponents of food factories. With the world population explosion, nothing but good can come from the entry of an alternate, high-potential source of food and water, providing *all* pertinent environmental and safety factors are understood.

The Future of the Nuplex

There is no doubt that the successful planning, construction, and operation of a large Nuplex would be an enormous technical, financial, and social undertaking. Many have been quick to point out these roadblocks. In most cases, economics would demand that its nuclear reactors should be as large as practical for the locale. Ultimately, the Nuplex power plants should be breeders, which may not be commercially available in this country until 1985. The many hurdles ahead of us will require years to overcome.

An agro-industrial Nuplex would have to be located in one of the coastal desert areas of the world where climatic and soil conditions were suitable for the type of highly intensive scientific farming that would be carried on. Studies have shown that a vast amount of land falls into this category, but naturally most of it is not inhabited now. People would have to be brought there to work and to settle. And they would have to be a unique combination of highly skilled specialists, semiskilled people, workers of many types, and their families. In other words, an entire self-sustained community—probably international in nature—would have to be established in a remote area at first not too hospitable or desirable for human habitation.

We suggest an international community because the size and location of a Nuplex would probably require that a developed nation or combination of them contribute the major portion of the financial and technical support, while one or more of the developing nations contributed the land and much of the labor. Such a situation is often cited as another roadblock to the realization of the Nuplex because it would involve a multitude of international political arrangements.

Then there are those who would question the economic success of the agricultural and industrial production that might result, even though the technical, political, and social aspects of the project succeed. How would the products compete in the world market? Would the Nuplexes become self-sustaining and eventually profitable ventures? There are many other similar questions that are difficult to answer with confidence.

While admitting that many obstacles would have to be overcome, and the validity of the many questions that remain unanswered, we should also ask what the Nuplex concept might mean to the world if we could make it succeed. First of all, as Gale Young, Assistant Director of Oak Ridge National Laboratory, has pointed out, "A third of the world's land is dry and virtually unoccupied, while half of the world's people are jammed—impoverished and undernourished—into a tenth of the land area." With the help of the Nuplex a good portion of the 8 million square miles of our world's deserts—of that "dry and virtually unoccupied" land—could be made not only habitable but productive. This would open up whole new frontiers and opportunities for millions of people who now face a bleak future in their current locations. The initial Nuplex community located on the seacoast would also provide the basis for expanding maritime activities and industrial operations connected with the sea. Scientific sea farming or aquaculture, mineral and chemical extraction and processing, and possibly manufacturing using these resources, would provide a livelihood for a growing community. The combination of desert climate and seacoast, with the addition of the necessary fresh water and power, and with the other attributes these make possible, could also lead to the building of nearby resort communities that the Nuplex would remotely sustain and service. Once underway, the possibilities are limited only by the imagination.

But the Nuplex concept does not have to be tied to the desert seacoast or its agricultural basis. There is the possibility that someday, when huge breeder reactor complexes produce power at costs substantially below those of today, we will see entire industrial communities, highly automated and possibly remote from our living communities, as the major manufacturing centers of the nation. Such centers, with their nuclear-energy hearts working round the clock, would be closed-cycle production machines that would ingest a combination of natural raw materials and scrap (all the solid waste and debris we could feed it) and process and reprocess them into the new products required by the outside world. And at the same time that these giant Nuplexes supplied us with material goods, they might also be supplying us with clean power to run our cities. Such cities, able to export their solid waste to, and import their energy and goods from, a remote Nuplex, could be something of an environmentalist's dream. We assume, of course, that Nuplex operations would be extremely clean, with a minimum of wastes, and that there would be a minimized impact on the environment.

We agree that such blue-sky thinking about the potential of the Nuplex is many years—probably decades—from becoming a reality. But we also believe that achieving the nuclear technology needed to make it a reality may be the least of the roadblocks ahead that we shall have to overcome.

Synthetic Food

Dirt farming is arduous and a slow way to get rich. More and more, farming is being left to machines and those few who cannot bear to leave the land for the enticements of the city. As machines become more dexterous, farming could be almost completely automated. In principle, even the sun could be dispensed with, because food is only air, water, a few minerals, and energy. ORNL's food factory could compress many acres into a compact, three-dimensional plant if food could be produced without solar photosynthesis. A single "farmer" of the future may only have to watch dials and press buttons to feed a million people, just as a few operators control huge chemical plants today. Out would come pills and pastes unrecognizable to us who still shred flesh, seed, and stalk.

Yet, solar photosynthesis is so cheap that one wonders what could replace it. The first synthetic foods will probably take advantage of what the sun has already accomplished. Crude oil, coal, papermaking by-products, wheatstraw, corn stalks, and similar sun-made organic materials can be converted into digestible substances by yeasts and bacteria. As long as organic by-products are in good supply, this kind of food factory would allay world hunger to a limited extent. In effect, man would substitute more efficient chemical processing plants (run by nuclear power, of course) for cattle, swine, chickens, and other inefficient, natural converters of organic materials.

Ultimately, as Malthus maintained, the supply of people will require more food than solar photosynthesis and the limited arable area of the Earth can provide. Nature will then have to be augmented by large-scale artificial synthesis of organic materials.

Factory photosynthesis is one approach, of course. Any source of energy can be converted partly into light suitable for photosynthesis. However, the overall conversion efficiency is so low—40 percent for the heat-to-electricity step, 10 percent for electricity-to-useful light, 4 percent, overall—that practical engineers search for other ways to transfer energy from the fuel to the molecular structure of food. The problem, then, is really one of improving efficiency so that nuclear fuel can compete with "free" sunlight.

Bypassing the inefficiency of converting electricity into light, Norman Weliky, at TRW Systems, has been exploring ways to grow food without light. In a plan designed for use on spaceships, where sun-illuminated area is a scarce commodity, Weliky sees electrochemical energy as a likely replacement for sunlight. The basic idea consists of pumping electrical energy directly into the food synthesis reactions at points in the chemical process where the energy of solar photons normally does the job. Electrodes and a power source take the place of sunlight and the sun. Research results are encouraging, but farmers need not worry for a while.

To eliminate both sources of inefficiency mentioned above, a nuclear power source capable of radiating photons of visible light would be required. It is tempting to imagine new food synthesis reactions that could use reactor-produced heat—perhaps at 500° F to 1000° F—to grow some novel, edible form of life in the dark recesses of a three-dimensional food factory. Since no such reactions are known, we look next to thermonuclear fusion as a possible source of photons equivalent to those emitted by our 10,000° F sun. Perhaps the waste heat from a fusion power plant—a magnetohydrodynamic (MHD) version that extracts electrical energy directly from the plasma—might be at a temperature suitable for artificial photosynthesis. There are many promising nuclear byways that have not yet been explored.

Tracers in Agriculture

It is easy to become so engrossed with the future that the significant but silent contributions to agriculture already made by the atom are overlooked. Tracers are not controversial and make no headlines. Nevertheless, they have given the world more and better food as well as a new understanding of plants and animals.

Tracer applications in agriculture number in the thousands. A list would be too voluminous, but a categorization and a few examples of practical results will prove the point.

Category 1. Plant Nutrition and Metabolism

· Phosphorus-32 experiments have demonstrated that 50–70 percent of a plant's phosphorus is absorbed from fertilizer during the first two or three weeks of growth.
· Other experiments showed that fertilizer can be absorbed by foliage and bark as well as roots, leading to the use of nutrients in foliage sprays.
· The intricacies of photosynthesis have been partially charted.
· For best results, fertilizer should not be mixed throughout the soil but injected in a small area about two inches below the seed.

Category 2. Plant Diseases and Weed Control

· Spore uptake of fungicides and other chemicals has been traced, leading to more effective treatment techniques.
· The ways in which herbicides work and their effects on different plants have been explored for the first time.

Category 3. Animal Nutrition and Metabolism

· The real nutritive value of various foodstuffs; e.g., the calcium in fodder fed to cows, has been checked with tracers and found to be different from that measured chemically. (Ordinary chemical tests could not distinguish between elements in the food and those temporarily withdrawn from the bones and other body reservoirs.)
· The role of the thyroid gland in milk and egg production has been studied with tracers. Thyroid activities may prove to be a

good indicator of the future milk-producing capabilities of calves.

Category 4. Insect Watching

· By tagging insects, their travels have been charted, aiding formulation of schemes to control them. Mosquito larvae and grasshoppers are typical subjects for tagging experiments. The effectiveness of insect predators, such as the praying mantis, can also be checked.
· Plants have been injected with a radioactive tracer, which appears in the pollen within a few days. By checking flowers in nearby fields for radioactive pollen, the distances insects carry pollen have been measured—a fact of great interest to seed growers trying to maintain pure strains of plants.

The science and engineering of agriculture have been built up from experimental observations like those above. Understanding how plants and animals function chemically and physically is the first step in making them more useful. The high productivity of agriculture in the advanced countries attests to the value of such research.

Radiation and Food

Too much radiation kills; a lesser amount sterilizes; still less reshuffles genes and introduces mutations. All three phenomena have beneficial effects in agriculture.

Radiation-Induced Sterility. The almost total eradication of the screwworm fly from the United States is one of the atom's greatest success stories. The female screwworm fly lays its eggs in open wounds of livestock, including the navels of newly born animals. The burrowing maggots almost always kill their victims. Before the atom came to the rescue, damage in the southeastern United States amounted to $15 million to $25 million annually.

The essence of the technique is the saturation of the infected area with male flies rendered sterile by exposure to at least 2500 roentgens of gamma rays from a cobalt-60 source. The sterile males seem to be just as attractive as their unirradiated brethren to the wild, fertile females. The females breed only once, and very likely with a sterile

male under the conditions established. With a screwworm fly generation lasting but three weeks, fly population plummets to near zero in
a couple of months, as the area is kept flooded with sterile flies.

During 1958 and 1959, after demonstration tests on a Caribbean island, a fly-breeding plant was set up in the United States. Flies were
raised to the pupal stage, irradiated with cobalt-60 gamma rays, and
permitted to mature. During the attack, some 50 million sterile flies
(both male and female) were released per week from airplanes over
Florida, Georgia, and Alabama—roughly 2 billion total during 1958
and 1959. The chances of native, fertile males and females mating became very low indeed. By 1960, the pest was eradicated from this region of the United States. Currently, the United States releases 125
million sterile flies each week in Mexico and along the Mexican bor-
der to prevent new invasions of the screwworm fly.

The sterilization technique is applicable to many species of agricultural pests. Desirable conditions are: easy rearing, easy dispersal, and
short generations. Some potential targets are the coddling moth, cotton pink bollworm, European corn borer, mosquito, tsetse fly, rice
stem borer, gypsy moth, and a long roster of other agricultural criminals. Successful tests have been carried out on the island of Capri and
in Nicaragua with another well-known pest, the Mediterranean fruit
fly. Even trash fish and some obnoxious animals are considered possible targets, although the ecological effects would have to be well understood before undertaking such an endeavor. It is a rare species
that totally lacks good points.

Breeding New Plants and Animals. Radiation doses falling well short
of the sterilization level cause genetic mutations. Generally, these mutations are harmful, but occasionally a mutation appears that is useful
to the survival of the species or to man. Natural radiation helps maintain evolution's progress (if such it is!)—for both good and bad species. Artificial radiation gives man the power to accelerate evolutionary rates in directions useful to him for certain species.

The effects of artificial radiation on mutation rates were first shown
by H. J. Muller in 1927 in experiments with fruit flies. Muller and
others soon demonstrated that the induced mutation rates were doubled when the radiation doses were doubled. Experimentation with
fruit flies (*Drosophila*) became scientifically very popular. During the
1930s and 1940s, many laboratories bombarded fruit flies with radia-

tion, producing an immense number of aberrant varieties. The results were curious but not frivolous because the *Drosophila* work helped lay the foundations of modern genetics.

When plants and animals are exposed to light doses of radiation, their progeny also exhibit strange, usually undesirable, characteristics. What is desirable for the natural survival of a species may be different from what man desires in a species. Man can impose artificial criteria for mutant survival, such as plant compatibility with harvesting machinery or shorter height, disease resistance, yield, and even attractive fruit color. Exposure to radiation will not guarantee the appearance of any specific desired trait; it is a "shotgun" technique. When desirable features do occur, man can select the mutants for further breeding and intensification of the valuable trait.

Radiation-induced mutations are particularly useful to plant breeders in two ways:

1. When it is desired to alter a single, specific trait, such as grain color or straw length, without seriously disturbing the good features of the species.
2. When breeders wish to induce mutations in species which reproduce vegetatively rather than by seed.

In the mid 1950s and early 1960s, radiation seemed to be a key that would unlock the door to agricultural riches; that is, valuable mutations of staple food plants, ornamental plants, and plants used for fiber. But it was slow, disappointing work and disenchantment followed the euphoria. Eventually, however, hard work and perseverance paid off. By 1970, over 80 commercially useful mutants had been created by radiation, mostly by X rays. Following are some of the types of crops benefited:

Crop plants	Number of new varieties
Wheat (bread)	7
Wheat (durum)	3
Rice	4
Barley	11
Oats	4

Soybeans	4
Beans	7
Peas	1
Groundnuts	1
Peaches	1
Tobacco	1
Castor beans	1
Rape	2
Mustard	1
Lezpedeze	1
Ornamentals	28

The modern Green Revolution, which certainly ranks with the discovery of antibiotics in terms of lives saved, is primarily a tale of those two great staple foods of the world, wheat and rice. In Mexico, the Rockefeller Foundation has supported the development of the Sonora variety of wheat, which is high yielding and responds well to the application of fertilizer. This wheat was created without the application of radiation. However, when this wheat was introduced into India, Indians objected strongly to its red color. It would have taken years by conventional techniques to change the color without compromising the strain's good points. Instead, workers at the Indian Agricultural Research Institute employed gamma rays to create an amber-colored strain called Sharbati Sonora. In just three and a half years, the objectionable color was eliminated and the new wheat was available in quantity to Indian farmers and consumers.

In the Philippines, the story was similar. The Rockefeller and Ford foundations had helped create the new Miracle Rice, called IR-8, by conventional agricultural techniques. Although IR-8 demonstrated a high yield, it succumbed too easily to a leaf blight and a rice blast fungus prevalent in the Philippines. Scientists at the Philippine Atomic Reactor Center used gamma rays to build three new strains of IR-8. The new rices were more disease-resistant, had improved milling qualities, and also matured earlier. Fittingly, Philippine newspapers dubbed the new strains "atomic rice."

Just as food crops are susceptible to improved adaptation, through radiation-induced mutations, so are the rusts, fungi, and other organisms that yearly destroy so much food in the field. A naturally induced mutant blight—for example, the United States corn blight epidemic of 1970—can devastate a region if science cannot mount a counterattack in time. Using artificial radiation sources, scientists may someday beat nature to the punch and create such mutations of plant and animal diseases before natural radiation does. Suitable counterattacks on these radiation-bred species, which hopefully simulate future natural mutations, could then be marshaled ahead of time. This trick might even work for human diseases, such as the many mutant strains of the flu viruses.

Food Preservation Through Irradiation. Beyond levels of radiation sufficient for mutation and sterilization lie debilitation and death. Such fates are considered suitable for those bacteria, fungi, and pests that invade man's foodstuffs. In some areas of the world up to 50 percent of all stored food is consumed or spoiled by pests and microorganisms. Unfortunately, the less-developed countries, where food is scarce, bear the brunt of these attacks. Over the centuries, canning, drying, freezing, fermenting, and pasteurization have helped stave off food spoilage and destruction. The atom presents us with a new technique: radiation processing.

The possibility of preserving foods by radiation was recognized in the early 1900s as soon as the destructive properties of too much radiation were understood. However, radiation sources were not powerful enough in those days to carry out practical experiments. After World War II, radioisotopes became more plentiful and the AEC began to search for practical applications. During the early 1950s, both the AEC and United States Army (always interested in food preservation) began programs to test the idea of processing food with radiation.

Irradiation has proved to be a most useful tool, but also a rather controversial one. As demonstrated by AEC and army work, irradiation can be used in four ways:

1. Inhibition of sprouting in root crops (principally potatoes and onions) using very low levels of radiation.
2. Delay of ripening or maturation (bananas, papayas, mushrooms) through moderate doses of radiation.
3. Insect control or disinfestation (wheat, rice, oats, flour, dried foods) of stored food products by low to moderate doses of radiation.
4. Destruction of some or all microorganisms in foods.

Potatoes photographed 16 months after exposure to gamma rays. Potato at top left received no radiation; bottom left potato received 20,000 roentgens. Radiation doses were progressively larger. The last potato received 106,250 roentgens (obviously too much). In some countries, radiation processing is now being used commercially to inhibit the sprouting of potatoes. (*Brookhaven National Laboratory*)

    a. Extension of shelf life (called radiation pasteurization) of perishable foods (fish, strawberries, poultry) through moderate doses of radiation.

    b. Indefinite preservation (called radiation sterilization) of perishable foods (meat, poultry, fish, seafoods).

    c. Destruction of food poisoning bacteria or other microorganisms injurious to public health (eggs and egg products, meat, poultry).

Both the AEC and the army have made intense efforts to explore the values and limitations of irradiated food. Some of the more important food irradiation facilities are listed in Table 6. The number of United States radioisotope-fueled irradiators is a measure of this effort. In the United States, several products have been brought to the brink of

Table 6
AEC Food Irradiators

Irradiator	Purpose	Location	Description
Research irradiators (4 total)	Immediate irradiation service for contractors in food irradiation program	M.I.T.; U. of Calif. (Davis); U. of Wash.; U. of Fla.	35,000-curie cobalt-60 source; capacity: 75 pounds per hour at 1 megarad dose, underwater irradiation in closed containers
Marine Products Development Irradiator (MPDI)	Semicommercial seafood irradiation; cooperative industry program	U.S. Dept. of Interior, Technological Lab, Gloucester, Mass.	250,000-curie cobalt-60 source; capacity: 2000 pounds per hour at 250,000 rads
Grain Products Irradiator	Disinfestation of bulk grain or packaged products	U.S. Dept. of Agriculture, Entomological Research Center, Savannah, Ga.	25,000-curie cobalt-60 source; capacity: 5000 pounds per hour bulk grain or 2800 pounds per hour packaged product at 25,000 to 50,000 rads
Mobile Gamma Irradiator (MGI)	Large-scale demonstration of feasibility of irradiation processing of fruits at harvest locations	U. of Calif. (Davis) (headquarters)	100,000-curie cobalt-60 source; truck-mounted, 60-ton unit; capacity: 1000 pounds per hour at 200,000 rads
Shipboard Irradiators (3)	Seafood irradiation on fishing vessels immediately after catch	U.S. Dept. of Interior, Seattle; the other two units, which were at Louisiana State University and Gloucester, Mass., are presently on international loan.	30,000-curie cobalt-60 source; transportable 17-ton unit; capacity: 150 pounds per hour at 100,000 rads
Hawaii Development Irradiator	Semicommercial irradiation processing of tropical fruit	Honolulu	250,000-curie cobalt-60 source; capacity: 4000 pounds per hour at 75,000 rads
Portable cesium-137 Irradiator	Research support for food processing industries	On site at industry locations	170,000-curie cesium-137 source; portable, trailer-mounted, 18-ton unit; capacity: 200–300 pounds per hour at 200,000 rads

commercial exploitation, but like many other nuclear programs, food irradiation has teetered on the brink a long time.

For all the success of food irradiation programs, the American housewife cannot find irradiated food on the grocery shelves. Food offered for sale in the United States must first be approved by the Food and Drug Administration, which passes on its wholesomeness. Radiation may affect the wholesomeness of food in three ways: (1) change of flavor and appearance, (2) changes in nutritional value, and (3) changes in the chemical constitution (as in radiation chemical processing). Flavor and appearance primarily determine consumer acceptance. The flavors of some foods are definitely altered by doses of radiation sufficient for preservation, but many irradiated meats and vegetables are as palatable as ever in both taste and appearance. Nor are food values affected unduly. In other words, many irradiated foods are perfectly acceptable to consumers. Why, then, are they not on the market?

The third category above has proven the most troublesome. Irradiated food is in the difficult position of being suspected of containing toxic and/or carcinogenic chemicals as by-products of irradiation. In the United States, its innocence must be proven beyond doubt before the FDA will permit its general consumption. So far, the FDA has not been completely satisfied with the tests conducted on many irradiated foods.

In 1963, the FDA did approve the sale of irradiated canned bacon, wheat flour, and white potatoes. These foodstuffs have not yet been marketed commercially within the United States, although the army has used some. In 1968, the outlook for irradiated food worsened when the FDA rescinded its 1963 approval of canned bacon on the basis of feeding experiments with rats. The AEC and the army were naturally discouraged at this turn of events. Some controversy also exists over the validity of the experiments. But, until the complete safety of irradiated foods is proved beyond a shadow of a doubt, the FDA will not release irradiated food for general consumption within the United States. Some foreign countries, particularly the USSR, have approved a wide variety of irradiated foods for domestic consumption. The United States will probably follow suit eventually. Whatever happens in America, the real impact of food irradiation will be in those less-developed countries where nonhuman mouths now consume almost as much of a harvest as do the human sowers and reapers themselves.

Chapter 5

Old Cities/
New Cities/
No Cities

The greatest pollutant on Earth is man. A thin film of humanity can be accommodated on this globe's surface; but as people congregate story-on-story and freeway-on-freeway the acidity of the human solution rises catastrophically. Nature is etched away bit by bit; soon cities become indelible scars on the face of the planet. Not only is nature destroyed in the heart of the modern city, but man's own wastes —smoke, garbage, sewage—threaten to asphyxiate and poison their maker.

This view of the city is the "apocalyptic" interpretation of modern urban trends. The future is not really so bleak. With imagination and a liberal dose of optimism, we can foresee our old cities remade in more human molds and new cities built with man in mind. The cities of the future could, *if men insist,* be the utopian centers of culture and inspiration we have read about ever since men began to write. Perhaps we would not care to dress for dinner every night or wear top hats perpetually as the city dwellers do in Bellamy's *Looking Backward,* but we do like to congregate with our kind. Given these social instincts, there is no technical reason why these places of congregation—the cities—cannot be stimulating, healthy, and aesthetically pleasing as well as economically sound investments.

What are the problems? The main problem is that 50 percent of the people in the United States live on 1 percent of the land. More move to the cities every day. It is already worse in some foreign countries. Rome's traffic is nearly impossible; Tokyo is a solid mass of humanity. The symptoms of overpopulation are air pollution, water pollution, high crime rates, short tempers, and encroaching ugliness, to name a few. Technology has been left off the list intentionally. As presently applied, technology leads to big garbage dumps, foul air, and cesspoollike lakes; but *it does not have to be this way.* Industrial wastes can be treated and smokeless fuels can be developed, to give just two examples. Technology can be turned easily to cleaning up the urban mess originally created partly through the misuse of technology and partly through gross underestimation of humanity's capacity to breed and consume, and to expel wastes. In fact, technology properly used may be the only short-term answer to the city's problems because it will take time to check population growth. More significant than old cities in the long run are the brand-new cities that are now possible, cities in which man and machine are no longer at each other's throats. Even the wastes of the city will one day become valuable lodes of minerals and chemical compounds.

Aspirin may relieve a headache and bring down a fever; technology can treat the symptoms of urban ills. The patient, however, is still sick in both cases. The disease is simply too many people in areas that are too small. Urban renewal cannot solve this problem; it is merely aspirin. Athelstan Spilhaus has said:

> The overgrown urban complex must be selectively dismantled and dispersed if we are to cure the ills of the megalopolis.*

Building brand new cities, the "minilopolis" instead of the megalopolis, is a good intermediate solution, possibly akin to substituting sulfa drugs for aspirin. The penicillin for urbanitis, the sure cure, though, is either population reduction or the complete elimination of the city. Why cannot people live wherever they wish and congregate electronically? Sight, sound, the sense of touch,† and, in the near future, even the sense of smell, can be transmitted anywhere in the

* From his article, "The Experimental City," *Science* 159, February 16, 1968, p. 710.

† Research by James Bliss, at Stanford University, has shown how pressure (touch) sensations can be reproduced by fingertip-sized clusters of dozens of tiny air jets.

world. Many of the business and cultural advantages of the city can be re-created equally well in a study high in the Rocky Mountains or in an artist's studio out on Cape Cod. Thus, the title of this chapter spans the spectrum: from old cities refurbished to brand new cities to no cities at all.

The atom plays its familiar roles in the old city/new city/no city production. It is first of all a clean source of cheap electric power and heat. Radioisotope tracers and high-temperature plasmas are also important players. The atom obviously cannot solve all urban problems. Electronics, computers, automation, chemistry, and many other engineering disciplines are in the cast, too.

Clean Power/Clean Cities

Each human being on this planet can claim the air above about forty acres of the Earth's surface as his own share of the atmosphere above him. This air has never been pristine pure. Even the crystalline air of the far north has always been laden with volcanic ejecta, particulate matter from forest fires, and a portion of the hundreds of tons of micrometeorites that filter down through the atmosphere from outer space each day. When Spanish explorers first landed in today's Los Angeles area in the sixteenth century, they noted layers of smoke from the Indian's fires hanging thickly in the air. And the famous "dark days" of New England, such as the one on November 19, 1819, are generally explained as due to smoke from massive forest fires to the west blotting out the sun.

Modern cities do not experience darkness as total as that of November 19, 1819, but the days are often hazy when winds fail to sweep away the exhalations of technology. In the Los Angeles basin today, the Indian's smoke has been replaced by the efflux of millions of automobiles and thousands of industries. Smog alerts are common. The eyes smart and water; respiratory and cardiovascular diseases increase. Just a few parts per million of sulfur dioxide makes air harder to breathe. In London and other English cities, sulfur dioxide from the combustion of low-grade coal is turned into sulfuric acid mist in the air that eats away at the facings of limestone buildings. In fact, the famous London pea soupers also date from the introduction of coal as a fuel. During World War II, the British encouraged smoky fires to increase the palls suspended over urban sections of the coun-

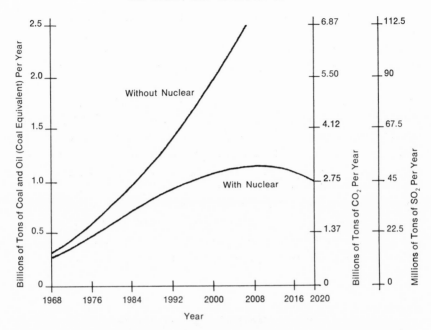

Projected sulfur dioxide (SO_2) and carbon dioxide (CO_2) emission by the electric power industry, showing the reduction expected as nuclear power plants assume a larger fraction of the load.

try to shield targets from German bombers. Air pollution, however, proved to be deadly, too, just as the aerial bombs were. The Committee on Pollution of the National Research Council has summed it up this way:

> It is shocking that so much is known about the possible injurious effects of airborne chemicals and so little public concern [is felt] over the potentialities of damage to health.°

Each city brews its own special variety of polluted air. Los Angeles creates a potent layer of photochemical smog stretching from the sea inland to the mountains that rim the natural basin. New York City is often overlain with thin inversion layers 60 to 100 feet thick, containing many varieties of aerosols, sulfur compounds, and carbon dioxide.

° Committee on Pollution, National Academy of Sciences/National Research Council: *Waste Management and Control*, Publication 1400, 1966, p. 82.

Roughly 125 million tons of waste matter are injected into the air over American cities each year. Table 7 shows the nature of these pollutants and their sources. Without question, the principal offender is the internal combustion engine, followed by the fires of industry and the combustion of fossil fuel in electricity generating plants. Very roughly, man's machines breathe ten times as much air as man does himself. And the situation worsens. Even if half the electricity in the United States comes from nuclear plants, in the year 2000 the new fossil-fueled plants will have quadrupled their contribution to air pollution, unless strict pollution abatement controls are enforced.

A city acts like an artificial hill: the continental air masses often split and flow around it rather than stream through the artificial canyons, cleansing them. In addition, the outpouring of heat from the city's engines sets up internal circulation patterns that concentrate the polluted air. If nature will not or cannot sweep away the dirty air

Table 7
U.S. Air Pollutants and Their Sources *

The Pollutants †		
		Percentage
Carbon monoxide	65,000,000 tons/year	52
Oxides of sulfur	23,000,000 "	18
Hydrocarbons	15,000,000 "	12
Particulate matter	12,000,000 "	10
Oxides of nitrogen	8,000,000 "	6
Other gases and vapors	2,000,000 "	2
	125,000,000 tons/year	
Their Sources		
Transportation	74,800,000 tons/year	59.9
Industry	23,400,000 "	18.7
Electric plants	15,700,000 "	12.5
Space heating	7,800,000 "	6.3
Refuse disposal	3,300,000 "	2.6
	125,000,000 tons/year	

* National Academy of Sciences, *Waste Management and Control*, Publication 1400, Washington, 1966, p. 11.

† Some of these pollutants are much more harmful than others on an equal-weight basis.

and replace it with clean air, the only remaining solution is the reduction of pollutants at their sources: the automobiles, industrial plants, and electric power plants.

Electrostatic precipitators, gas scrubbers, and filters will remove large fractions of the more obnoxious pollutants emanating from furnaces and automobiles. However, they will not eliminate all pollutants, particularly those from private automobiles. Neither will they remove carbon dioxide (CO_2), which is not even listed as a pollutant in Table 7, but which, nevertheless, contributes 6 billion tons each year to the air over America. Carbon dioxide is the primary efflux of our machines, our fires, and man's own metabolism. It is innocuous in small quantities, and green plants remove some of it from the atmosphere during photosynthesis. In the long term, however, man is producing more carbon dioxide than nature assimilates. Carbon dioxide levels have increased markedly—perhaps doubled—in concentration since prehistoric days. The fear rises that the increased concentration of carbon dioxide in the air will cause a planet-wide "greenhouse effect," wherein the infrared radiation emitted by the Earth's surface will be intercepted by the atmosphere's carbon dioxide. In theory, without the cooling effect of infrared radiation to outer space, the planet's temperature would rise. A few degrees average temperature rise on a global basis would begin to melt the polar caps, threatening low-lying coastal cities with inundation. Such flooding would certainly extinguish the smoky furnaces of New York and Los Angeles—a sort of natural catharsis. To some extent, however, the greenhouse effect is offset by the increased reflection of sunlight from the increased cloud cover stimulated by pollutants.

The atom, of course, does not "combust"; it splits or fuses. Its radioactive "ashes and smoke" can be contained and buried where they can do no harm. Atomic power, therefore, may be labeled "clean" power. Cleanliness is relative, though, and considerable attention was devoted to the matter of nuclear safety and waste disposal in Chapter 2. An important point here is that the nuclear industry has always been pollution-conscious and safety-conscious. Federal regulations and technical advances have always kept the radioactive hazard problem under control and open to public debate and revision. This has not been the case with most other technical innovations. Radioactivity is inescapable, but the bulk of this environmental radioactivity is generated by nature. Natural radioisotopes have always been mixed

with every cold air mass that pushes south from the Canadian wilder-
nesses. Even our foods are naturally radioactive to some slight degree. Nuclear power plants are designed so that they do not add significantly to the radioactivity already created by nature. The great
bulk of their radioactivity is contained. Fossil-fueled automobiles and
furnaces, on the other hand, have no choice; by nature they must
dump most of their voluminous efflux into the atmosphere.

In principle, the new "planned cities" on the drawing boards could
rise from wilderness tracts without the usual palls of polluted air
hanging over them. There need be no rumble of coal-laden freight
trains, and no oil tankers need ply the waterways. Inhabitants of old
cities could breathe freely again. It all sounds too idealistic, something like the excerpt from H. G. Wells quoted in Chapter 1. Nevertheless, properly applied technology, including nuclear technology,
could greatly improve our environment.

Pollution Sleuths

Nuclear engineers have been attacking environmental pollution for
some time—long before it became a national campaign. The sensitivity, versatility, and high specificity of isotope-based techniques have
proven to be of great value to the environmentalist and the sanitary
engineer. With nuclear methods they can measure the concentrations
of pollutants and potential pollutants in the atmosphere and hydrosphere. With stable and radioactive tracers they can learn the origins,
physical and chemical behaviors, dispersion paths, and the ultimate
fates of many kinds of pollutants.

Mercury pollution of our lakes and waterways is a serious current
problem. The AEC's interest in mercury toxicology dates back to the
early 1950s, when the University of Rochester Atomic Energy Project
began extensive studies of the biological effects of mercury. Conventional chemical techniques were used in the early work. Only later
did neutron activation analysis arrive on the scene. Over 150 million
pounds of mercury have been used within the United States during
this century, and a large fraction of this huge amount has gone directly into the biosphere. Despite the known toxicity of mercury, this
element has just recently been recognized as a serious hazard in our
environment. The AEC's laboratories have been able to contribute
greatly to our understanding of the mercury problem through neutron

activation analysis. Some examples of these studies of the biosphere follow:

1. In 1969, the Pacific Northwest Laboratory determined that rainbow trout from the Columbia River concentrated mercury in certain organs and tissues to levels above one part per million even though the mercury concentration in their food was only one tenth this value. At about the same time, scientists at Argonne National Laboratory helped place in perspective the buildup of mercury concentrations in the fish and bird life of the Great Lakes area.

2. In 1969, in cooperation with the United States Department of Agriculture, Oak Ridge National Laboratory employed neutron activation analysis to show that corn grown from seeds treated with a mercury-based insecticide had fifteen times as much mercury as corn grown from untreated seeds.

3. At the request of the State of Idaho, the AEC's National Reactor Testing Station measured the amount of mercury in approximately 1000 pheasants and in other plant and animal samples collected throughout the state. The amazing sensitivity of activation analysis can be seen in the fact that mercury was measured down to 0.018 part per million in these tests. Some 90 percent of the game birds sampled showed detectable amounts of mercury, and about 25 percent of them exceeded the allowable limit of 1.0 part per million. Consequently, Idaho health officials decided to inform the public of the mercury problem but to keep the hunting season open, provided certain precautions were taken. Idaho is now considering legislation that will control or eliminate the use of certain fungicides containing mercury.

Thus, neutron activation analysis has been instrumental in sketching the dimensions of the mercury problem on both local and national levels. An important feature of activation analysis in environmental work is that it is nondestructive; that is, the original sample is not changed in any way and may be kept for legal purposes or additional analysis.

Isotopic tracers are solving environmental problems, too, in a surge of new applications paralleling the growth of activation analysis. For example, Brookhaven National Laboratory has developed a stable-isotope-ratio technique to follow the dispersion and ultimate fate of the sulfur dioxide in coal and oil smoke. More than 20 million tons of sulfur dioxide spew forth from American smokestacks each year, but no one seems to know just what happens to it. The Brookhaven technique measures the ratio of sulfur-32 and sulfur-34 with a mass spectrometer in conjunction with a method for tracing stable sulfur hexafluoride developed by the National Air Pollution Control Administration. Together, they help scientists study the chemistry of sulfur oxides in the smoke plumes from fossil-fuel power plants. The sulfur-32/sulfur-34 ratio technique shows how the sulfur oxides are converted to sulfuric acid and sulfates. The sulfur hexafluoride tracer reveals the dispersion of the plumes. This work and other studies like it will help form the basis for setting sulfur emission standards.

Radioisotopes also help control waste disposal operations. Over 300 gamma-ray density gauges are now in use in sewage treatment plants in the United States. (In these gauges, the density of sewage is measured by the fraction of gamma rays absorbed in the sewage. See Chapter 9.) Most gamma-ray density gauges are used in routine sludge transfer operations to determine the best solids concentration for maximum plant efficiency.

Neutron activation analysis, tracers, density gauges: these are just a few of the tools nuclear technology provides to the environment engineers, who must know where wastes come from, where they go, and how they can be efficiently reclaimed.

Smokeless Fuels

Well over 90 percent of the energy that keeps a city's pulse beating today comes from the combustion of fossil fuels. Even the best coal and purest oil will foul the air with their manifold combustion products. Fortunately, there are chemical fuels far superior to those based upon petroleum. Hydrogen and ammonia come to mind; methanol is another possibility; and, of course, battery chemicals are really fuels that can be regenerated by replacing the extracted energy.

Why bother with exotic, "far-out" chemical fuels at all? Electric houses and electric trains have been on the scene for decades and are noted for silence and cleanliness. An all-electric city, with its electricity generated well outside of its boundaries, is certainly appealing and may be a reality some day. However, the world today moves on wheels. The problem of connecting over 100 million vehicles in the United States to busbars via sliding contacts is one that could not be solved without rebuilding every city from the ground up. The planners of the brand new cities can dream of fully automated electric highways, but interim solutions will be applied first. Battery-powered cars, or cars with engines that burn clean, energy-rich chemical fuels, offer the most promise for abating pollution and at the same time permitting the continued luxury of private, individually controllable vehicles.

Before too many years have passed, corner gasoline stations may be replaced by ammonia or methanol stations or battery-recharging terminals. Notwithstanding the new oil fields in Alaska, some sort of synthetic fuel is inevitable. The urban pollution problem merely accelerates the changeover, a changeover that will alter the basic fabric of American industry and—hopefully—cleanse the atmosphere, too.

The first internal combustion engine appeared in 1859, when the Frenchman Étienne Lenoir designed an engine that exploded coal gas behind a piston. The internal combustion engine was soon adapted to inexpensive petroleum derivatives, such as naphtha, benzene, and gasoline, but experiments showed that alcohol or hydrogen, indeed, almost any highly flammable fuel, would also suffice. For example, modern tests with engines utilizing the synthetic fuel ammonia indicate that very few oxides of nitrogen are formed, and that traces of ammonia vapor even help dispel smog. A well-designed engine, however, would emit no ammonia fumes. Gas turbines are even more tolerant. Thus, if a synthetic fuel could be manufactured cheaply enough and if it could be stored and transported with safety, the modifications to our transport industry would be rather minor.

In the United States, gasoline has always been abundant. Despite this fact, the United States Army seriously considered "energy depots" in the 1950s. The basic objective was simplification of fuel logistics for combat vehicles by switching to fuels that could be synthesized on the spot from air and water. Hydrogen and ammonia were the primary fuels considered. Hydrogen was to be derived from water through electrolysis, using electricity generated in a mobile station built around a small nuclear reactor. Liquid hydrogen would be used directly as a fuel or, more likely, converted into ammonia, which is easier to handle. The fact that the "energy depot" and the host of vehicles it supported would not pollute the atmosphere was not an important consideration.

The modern version of the energy depot would be useful in two ways: first, because it *would* greatly relieve urban pollution and, second, because *eventually* gasoline will have to be replaced as our primary vehicle fuel regardless of environmental considerations. Admittedly, fossil fuels are still cheap and abundant. When petroleum becomes scarce, perhaps half a century from now, nuclear heat can be employed to gasify coal and further extend the sway of fossil-fueled internal combustion engines and their turbine counterparts. But speaking in ultimate terms, as we so often do in this book, synthetic fuels are as inevitable as, say, the replacement of gas lamps by Edison's incandescent lamp.

Energy from the nucleus, from coal, or from the sun would be equally suitable for operating a synthetic fuel plant. All that is needed is a source of electricity to electrolyze water into hydrogen and oxygen. Nitrogen, if needed, would be taken directly from the atmosphere. If an easily handled fuel, such as ammonia, is synthesized, the fuel plants could be located well away from city centers. This would be consistent with nuclear and chemical safety objectives. To reduce the possibility of releasing ammonia into the environment, underground pipelines would carry the fuel from the plant to distribution centers in the city and thence to individual consumers, each still the proud possessor of a private vehicle. The scheme would work equally well, of course, for public vehicles, particularly in the more highly organized cities that have been proposed where privately owned conveyances are prohibited.

Carry the synthetic fuel concept one step further: electric power lines are energy carriers. They are also costly and not aesthetically appealing. Underground pipelines carrying anhydrous ammonia or some other synthetic fuel could transport life-giving energy into the city from satellite plants more cheaply than above-ground electrical transmission lines.° Once in the city the fuel could be used for vehicles and space heat directly and also converted back into electricity for home and industrial use in highly efficient fuel cells. In passing, it should also be noted that chemical transmission of energy, with its local holding tanks and reservoirs, would eliminate the possibility of regional electric power failures. Each city would have several days' supply of fuel stored on its outskirts. Further, energy generated dur-

° L. Green, Jr.: "Energy for an Inland Agro-Industrial Community," American Society of Mechanical Engineers Preprint 68-WA/ENER-12, 1968.

ing off-peak hours could be easily stored in the form of synthetic fuel for later use—something impractical in current power practice.

Most of us think of ammonia as pungent and rather disagreeable. Hydrogen has a reputation for being explosive and dangerous. The chemical and space industries, however, have tamed both fuels in recent years. In some ways, anhydrous ammonia is just as safe to handle as gasoline; and liquid hydrogen is becoming common as a high-performance rocket fuel. Ammonia is most often encountered (as far as the nose is concerned) in household cleaners. It is less well known that fully 80 percent of the world's fertilizer requirements are met by synthesizing ammonia from natural gas and steam. Roughly 40 million tons of ammonia are consumed annually in agriculture. Consumption increases almost exponentially. Thus, we can conceive of ammonia production plants that will "fuel" *both* farm and city. However, ammonia, methanol, or hydrogen will probably never replace petroleum completely. But the profligate burning of petroleum products seems a great waste of those remarkable petrochemicals that could otherwise be turned into lubricants, synthetic fabrics, drugs, and a host of other useful products.

Currently, the production of ammonia from electrolytic hydrogen cannot compete economically with hydrogen generated by the steam-methane reforming process or steam-naphtha reforming. If extremely low-cost power were ever to become available from large nuclear power plants, electrolytic hydrogen would become competitive. Studies at the AEC's Oak Ridge National Laboratory indicate that this turning point might come as soon as 1980 in areas served by large nuclear plants where hydrocarbon feedstocks would otherwise have to be imported for ammonia synthesis. The same studies also predict that by 1980 advances in the nuclear and electrolytic technologies will give ammonia derived from their application a competitive edge over naphtha- and methane-derived ammonia.

That gas station on the corner will not begin dispensing ammonia instead of gasoline for a decade or two, perhaps longer if we discount the usually overoptimistic predictions of the engineers. Synthetic sources of energy are, nevertheless, an inevitability. The adoption of an ammonia fuel economy would represent a revolutionary change in a world of more than 200 million petroleum-powered vehicles and tens of millions of oil-burning homes. The smog and smoke hanging over our cities and our passion for private transportation may give us no other choice except such a revolution.

Electric Transportation

The newspapers are full of pictures of small personal electric cars and high-speed electric trains that supposedly will help solve the urban transportation problem. Electric vehicles seem perfect for the city; they are clean, quiet, and consume power only when they are moving. The power that keeps them moving, however, must come from somewhere. If that "somewhere" is a fossil-fuel power plant, we have merely removed the air pollution problem to another locality; possibly somewhere away from the heart of the city, but where people live nevertheless. The reasons for using nuclear plants to generate electric power for vehicles are the same as they are for any commercial application of power: low cost, cleanliness, and reliability.

In addition to saying merely that the atom is a "natural" for supplying energy to electric vehicles, we should attempt to acquire some insight into the workings of these proposed vehicles. It may be that an all-electric vehicle economy would be superior to the ammonia or synthetic fuel economy just described. Either approach, however, would change urban transportation as deeply as the introduction of the first automobiles. Indeed, one of the central problems with modernizing urban transportation is that the best solutions are radical and not evolutionary. Tomorrow's cities will probably not be compatible at all with gasoline-powered, individually driven vehicles traveling upon relatively uncontrolled streets and highways.

An electric vehicle could acquire needed power from a portable source, most likely a battery or fuel cell, or from a guideway mounted overhead or fixed in the roadbed. Or, by doing away with separate vehicles completely, passengers and goods could be carried on electrically powered moving belts or pallets. None of these ideas is new; all have been built and proved in limited applications. From the power plant standpoint, all schemes consume electricity. The battery-powered vehicles, however, display an interesting economic advantage; the batteries would draw energy from the power plant only while being recharged. Since the recharging could be done when most of the city sleeps, the power plant problem of varying loads would be alleviated. In contrast, moving belts and vehicles using guideways draw power just when industry also needs it. Of course, if the energy were conveyed into the city via ammonia pipelines, this peaking of the power-demand curve would be no cause for concern, because ammonia can be stored to meet demand peaks whereas electricity cannot.

Within the city proper, travel during working hours usually consists of excursions of less than 1.5 miles. People would be healthier if they walked such short distances (particularly if the city air had been cleaned up), but experience has shown that they generally will not walk if the distance exceeds a quarter mile. For this kind of travel, moving belts or some other kind of "continuous-capacity" system looks very attractive if it is fast enough and people can get on and off it safely. Many variable-speed belts and platforms are now being examined in this country and abroad. When the day is over, however, workers in present-day cities head home. Earlier in this century, mass transit systems—buses, trolleys, commuter trains—carried most of this traffic. As the automobile came within financial reach of more and more people, Americans became psychologically attuned to private conveyances, regardless of the burden they placed upon the city's capacity to digest and disgorge them morning and night. Although the cities of the future may be so pleasant to live in that the 5:00 P.M. mass exodus is partially stemmed, there will always be substantial traffic flow in and out of any metropolis. Many concepts have been proposed to handle this longer-range traffic. The United States Department of Transportation has sponsored several studies of various concepts. Table 8 summarizes the findings of a survey concluded by TRW Systems in 1968.

It is impossible to tell at present just which of the transportation schemes presented in Table 8 will prove superior in actual city use. Only the last two entries can be considered *radical*, that is, revolutionary in the same sense as an ammonia fuel economy. Air-cushion vehicles, tube vehicles, and the others are simply better ways of doing the same things we do today in urban transportation. The authors lean strongly and intuitively toward the radical solutions. This view is based on the premises that if one must travel in a vehicle, the trip should: (1) demand little or nothing from the traveler (after all, men should not have to serve machines), and (2) not infringe upon the traveler's "right" to go where he wishes, when he wishes. It is in the spirit of modern technology to step into a private vehicle, dial an

Table 8 Comparison of Various

	Class	Characteristics	Speed Range (mph)	Guideway
Track levitated vehicles	Tracked air cushion vehicle	Vehicle guided along track and supported by air cushions	150–300	Flat concrete horizontal surface for support and vertical surface for guidance. Inverted "T," box, and "U"
	Electro-magnetic suspension	Vehicle guided along track or enclosed tube and supported electromagnetically	250–500	Passive aluminum loops buried in guideway
Rolling-support systems		Vehicle guided and supported by either conventional surface rails or monorail	150–300	Conventional rail roadbeds; elevated structures
Tube vehicle systems		Vehicle guided and supported by enclosed guideway or subterranean tube which may be evacuated	300–450	Concrete or steel tubes, either at atmospheric pressure or evacuated, above or below ground or on the surface
Multimodal systems		Vehicle using both conventional surface routes and new automated guideways for intercity portion of trip	80–150	Suspended and over-running

° Adapted from a 1968 study by TRW Systems for the Department of Transportation.

High-Speed Ground Transportation Systems *

Suspension	Propulsion †	Potential Advantages	Disadvantages
Air cushions pressurized by centrifugal or axial compressors	Linear electric motor with reaction rail in guideway; propeller driven by gas turbine or rotary electric motor	Guideway may be cheaper to build and maintain; smoother ride at high speeds than rail system; no wheel hop or traction limitations	Power to support weight of vehicle is high. Air cushions may be noisy. Switching is difficult
Electromagnetic forces generated by super-conducting magnets on vehicle	Linear electric motor or propellers	No apparent speed limitation imposed by method of support. Power needed for support potentially less than for air cushion	Intense magnetic fields may affect passengers, subsystems; vehicle may require heavy shielding
Steel wheels on welded steel rail; rubber tires	Rotary electric motor; gas turbine engine (both with drive through wheels); linear electric motor	Wheel support requires no power expenditure beyond friction effects; extends conventional technology to a higher speed range	Guideway maintenance costs may be high; traction falls off at high speeds; monorail poses switching problems
Wheels; electromagnetic; aerodynamic— large air cushion	Linear electric motor; neumatics and gravity; linear turbines; propellers	Very high speeds without disturbing environment or corridor community. Attractive for highly urbanized areas such as Northeast Corridor	Existing power pick-up devices are unsuitable for high speeds; tunneling costs are high
Steel wheels or rubber tires; magnetic; air cushions	Linear electric motor; rotary electric motor; internal combustion engine	May offer shorter door-to-door travel time. Retains advantages both of private auto and high-speed mass transit. Possibly compatible with urban systems	Vehicle unit costs per passenger are higher than for conventional auto or mass transit; maintenance of privately owned vehicles must be verified before use on public guideway

† Combustion engines could be fueled with petroleum or the "clean" fuels discussed elsewhere in this chapter.

Table 8

Class	Characteristics	Speed Range (mph)	Guideway
Autotrain systems	Conventional autos, along with drivers and passengers, are loaded on a carrier vehicle and transported over the high-speed link	100–150	Standard gauge for lengthwise loading and 17-ft width for cross-wise loading of autos
Automated highway systems	Conventional autos and highways are modified to provide automatic control of traffic flow on the high-speed link of intercity trips	Auto speeds	Conventional concrete highway, special purpose or modified to accommodate appropriate control system
Continuous capacity systems	Transportation is continuously available to passengers at a given point without regard to demand. Employs variation of endless-belt principle	15–50	Enclosed belts, elevated or subsurface

° Adapted from a 1968 study by TRW Systems for the Department of Transportation.

(Continued) *

Suspension	Propulsion †	Potential Advantages	Disadvantages
Steel wheel on welded steel rail	Rotary electric motor; gas turbine engine; diesel-electric locomotive	Offers door-to-door service. No parking problem at terminal	Access to urban terminal could pose problems; flexibility of loading for different destinations is poor
Automotive	Internal combustion engine, rotary electric motor	Appears safe and offers increased density over existing highways. Driver becomes backup controller. Door-to-door service. No terminal interface required	Vehicle maintenance may be beyond control of system operator; merging for entry, exit, and landchanging requires complex central control system
Rollers; wheels; air	Rotary and linear electric motors; air pressure	Offers uninterrupted, continuously available service to many passengers	Passenger acceptance is not widespread. Slow. Not suited for growth or intercity speeds

† Combustion engines could be fueled with petroleum or the "clean" fuels discussed elsewhere in this chapter.

address, and be whisked off to your destination at a rapid clip. Transportation should be as painless and as automatic as dialing a telephone number or selecting a television station. The foregoing philosophy has little to do with the atom specifically but a lot to do with the way technology approaches urban problems.

Gold to Garbage and Back Again

If the pall of polluted air permits, one of the first sights to greet a motorist or air traveler when he approaches a large city is often garbage or waste (a cleaner name for it), smoking acres of it; redolent sewage disposal plants; automobile carcasses stacked in high piles. From aesthetic considerations alone, this garbage is offensive enough; it is much worse from the standpoint of public health; when it comes to the conservation of natural resources, it is perilous for coming generations.

Each pound of discarded metal in the rubble heaps came from nonrenewable mines that will play out some day. The cellulose products (paper, plastics, etc.) that are thrown away or incinerated are renewable to some degree, but they are synthesized from raw materials won from fields and forests increasingly needed for food production, home construction, and recreation. Because nature's power to recycle man's wastes is very limited, the human race must learn to use these resources over and over again if it is to be a permanent fixture on this planet. Dandridge Cole, in his prophetic book *Beyond Tomorrow*, puts it this way:

> The closed-cycle society is not an optional human development. It is not a way of life which may or may not be adopted, depending on whether people decide for or against it, whether some master salesman convinces people that they should have it, or whether people are strongly opposed to it. It is going to happen whether we like it or not and whether we decide to do anything about it or not—if the population explosion continues. It will probably happen even in the unlikely event that the population explosion is halted in the near future, since it represents greater efficiency, less waste, greater control over the environment, and greater security.

If our civilization chokes on its own wastes, archeologists (perhaps not human) of the distant future will be perplexed as they poke through our refuse heaps: How could it have happened amidst such wealth? The atom, with its potential for producing cheap electricity, extremely high temperatures, and sterilizing radiation, can help us move toward that closed-cycle society that Cole considers inevitable.

The United States Public Health Service estimates that 95 percent of the 300 million people residing in this country in the year 2000 will

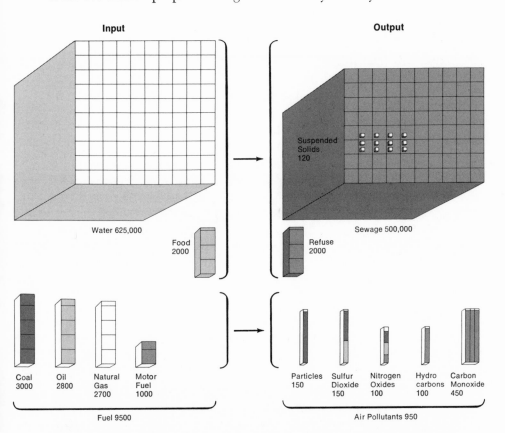

The metabolism of a typical city shown in input-output form. Figures are in tons per day per million inhabitants. About 20 percent of the water input of 1200 pounds per person per day is diverted to lawns and other nonrecoverable uses. Carbon dioxide, which is not considered a pollutant in this chart, may also have long-term effects. (*Adapted from A. Wolman, "The Metabolism of Cities,"* Scientific American 213 (Sept. 1965):80. Reprinted with the permission of Scientific American.)

live in urban areas. Each one will discharge an average of 130 gallons of liquid waste per day through the municipal sewers. Added to this, the city will have to deal with the more voluminous liquids (five to ten times the volume of household wastes) discharged by the city's industries. Both waste streams will *have* to be cleaned up because, by the year 2000, the total volume of liquid wastes will roughly equal the total stream flow within the continental United States. Power plants, farmers' fields, and other water users wait for clean water downstream from every city. One wag has remarked that only one law would be sufficient to encourage cities to clean up their waste water: Make them place their water intakes downstream and discharge their wastes upstream. This would amount to a nearly closed water cycle.

Radiation purification of waste water has been studied by UMC Industries, Purdue University, the University of Vermont, Battelle-Northwest, and the city of Chicago. In various pilot experiments, radiation has displayed its sterilizing potential and the ability to induce oxidation of organics. None of these techniques has been employed on a large scale as yet. Identification of their proper places in the spectrum of waste treatment processes will have to wait until comprehensive comparative studies are completed by the government.

W. F. Schaffer, Jr., and his colleagues at Oak Ridge National Laboratory have studied the impact of low-cost electricity upon the treatment of waste water. Some likely processes that would benefit from cheap electricity are electrolytic sewage treatment, salt removal through distillation, wet air oxidation of sewage or sludge, and ozone treatment. Again, we do not yet know how these processes compare economically with competing chemical and thermal schemes.

Rivers foam from detergents and lakes "bloom" with green algae due to the unwise dumping of liquid wastes, but the plethora of solid wastes is even more obvious. In 1920, each American discarded about 2.75 pounds of solids per day—paper, cans, Model Ts that finally wore out. Today, he throws away almost five pounds of more sophisticated trash—aluminum beer cans, defunct television sets, plastic containers. The greater the national product and the higher the standard of living, the more each American throws away and the harder it is to dispose of his refuse (Table 9).

Municipal trash is usually collected by individual trucks that carry it to centralized incinerators and dumps. Sometimes, there will be

Table 9
Average Composition of Municipal Refuse

Rubbish (64 percent)	Percentage by weight
Paper, all kinds	42.0
Wood and bark	2.4
Grass	4.0
Brush	1.5
Cuttings, green	1.5
Leaves, dry	5.0
Leather goods	0.3
Rubber	0.6
Plastics	0.7
Oils and paint	0.8
Linoleum	0.1
Rags	0.6
Street refuse	3.0
Dirt, household	1.0
Unclassified	0.5
Food Wastes (12 percent)	
Garbage	10.0
Fats	2.0
Noncombustibles (24 percent)	
Metals	8.0
Glass and ceramics	6.0
Ashes	10.0
	100.0

hand or machine sorting of reclaimable materials such as paper, rags, and useful metals. Undoubtedly, the sorting process could be further mechanized, but technology ought to do better than this. Most of the advanced waste processing techniques depend upon very high temperatures. For example, the so-called Melt-Zit Destructor, developed by the American Design & Development Corp., operates at about 3000° F. At this temperature, about 95 percent of all the refuse introduced is completely incinerated; the residue emerges in a stream resembling molten lava. Upon cooling, the lava is found to be a sterile metallic silicate that is suitable for road foundations, fiberboard, and other applications.

If high incinerator temperatures are obtained from a jet engine or other fossil-fuel burners, there will be large volumes of gas discharged into the atmosphere from both fuel and refuse. This hardly alleviates air pollution. Nuclear reactors cannot sustain temperatures of 3000° F for more than a few hours, so that direct incineration within a reactor is out of the question. An electric furnace supplied with low-cost nuclear electricity is a possibility, however. In summary, high-temperature incineration seems to be a step in the right direction, turning all refuse indiscriminately into gases and solids.

Despite the attractive features of these relatively high-temperature processes, they assume *waste disposal* rather than *waste recovery*. The atoms of iron, chromium, silver, copper—essentially all the metals—are often hard won from dwindling ores. To extend man's domain on this planet, at least the rarer elements must be rescued from solid wastes.

Trash collection in the future will doubtless be silent and more hygienic. Battered, smelly garbage cans and traffic-blocking trucks will probably be replaced by either a continuous, underground collection system or a fully automated army of computer-controlled robot vehicles that collect standardized, disposable containers. The latter technology is already being developed for use in automated warehouses and containerized ships. Once waste arrives at the recycling plant, it could be reduced, container and all, to its chemical elements in the fusion torch described in Chapter 3. In principle, all kinds of waste could be consumed in the multimillion-degree torches. The operator of the recycling plant could, at his discretion, do one of three things with the hot mixture of resultant ions:

1. He could separate the ions into elements of high purity.
2. He could "quench" or cool the plasma quickly to encourage the formation of simple, usually gaseous compounds, or he could quench the plasma slowly to promote the synthesis of stable solids.
3. He could hold the plasma at that specific temperature and pressure favorable to recombination of desired compounds.

Other possibilities doubtlessly exist, for the plasma torch is an unexplored concept.

The essential ingredient in the above recipe for automated, flexible waste recycling is cheap energy. Once abundant energy exists, one can recycle garbage as well as go to the moon; energy plays no favorites.

Nuplex: An Integrated Energy Center

As optimists, we see the city of tomorrow as a place where people will meet to do business, exchange ideas, govern the land, perform scientific research, write music; in other words, a place for creative expression. An IBM advertisement puts it nicely: "Machines should work. People should think." Except for air conditioners, computers, and a few robot devices, there is no need for heavy machinery and processing plants in the city proper. In the future we are discussing, most industry will be automated. Those people who prefer urban life will collect in the city to apply their brains and artistic talents, primarily to those things more important to the human race than filling out forms in quintuplet or hauling garbage. Industry and agriculture (also automated) will concentrate around the energy centers located *away* from the city's heart. As noted earlier, we call these energy centers *Nuplexes* (for nuclear energy complexes); they do all the "dirty work."

By segregating industry outside the central city, architects can design cities for people instead of machinery. With the smokestacks, rail yards, and warehouses gone, we might yet see cities like those graceful creations of the future that artists are wont to draw when freed of today's urban conventions. We already see glimmerings of the architecture of tomorrow in the advanced buildings at the great world fairs.

Nuplexes will supply the city with energy, food, and manufactured goods. The city would send back wastes and work orders to the machines. The Nuplex might even be put underground and covered with fields and pastures. The city, then, becomes a large artifact supporting the higher aspirations of man. Men free to think and create individually or *en masse,* twenty-four hours a day, might indeed surpass all past human achievements.

Enough philosophizing; can nuclear power do all of the things implied above? We have already described the potential of the atom for:

Desalting water
Increasing food supplies
Improving extant water sources
Recycling wastes
Providing abundant, cheap energy as electricity or nonpolluting
    chemical fuels

All of these functions are essential to the well-being of future cities;
they are all intrinsic to the concept of a Nuplex; however, one vital
Nuplex function remains to be described. The true Nuplex is an
agro-*industrial* center, providing manufactured goods and products as
well as administering to the other vital needs of its nearby "master"
city.

The atom has already served industry for over two decades, although silently and without fanfare. The industrial atom is commonly
found in thickness gauges, level indicators, and other sensors widely
used on automated process lines. These applications are covered in
more detail in Chapter 9. Important as these applications are, the
greatest impact of the atom on industry will be felt when nuclear
heat and electricity are cheap enough to be used directly in the process industries.

Electricity is now used directly in some industrial processes, such
as electroplating, chemical electrolysis, and electric furnace operations. Huge quantities are required to win aluminum from its ore. Because there is no intrinsic difference between nuclear electricity and
electricity from any other source, the primary criterion is cost. With
process heat, however, it is a different story. Besides the comparative
costs of nuclear heat and heat from other sources, we have to take
into account the fact that nuclear heat is relatively low-temperature
heat and available only in very large quantities. Some industrial processes require heat at 2000° to 3000° F, whereas most nuclear reactors
can offer only about 1000° F on a sustained basis. (The High Temperature Gas-cooled Reactor can be adapted to produce process heat at
about 2000° F.) Some industrial processes can utilize 1000° F heat,
but they rarely need enough of it to justify the operation of a large
reactor. However, the Nuplex envisions the use of this low-temperature heat and, in addition, sources of high-temperature heat. In a
Nuplex, there will be large quantities of relatively *low-temperature*

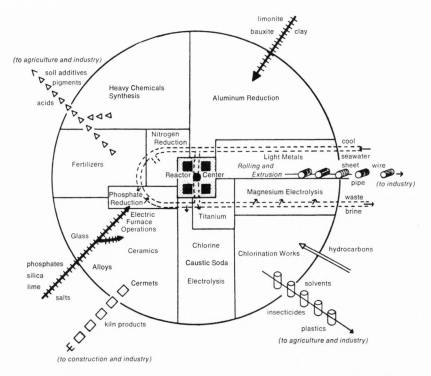

An early Nuplex concept. Originally presented in the 1956 edition of R. L. Meier's *Science and Economic Development,* the caption is reproduced unaltered. "This is a hypothetical map of an industrial estate that might naturally grow up around a million-kilowatt nuclear reactor. The cluster of industries represented here would use the off-peak power to advantage. The processes integrate very nicely with each other so that marked savings can be made in freight costs and waste disposal as well as line losses for power transmission. The manufacturing methods are almost fully automatic so there is little danger to humans in case of reactor failure. Although there may be only a few thousand workers regularly employed in this billion-dollar complex, the fabrication and handling of the products should provide productive work for hundreds of thousands of workers in construction, other manufacturing, agriculture and services." (*Reprinted with the permission of the M.I.T. Press*)

waste heat from the central sources of nuclear electricity. Processes, such as water distillation and those using process steam, will be integrated into the Nuplex to extract as much energy as possible before the heat is released to the environment (probably for agriculture and local climate control). In addition to waste heat, the Nuplex will supply electricity so cheaply that it can be converted economically to *high-temperature* process heat.

Each Nuplex would be different, molding itself to the local terrain and resources. At the heart of each Nuplex would be one or more large nuclear reactors. Heat and electricity would flow from the power plants to the component industries. Besides raw materials, the Nuplex influx would include waste materials for reuse as part of a nearly complete recycle economy. Within the Nuplex, pipes, conveyor belts, and automated vehicles would carry raw and processed materials between various sections of the complex. Out would flow finished goods and products for the "master" city and for export.

The nuclear energy center has been studied in depth at the AEC's Oak Ridge National Laboratory. A key and probably valid assumption in the studies has been the inevitability of very cheap electricity from large nuclear power plants. The cheaper the electricity, the greater the number of industrial processes that become candidates for Nuplex operations. Oak Ridge singled out ammonia, phosphorus, aluminum, chlorine, and caustic as the products most likely to succeed in the first Nuplexes. Looking at various product mixes and plant sizes, Oak Ridge engineers summarized their results as in Tables 10A and 10B. Although the Oak Ridge study confined itself to only the process industries, ignoring waste recycling and manufactured goods, the outlook is very favorable. Product Mix III, for example, shows a good rate (12.7 percent) of return for a domestic energy center. The rate of return for the same energy center located in a foreign country is even higher (16.6 percent). In fact, the foreign country might profitably export the phosphorus and aluminum.

The Oak Ridge work has been criticized from several standpoints:

1. Very cheap nuclear electricity is a dream only.

2. Even a Nuplex of the limited sort proposed by Oak Ridge would cost a billion dollars to set up.

3. The "people" problems would be enormous, especially in the less-developed lands where local residents do not have the technological skills to help build or operate an energy center.

Yet, most critics admit there is great promise in the energy center idea. Industrial complexes coalesce naturally around oil refineries and big hydroelectric plants. The Tennessee Valley Authority (TVA) introduced technology and a higher standard of living to a whole region of the United States through the medium of cheap electricity.

The Oak Ridge agro-industrial center outlined above is, of course, only a primitive Nuplex. A city would be built up around early Nu-

Table 10A
Summary of Oak Ridge Energy Center Studies *

	Product Mixes Considered					
Commodity production	Mix I	Mix II	Mix III	Mix IV	Mix V	Mix VI
Ammonia (tons/day)	3000	0	0	0	0	3080
Phosphorus (tons/day)	1120	1120	1150	1180	1280	1500
Aluminum (tons/day)	514	514	685	685	342	
Chlorine (tons/day)	1000	1000	0	0	1000	2000
Caustic (tons/day)	1130	1130	0	0	1130	2260
Electric power consumption (megawatts)	2048	1050	1038	1050	1021	2026

° Extracted from: Oak Ridge National Laboratory: *Nuclear Energy Center, Industrial and Agro-Industrial Complexes, Summary Report*, ORNL-4291, 1968. For the summary table of agricultural potentialities, see Table 5, p. 130.

Table 10B
Product Mixes Considered in Oak Ridge Energy Center Studies

		United States		Foreign		
Product Mix [a]	Industrial plant power (megawatts)	Capital investment (millions of dollars)	Internal rate of return (percent)	Capital investment (millions of dollars)	Internal rate of return percentage Domestic prices	Export prices
I	2048	$812	11.4	$890	16.1	7.7
II	1050	628	12.1	693	16.6	7.8
III	1038	699	12.7	755	16.6	8.9
IV [b]	1050	508	18.7			
V	1021	555	11.4	612	15.1	5.5
VI	2026	521	13.1	592	16.3	4.5

[a] Product output scaled to power rate. Product mixes defined at top of table. Light water reactors assumed.

[b] Florida location near phosphate rock deposits; aluminum made into ingots only.

plexes to provide workmen and to take advantage of cheap power. The advanced Nuplex we visualized at the beginning of this section would be fully automated, leaving men free to pursue other objectives in the city it serves. We expect, however, that progress toward this goal will be via intermediate steps, particularly in areas of the world that still till the land with beasts of burden and even now spin by hand.

It is a fascinating coincidence that Milton Burton, former head of the Radiation Chemistry Section at what is now Oak Ridge National Laboratory, spoke the following words back in 1947, when there seemed little chance that the atom would ever generate electric power or become an instrument of social progress:

> . . . a municipal atomic energy plant may be used in the future as the principal unit in the city's sanitation system, purifying its water supply, sterilizing its waste, and producing new products at the same time that it produces power. (*Scientific American,* March 1947, p. 124.)

History is now proving that even the wildly optimistic dreams of more than two decades ago are conservative.

A Longer View

Two decades from now, we will probably look back and find that we too were technological conservatives. No matter how bold the forecast, unexpected discoveries will sweep technology and society into more exciting and potentially rewarding times.

Some of the new cities, or whatever the people of the future call the places in which they congregate, will be located in spots of great natural beauty. With closed-cycle cities, good harbors and favorable natural resources will no longer be critical to urban sites. Furthermore, the future cities can be designed with great architectural freedom; they will complement the natural surroundings rather than shroud them in smoke and smog. Then, too, closed-cycle cities will be built in the new frontier lands for the more adventuresome. We can conceive of floating cities far at sea; cities on the sea floor; perhaps even subterranean cities, where the man-made sun—a nuclear reactor, of course—never sets. Moon colonies and huge orbiting complexes may be the frontier towns of the next millennium. Adventure would not be the sole justification for living in these frontier cities; population pressure, new economic opportunities, just the urge to "move on" might stimulate some.

Man may not follow such a fantastic trail. He may not care to carve out a wider physical empire in space or under the sea. The world of our children's children may be reached through reversal of the population growth rate and through the dissolution of the cities. As intimated earlier, the greatest technical advances may be those that project across great distances man's senses and his unique ability to manipulate things. There will be no need to build cities on a planet where every person can contact every other person with all his senses at the speed of light, or when the world's knowledge can be spread out before one on a video screen. The real city of the future may cover the globe, not with concrete and steel but with radio pathways that pass invisibly over fields, prairies, and forests. Only occasional remnants of the great cities, even now decaying, would then remain as museums of a once overly gregarious culture.

Chapter 6

Planetary
Engineering

A Slightly Flawed Planet

The Earth is a unique and generous home for man, but is not perfectly suited to his aspirations. A lack of good harbors stifles development of coastal Peru and Australia. Valuable natural resources are overlain by hundreds of feet of rock; huge deposits of oil and gas are trapped, immobile in fine-pored rock. The weather is not what we would like it to be, despite millennia of sacrifices to the gods and many tons of dry ice and silver iodide crystals. All of humanity's efforts to restore the Garden of Eden have been futile so far. Man's machines have not been powerful enough to compete with the forces of nature. Not that man has not altered nature (remember the Dust Bowl?); rather, he has not always changed it the way he wishes.

The ancient Egyptians lost an estimated 120,000 men trying to carve a canal from the Nile to the Red Sea. Today we could accomplish the task within a time span of a few years and with a high degree of safety with nuclear explosives. The atom, as a source of heat and explosive power, is at last making the discipline of planetary engineering a practicable one. Technology now has enough muscle to move mountains, possibly change the climate, and extract natural resources previously locked tight far below the ground. The real questions are whether man will seize the opportunities and whether he will use them wisely.

Most of the world is classified as "underdeveloped." Hundreds of millions live on the brink of starvation while surrounded by great natural wealth—minerals, fertile land, the rich ocean—all that is needed to synthesize food and the other commodities of an affluent society, except energy. This wealth cannot be tapped until nature is nudged a bit; and a nudge on nature's scale means releasing energy far beyond preatomic capabilities. There are hazards involved, but often they can be made small in comparison with those now forming in a world grinding toward deeper poverty and more starvation.

Atomic Underground Engineering

Atomic underground engineering comes under the aegis of the AEC's Plowshare Program which, as we have seen in Chapter 3, has the responsibility for applying the colossal forces produced by the nuclear fireball to peacetime purposes. Nuclear explosives can blast away massive layers of soil and rock (overburden) more quickly and more cheaply than conventional explosives. Other Plowshare techniques are less conventional but more acceptable to conservationists who wish to preserve scenic values. Nuclear "dynamite" can, for example, shatter huge volumes of rock far beneath the surface, releasing oil and gas trapped for eons in porous rock of low permeability. This technique is termed "stimulation," because the explosion (the stimulus) forces the rock to yield oil or gas it would otherwise retain. More ingenious is the thought of employing the underground column or "chimney" of shattered rock created by the nuclear detonation as a gigantic chemical retort hundreds of feet high. We will next describe some Plowshare projects aimed at capitalizing on these notions. Atomic underground engineering, like space travel, requires a fundamental shift in perspective. Of course, regulatory control must ensure that the gas, oil, and other minerals produced do not endanger the consumer with residual radioactivity resulting from the nuclear explosion. The development of clean nuclear explosives will help attain this goal.

Releasing Trapped Natural Gas. Large areas of Wyoming, Utah, Colorado, and New Mexico are underlain by gas-bearing strata that oil and gas men call "tight"; that is, the rocks are relatively impermeable and do not release their natural gas readily. "Stimulation" of reluctant gas wells usually involves detonating nitroglycerine in the rock or forcing fluids into the rocks under high pressure (hydraulic fracturing). When the Plowshare Program commenced, the petroleum industry was quick to react. In 1958, the El Paso Natural Gas Company wrote to the Lawrence Radiation Laboratory, technical director of Plowshare for the AEC, to inquire about the possibility of nuclear stimulation.

Strong commercial interest is understandable: there are over 44,000 United States gas wells drilled into "tight" strata that could benefit from stimulation. Nuclear stimulation of these wells might double the country's natural gas reserves. On an engineering basis, one might expect stimulated wells to yield 300 trillion cubic feet of otherwise unattainable gas. This gas would add more than $8 billion in royalties alone to the United States Treasury and save consumers billions more. Nuclear stimulation could reverse the current trend wherein natural gas is becoming scarcer. Prices for natural gas are soaring because of the diminishing supply. Soon, it will no longer be an economically acceptable fuel unless drastic measures, nuclear or otherwise, are taken. Economics aside for the moment, natural gas is a relatively clean fuel which is in great demand in pollution-conscious cities. Nuclear stimulation of gas wells would forestall switching back to the dirtier fossil fuels.

Gas stimulation is relatively simple and straightforward. In addition, the prime areas are sparsely populated. Thus, it is not surprising to find this phase of Plowshare well-advanced. The first two industrial experiments carried out under Plowshare auspices were Gasbuggy and Rulison, both gas-stimulation experiments (Table 11). Gasbuggy was a joint experiment of the El Paso Natural Gas Company, the Department of the Interior, and the AEC. The principal objectives were: (1) to determine the effectiveness of nuclear stimulation; (2) to measure the radioactivity of the gas evolved; and (3) to record the seismic motion.

The shot site was located about 175 miles northwest of Albuquerque, near Farmington, New Mexico. The Gasbuggy nuclear charge, equivalent to 29,000 tons of TNT, was lowered into a drill hole to a depth of 4240 feet, just 40 feet below the bottom of the 287-foot-thick gas-bearing Pictured Cliffs Formation, a rock stratum named by some poetic geologist who viewed it where it had been exposed by erosion. Right after the detonation, which took place on December 10, 1967,

Table 11

Plowshare Gas Stimulation and Storage Projects

Code Name	Date	Size of Explosive	Location	Depth	Results
Gasbuggy	Dec. 10, 1967	29 kilotons	New Mexico	4240 ft.	Highly successful
Rulison	Sept. 10, 1969	30 kilotons	Colorado	8500 ft.	Highly successful
Dragon Tail	Withdrawn	20 kilotons	Colorado	2950 ft.	
Wagon Wheel	Design phase	Undetermined	Wyoming	9–15,000 ft.	
WASP[a]	Design phase	Undetermined	Wyoming	9–15,000 ft.	
Rio Blanco	Design phase	60–100 kilotons[b]	Colorado	5– 7000 ft.	
Ketch	Withdrawn	24 kilotons	Pennsylvania	5– 7000 ft.	

[a] Wyoming Atomic Stimulation Project (not to be confused with any sociological acronyms).
[b] Two 30–50 kiloton explosives.

drilling began toward the column of broken rock (the so-called "chimney"). On January 10, 1968, the drill broke through into the top of the chimney at 3907 feet. This inferred an apparent chimney height of at least 333 feet. Subsequent exploration showed the chimney to be roughly 160 feet in diameter, with fractures extending out to about 400 feet from the point of detonation.

The Gasbuggy explosion was strong enough to stimulate high gas-flow rates. By November 1969, nearly 300 million cubic feet of gas had been withdrawn from the chimney region. In contrast, a conventional well located only 400 feet from the shot point had yielded only 81 million cubic feet in the previous ten years. The level of radioactivity in the released gas has been decreasing rapidly as fresh gas flows into the chimney from the fractured rock. Soon, the gas will be safe enough for general use, although there are no plans for commercial use.

The second gas stimulation shot, Project Rulison, occurred on September 10, 1969, some twelve miles southwest of Rifle, Colorado.

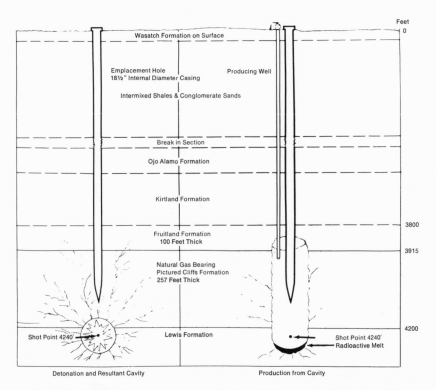

Underground effects from the 29-kiloton Project Gasbuggy explosion near Farmington, New Mexico, on January 10, 1968. (*El Paso Natural Gas Company*)

The Project Rulison 30-kiloton nuclear charge is shown being lowered into the shot hole near Rifle, Colorado, in 1969. (*U.S. Atomic Energy Commission*)

Working with the AEC and the Department of the Interior were the Austral Oil Company and CER Geonuclear Corporation. The purpose of Rulison was primarily to extend Gasbuggy experience to greater depths and different types of rocks.

Considerable opposition developed to Project Rulison during the summer of 1969 prior to the detonation. Some scientists and conservationists objected to the dangers from possible vented radioactivity and seismic effects. No radioactivity was released by the explosion, and except for the inevitable minor shock at detonation, no earth-quakes or large aftershocks were triggered. Pressure buildup at the well head has been very encouraging so far, and tests made on the natural gas burned ("flared") at the well head indicate that the ra-dioactivity content is very much less than predicted before the experi-ment. Production flow tests of Rulison gas were well underway in late 1970. As with Gasbuggy, the gas is not being distributed commer-cially.

A nuclear explosion could also provide a means of manufacturing a large reservoir for natural gas in rock far underground. Given a stratum of impermeable rock sufficiently thick, a nuclear-created chimney would be an extremely cheap way to store gas (extracted elsewhere) at high pressure. Another plus: there would be no offen-sive storage tanks above ground. The first such proposal, Project Ketch, involved the AEC, the Columbia Gas System, the Bureau of Mines, and Lawrence Radiation Laboratory. It was expected that the Ketch chimney, to be created in shale some 3300 feet down, would store about 465 million cubic feet of natural gas at 2100 pounds per square inch pressure. In July 1968, because of political pressures from concerned people, the Columbia Gas System withdrew its request to the State of Pennsylvania to use state forest land. The Columbia Gas System is now looking for another site in the Appalachian area.

Subterranean Oil Retorts. Three immense North American reservoirs of petroleum reside tantalizingly just beyond the reach of economical recovery methods:

1. Roughly 200 billion barrels remain behind in oil wells no longer economic to pump.
2. Possibly 450 billion barrels reside in the Athabasca oil sands of Alberta, where the viscous oil cannot be pumped out of the dense sand.
3. The greatest untapped oil reservoirs of all exist under the Uinta, Green River, and other geologic basins of Colorado, Utah, and Wyo-ming. Layers of oil shale up to 2000 feet thick contain an estimated 2 trillion barrels. The hydrocarbons are in the form of waxy kerogen, which cannot be pumped out by the usual methods.

To win the petroleum from depleted wells and from oil-shale deposits, two nuclear techniques have been proposed: (1) stimulation by blasting; and (2) fracturing followed by *in situ* retorting. No good method has been found for extracting oil from the tar sands.

The rate of oil flow from a conventional well is proportional to the permeability of the oil-bearing stratum and the difference in pressure between the oil below and the pump at the well head. Thus, stimulation of oil wells by hydraulic fracturing or nuclear explosives should work for the same reason it works in gas wells: increased permeability. The intense heat during the explosion would have a positive effect, too.

Oil shale is a silty carbonate rock or marlstone packed with waxy kerogen (up to twenty-five gallons/ton). When heated above about 700° F, the kerogen decomposes into gaseous and liquid hydrocarbons plus some free carbon. In retorting experiments aboveground, the combustion of some of the gases and the residual carbon has been sufficient to sustain high temperatures and kerogen decomposition, leading to the production of recoverable oil. Mining and subsequent retorting of the oil shale is thus technically feasible. It is much cheaper, however, to pump out oil from underground reservoirs as long as they hold out. Furthermore, one objects to pictures of Utah and Wyoming, states of great natural beauty, defaced by piles of retorted shale.

A recent indication of interest in Plowshare services from the petroleum industry led to Project Bronco. In 1967, the CER Geonuclear Corporation proposed, on behalf of a large group of oil companies, the *in situ* retorting of oil shale. This cheaper and more aesthetically pleasing method of extracting oil from the shale would consist of blasting a subsurface chimney full of fractured oil shale with a nuclear explosive, igniting the shale, and keeping the combustion process alive by pumping air down into the area. Retorted liquids and gases would be pumped up as they collect in the chimney. A fifty-kiloton explosive should blast about a million tons of shale into fragments suitable for retorting. Impermeable strata above and below the thick shale layer would ensure a tight effective natural retort. Above ground, there would be no torn-up landscape like that accompanying most mining endeavors.

As in the case of natural gas, dwindling oil resources should hasten the development of additional sources of supply. Nuclear methods again offer a possible solution.

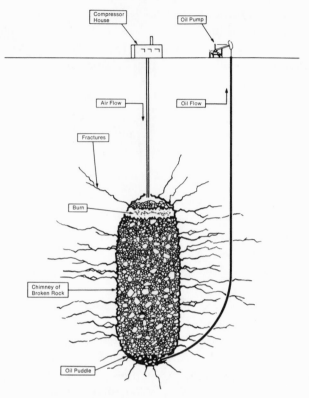

Oil can be recovered from oil shale *in situ* by burning (retorting) the shale in a chimney created by an underground nuclear explosion. (*CER Geonuclear Corporation*)

Mining Operations. The United States is fast using up its resources of such critical metals as copper, nickel, gold, silver, and titanium. Mines play out not because every last atom of their valuable minerals has been extracted but because the ore is so hard to get or of such low quality that further operations are uneconomic. The same situation that exists with oil sands and shales extends to other minerals, copper in particular. While one cannot "stimulate" a copper deposit with a nuclear explosion, nuclear fracturing of the rock followed by *in situ* retorting constitute a possible approach.

The Kennecott Copper Company, in cooperation with the AEC and the Bureau of Mines, is engaged in evaluating the concept of Project Sloop. The concept calls for exploding one or more nuclear charges equivalent to about 20 kilotons of TNT in the Arizona copper-bearing stratum. A chimney perhaps 440 feet high would be formed. After waiting about eight months for the radioactivity to decay to levels that would not interfere with chemical operations, holes would be drilled down to the chimney. Sulfuric acid pumped down one hole would leach out the contained copper in the form of copper sulfate. Brought to the surface from the chimney bottom by other pipes, the copper would be precipitated out and the acid replenished and re-cycled. A neat approach, *in situ* leaching would leave no piles of tailings or otherwise mar the landscape permanently. Project Sloop is still in the preliminary stages, but its general approach may find ap-plications in other localities and with other minerals where conven-tional mining does not pay off.

In block caving, the central idea is extensive fragmentation of the mineral-rich rock by explosives. Tunnels are then dug under the frac-tured region, and the rock is loaded and transported away as it falls into the tunnels. Nuclear explosives have proven themselves excellent rock crushers; in fact, they crush rock much more cheaply than any chemical explosives.

Large-Scale Nuclear Excavation

The first planetary engineering of any consequence was consummated in fiction; for example, the time Paul Bunyan and Babe, his Great Blue Ox, straightened out the roads of North America with a mighty tug. Babe's hoofprints can still be seen as lakes in the Northland. Great works of engineering represent a natural human dream; other-wise they would not appear so often in fiction. Nuclear explosives, of course, greatly magnify man's ability to build canals, artificial har-bors, and other large-scale construction projects. The great advan-tages of nuclear explosives over conventional earth-moving equip-ment are speed and economy. The dangers of radioactivity from the new, *nearly-fission-free* explosives are small, as mentioned in Chap-ter 3. Of course, each projected nuclear construction project is studied in great depth to identify and minimize any possible adverse fallout and seismic effects on man, his structures, or his natural surroundings.

Some Problems. A serious geopolitical issue that must be resolved be-fore any large-scale nuclear excavation projects can begin involves the restriction imposed by the Limited Test Ban Treaty of 1963. Of the nuclear weapons powers, the United States, the Soviet Union, and Great Britain are signatories to the treaty. Under the treaty, they have agreed that they will not set off any nuclear explosion, regardless of purpose, that causes radioactive *debris* to be present outside the country in which the explosion occurs. The methods of detecting radioactivity are so sensitive that a few atoms of radioactive material in a roomful of air can be identified. Even a "clean" nuclear explosive could release a harmless but measurable amount of radioactivity that might drift into another country. If we take into account these almost unbelievably sensitive detection methods, all nuclear excavations, except very small ones, might be ruled out. A more realistic approach to the interpretation of the treaty would permit nuclear excavation experiments with the proviso that the "radioactive *debris*," to use the words of the treaty, would be present in no more than *de minimis* (that is, very minute) quantities. However, even with this approach, large excavation projects might require an amendment to the treaty, permitting these projects to be undertaken when the level of radioactivity is well below that which would be harmful to living things.

The Non-Proliferation Treaty of 1969 contains a feature (Article V) under which the signatories (again the United States, the Soviet Union, and Great Britain) are obligated to furnish nuclear explosion services on a nondiscriminatory basis to the other signatory countries which have agreed to refrain from developing and building nuclear weapons. Thus, one international treaty inhibits the use of peaceful nuclear explosions and the other encourages them.

Another serious deterrent to the use of nuclear explosives for excavation is public concern that even small amounts of radioactivity released by the explosions can have adverse effects on the health of people living in the general area of the project. The development of even cleaner nuclear explosives and proper regulatory control should insure protection of the public. However, as in the case of the radioactive effluents released from nuclear power plants, there is much misunderstanding and fear.

Canal Construction. Even before the Panama Canal was opened for traffic in 1914, the need for a second waterway across the narrow waist of Central America became apparent. The present canal is too narrow for large, modern cargo ships and already operates at about

80 percent of capacity, handling more than 20,000 ships per year. By 1990, a new canal will be imperative.

The plans for a new canal have been plagued by many of the same kinds of difficult political and administrative problems that affected France and then the United States decades ago. The French began excavation in 1881 and the United States did not finish the canal until thirty years later. In between were many changes of plans and over 6000 deaths. These tragic consequences could be avoided in the construction of the next canal, when a series of a few hundred well-placed nuclear detonations could blast out a trench from coast to coast in a fraction of that time—and without loss of life. In this manner a wide, deep, sea-level canal, eliminating the need for locks, could be excavated.

The picture, however, is not quite so simple. Nuclear excavation might entail the temporary removal of several tens of thousands of people at the very least. Such operations, combined with fears of radioactivity and of induced earthquakes, could lead to considerable apprehension, regardless of the great saving to world commerce and the substantial benefits a canal would bring to the area.

The effect such a canal might have on marine biology is a major concern to ecologists. Professor John C. Briggs, Chairman of the Zool-

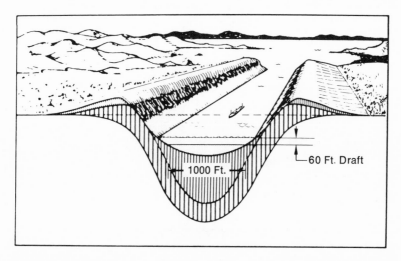

Sketch of a sea-level canal excavated by a row of nuclear charges. (*U.S. Atomic Energy Commission*)

ogy Department of the University of Florida, has labeled a sea-level canal a "potential biological catastrophe." Briggs feels that the free migration of Atlantic and Pacific marine species from one ocean to the other would radically upset the relatively stable ecologies that now exist on both sides of the Isthmus. The invasion of the Great Lakes by the sea lamprey through the St. Lawrence Seaway is cited as an example of biological havoc. Other scientists, such as Robert W. Topp, a marine biologist with the Florida State Board of Conservation, disagree with this view, claiming that biological changes will be minor, with faunal enrichment rather than deterioration distinctly possible. Here again, tempers are short and arguments long, and the great advantages of a new canal to the human species are often underestimated or forgotten in the heat of conflict.

Studies of a Plowshare-excavated canal have gone ahead. During 1959 and 1960, the data from the 1947 Isthmian Canal Study were reviewed. At that time five of the thirty likely canal routes examined in 1947 were identified as being feasible for nuclear excavation in the light of experience with nuclear explosives. The conclusion was that nuclear excavation appeared to be far cheaper than conventional techniques.

Canal excavation would consist of planting strings of nuclear row charges along the route. Detonated over a period of two or three years, each string would blast out a wide, deep channel that would require no further excavation by machinery. The charge sizes would be gauged by the amount of material to be removed. The greatest cost saving appears in this part of the project. For example, cutting through the 1000-foot-high Continental Divide along the Sasardi-Morti route in Panama would take years of time with thousands of laborers, even with modern equipment. Nuclear explosives would make long terrestrial incisions in less than a minute, slicing down 200 feet below sea level.

The primary concern in nuclear planetary engineering is the safety of the surrounding populace. AEC experience with many Plowshare tests indicates little risk, particularly with the newer explosives that derive most of their energy from fusion rather than uranium fission (Chapter 3). An additional safety factor along the Isthmus comes from the prevailing winds which would permit deposition of much of the airborne radioactivity—which is very small in any case—in a zone from which all inhabitants would be evacuated. The remainder would be swept out to sea rather than into populated areas.

The basis of the AEC's optimism regarding the effectiveness and safety of nuclear explosives derives from the many explosive tests and cratering experiments conducted at its Nevada test site. Some early shots, Danny Boy and Sedan, which put the first experimental foundations under crater scaling laws, are noted in Table 12. In 1968, three nuclear cratering experiments—Cabriolet, Buggy, and Schooner —were conducted. Cabriolet and Schooner were point shots utilizing single charges to check cratering laws, the release of radioactivity, and seismic effects. The Buggy experiment was the first detonation of a row of nuclear charges. Five 1.1 kiloton charges, spaced 150 feet apart and buried at 135 feet, simultaneously detonated, excavated a "ditch" 865 feet long and 70 feet deep in hard rock.

Evaluation of the above experiments and other advances in nuclear excavation technology were considered by the Atlantic-Pacific Interoceanic Canal Study Commission in its investigations during 1965– 1970. This commission was formed to make a complete study of the feasibility of excavating a sea-level canal between the Atlantic and Pacific, using the best means of construction, whether nuclear or conventional. Other considerations were national defense, foreign relations, intercoastal and interoceanic shipping, and total cost. Two of

Table 12
Major Plowshare Cratering Experiments[a]

Code Name	Date	Size of Explosive	Remarks
Danny Boy[b]	Mar. 5, 1962	0.43 kiloton	Low-yield shot; crater 214 ft. in diameter, 68 ft. deep
Sedan	July 6, 1962	100 kilotons	1200-ft. crater diameter, 320 ft. deep
Cabriolet	Jan. 26, 1968	2.5 kilotons	Point shot; crater 360 ft. in diameter, 120 ft. deep
Buggy	Mar. 12, 1968	Five 1.1 kiloton charges	First nuclear row-charge experiment; row 860 ft. long, 280 ft. wide, 68 ft. deep
Schooner	Dec. 8, 1968	35 kilotons	Point shot; crater 850 ft. in diameter, 240 ft. deep

[a] All shots fired at the AEC Nevada test site.
[b] A Department of Defense experiment with Plowshare participation.

188   Applying the Tools

the routes considered involved the use of conventional excavation
techniques only; four others would have employed nuclear or a combination of nuclear and conventional means.

The Canal Commission finished its work in December 1970. It concluded that, if a decision were to be made to construct a new sea-
level canal in the near future, nuclear excavation technology should
not be recommended because it had not advanced far enough. The
commission did comment, however, that a canal through Colombia,
excavated partially by nuclear methods, might someday be politically
acceptable, providing technical feasibility could be demonstrated be-
forehand. Therefore, the commission recommended that the United
States continue its development of nuclear excavation methods.

On the ecological front, the commission concluded that a sea-level
canal would not significantly harm the Atlantic and Pacific ecologies
if precautions were taken to restrict the flow of water through the
canal. But it also noted that long-term studies should be initiated
long before construction begins and that they should continue for
many years after the opening of the canal for proper evaluation of
ecological effects.

Proponents of nuclear excavation have also looked for other spots
where canals are needed but have not been built because of the high
costs of conventional excavation techniques. One possibility is a sea-
level canal through a narrow section of the Alaska Peninsula to avoid
the difficulties in circumnavigating the fogbound Aleutian chain. Another canal might cut across the Isthmus of Kra, on the Malay Peninsula, eliminating 1000 miles from the sea route between Japan and
India. The technical possibilities are many but so are the political
problems.

Instant Harbors. During the hydrogen bomb tests at Eniwetok in the
early 1950s, some of the explosions (Mike and Bravo) gouged nearly
circular holes more than a mile across out of the coral strand. There
are many seacoasts in the world where a hole this size would be very
welcome indeed. Good, natural harbors are rather rare. Even when
an indentation in the shore does exist, the anchorage area must often
be protected by breakwaters and seawalls. Large nuclear explosives
give us, for the first time, the capability to remedy nature's oversights.
Manifestly, nuclear-blasted harbors would not be acceptable on the
California coast, but several sparsely populated localities rich in nat-

ural resources do exist; they are ideally suited for nuclear excavation. Actually, these localities are sparsely populated today because good harbors do not exist. Historically, harbors are magnets for people and industry. Some promising candidates are:

—The Alaskan coast, on the Katalla River, for the purpose of tapping the huge coal deposits.
—Some of the Hawaiian Islands.
—The west coast of South America.
—The northwest coast of Australia, where large deposits of iron and other minerals are waiting for cheap transportation.

From 1957 to 1961, soon after the Plowshare Project was created, plans were laid to blast a harbor on a remote section of the Alaskan coast as a demonstration project. In this operation, called Project Chariot, four 20-kiloton and one 200-kiloton charges were to be detonated simultaneously. Eventually this project was dropped because the economic incentives were not great enough.

A recent harbor project studied by Plowshare engineers centered on Cape Keraudren, Australia. There is no suitable harbor on Australia's northwest coast near some of her greatest mineral deposits. In addition, this is a "typhoon coast," where unprotected harbors would be inadequate during storms.

One possible pattern of explosives suggested for the Australian harbor required the burial of five 200-kiloton charges 1100 feet apart along a line and 800 feet beneath the ocean floor. Detonated simultaneously, the explosions would have created an artificial harbor 6000 feet long, 1300 to 1600 feet wide, and 200 to 400 feet deep at the center. The crater lips, 200 to 300 feet above sea level, would have protected dock facilities and ships from typhoon seas and winds. As with the second Panama Canal, there were serious political obstacles to the implementation of this plan. The economic picture also changed. Ultimately, this excavation project, too, was dropped.

Soviet Accomplishments in Planetary Engineering

In September 1970, the Soviet Union announced that it had already harnessed nuclear explosions for practical, peaceful applications—not merely experiments. Although it had been known that the Soviet

Union was pursuing a program like the United States Plowshare effort, news of its actual use for engineering purposes came as a surprise. At least three different kinds of projects were carried out, apparently in the 1967–1970 period:

.1. The construction of a water reservoir in a dry riverbed for storing heavy spring runoff.

2. Stimulation of oil recovery from a geological formation that had been previously depleted.

3. The snuffing out of runaway oil and gas fires when they could not be controlled by ordinary means.

In the reservoir project, a single nuclear explosive with a yield of over 100 kilotons formed a crater over 1300 feet across and about 325 feet deep. Channels for diverting the flow of water into the crater were carved out by conventional earth-moving equipment.

The oil stimulation project involved the detonation of two 2.3 kiloton explosives in separate holes 4400 feet below the surface in the middle of the oil-bearing rock formation. Several weeks later, an 8 kiloton explosion was set off just outside the previously fractured region. Increases in oil flow at nearby wells ranged from 25 percent to 60 percent, stabilizing at these enhanced levels over the subsequent two- or three-year period that has elapsed since the last explosion. The site of the project was a little over a mile from a village of 2000 people and about ten miles from a town with a population of 24,000 and several five-story buildings.

On two occasions, raging fires that had ignited on the surfaces of oil and gas production fields were snuffed out by 30 kiloton nuclear explosions deep beneath the surface. The explosions sealed off the fissures which carried natural gas to the surface.

The Soviet Union is studying other applications. Experiments in gas stimulation and gas storage have been carried out. Several low-yield cratering tests have been made, including a row-charge test for future canal digging. A block-caving test for ore mining is planned. Probably the most unusual project under consideration would create a channel between the north-flowing Pechora River and the south-flowing Kama River, the latter a tributary of the Volga, which flows into the Caspian Sea. The objective is to increase flow into the

Caspian Sea, which has dropped nearly ten feet in the past thirty-five years.

The Soviet Union, with its huge expanses of sparsely populated terrain and enormous unexploited mineral resources, should make excellent use of peaceful nuclear explosives. It deserves credit for being the first to apply nuclear explosives to some of the world's practical problems in planetary engineering.

Climate and Weather Control

Weather changes come with each new cyclone and anticyclone that wheels across the continent. Climate changes, in contrast, are barely felt from one century to another. To assert that man cannot affect the weather or climate is incorrect; it is more proper to say that he cannot control weather and climate the way he wishes.

Man's long-term effects on climate are most evident in the Mediterranean Basin and the American prairie lands. Denudation of forests and heavily sodded grassland are commonly blamed for the presently arid lands around the Mediterranean and the Dust Bowl conditions in the United States of the 1930s. It is hard to prove such statements about climatic changes because of the long time scales involved.

A few calculations show that even with the atom we cannot manhandle the highs and lows on the weather map. The H-bomb is dwarfed by the smallest storm. A mature hurricane releases energy equal to more than a dozen Hiroshima-sized bombs *per minute*. We must find an Achilles' heel in natural weather processes to have any effect at all. Weather changers are forever buoyed by the hope that they can find pressure points where man-sized forces can manipulate continent-sized weather patterns.

The most effective tool of the weather changers is the silver iodide crystal. When these minute crystals are released from aircraft or wafted aloft from ground-based generators to high altitudes, they serve as condensation nuclei for raindrop formation. There is no longer any question that local precipitation patterns can be modified. The first United States cloud-seeding experiments were carried out in 1946 by Vincent Schaefer and Irving Langmuir, both with General Electric Company at the time. Originally, dry ice was used for seeding; silver iodide crystals were introduced a year later by another GE scientist, Bernard Vonnegut. Dry ice induces precipitation through supercooling while silver iodide crystals serve as artificial condensation nuclei.

Where does the atom come in? The base surges from big nuclear weapons tests at Bikini set off showers that lasted from twenty to thirty minutes. But clearly, this is not a practical way to make weather. Neither would it be conscionable to employ nuclear explosives after the fashion of the cannons and rockets that were part of the paraphernalia of the rainmakers of the last century. Impressed by the great power of a nuclear explosion, it was popular for a time to suggest stopping hurricanes just as the mariners of old attempted to break up waterspouts by cannon fire. The energy of the largest nuclear detonation is just too small to break up a storm several hundred miles in diameter. So far, meteorologists have discovered no "trigger mechanism" by which a bomb's relatively small contribution of energy could upset the stability of a hurricane's heat engine. And even with our cleanest Plowshare explosives, it would be foolish to search at random for a trigger. The reluctant conclusion is that none of our atomic tools seem to have much impact on local weather, but the search for "sensitive" spots in the weather machine goes on.

Climate control is a different matter. Several bold schemes that might use nuclear energy have been proposed. But before describing some of the more promising ones, it will be well to put them in perspective with the following quotation from a key report of the National Research Council:

> It can be stated categorically that there is, at present, no known way deliberately to induce predictable changes in the very large-scale features of climate or atmospheric general circulation. While man may attain the technological capability to induce perturbations sufficient to trigger massive atmospheric reactions, we cannot now predict with certainty all the important consequences of such acts. As long as our understanding is thus limited, *to embark on any vast experiment in the atmosphere would amount to gross irresponsibility.* (*Weather and Climate Modification,* NAS/NRC Publ. 1350, 1966.)

Thus, any climate-modification scheme would be both speculative and extremely risky.

Climate modification schemes involving nuclear energy fall into three classes:

1. The reversal of the supposed "greenhouse effect" as nuclear power plants replace fossil-fuel plants.
2. The addition of waste heat from nuclear power plants to rivers and lakes to keep them ice-free. The Soviets have even suggested opening up the Arctic Ocean in this way; climatic changes would surely follow this event.
3. Nuclear-aided engineering projects that would dam ocean straits, deflect ocean currents, and extensively modify watersheds.

Those in the last category are the most intriguing, and we shall concentrate upon them.

The great ocean currents that flow around and across most oceans carry immense quantities of heat or cold from one part of the Earth to another. The Gulf Stream carries the tropics as far north as North Carolina's Outer Banks; and the Labrador Current brings much cold fog to Newfoundland waters. These streams are really rather narrow rivers a score or so miles in width—almost within the realm of human engineering. In 1913, for example, plans were laid for building a jetty 200 miles long off Newfoundland to deflect the warmth of the Gulf Stream toward England. Engineers have occasionally contemplated blocking the Belle Isle Straits, which carry the icy waters of a branch of the Labrador Current southward toward New England. (The concrete causeway was also to have extended mainland rail service to Newfoundland.) Similarly, the U.S.S.R. has long desired to fill in the Tatar Straits, which carry the Arctic waters that chill Vladivostok to the point where the port is frozen shut several months each winter.

The most ambitious Soviet scheme—and the most controversial one —is the damming of the Bering Strait. This would be highly beneficial for Siberia because the cold Arctic waters bathing the eastern coast would be replaced by warmer Pacific water. Eastern Siberia might then be opened up to agriculture. But the cold Arctic water has to go somewhere, and Western scientists claimed that it would increase the southward flow of frigid water along the eastern edge of North America. While eastern Siberia basked in its new climate, Canada and the northeastern United States would be back in the Ice

Ages. It was proposed that in self-defense the North American countries would have to build a series of nuclear power plants to heat the
Arctic Ocean to compensate for the Soviet upsetting of the climatic
balance. Next, the Soviets suggested that the Bering Strait dam
would actually warm the Arctic Ocean and remove its ice cover permanently. By this time (1962), it became painfully obvious that accurate predictions could not be made, indicating that the admonishment
just quoted from the 1966 NAS/NRC report is well founded.

The role of nuclear energy, other than warming the Arctic Ocean,
might include dam building: for example, providing aggregate from
local rock formations or blasting cliffs that would slump and partially
fill in narrow straits. More specifically, nuclear blasts could help fill
in the Strait of Gibraltar, a feat which, according to its proponents,
would cause the Mediterranean to rise a bit and freshen to the point
where the Sahara could be irrigated. Of course, the advantages of a
verdant Sahara would have to be weighed against the loss of Venice
and other sea-level cities. We repeat these proposals primarily to stimulate thinking about both pros and cons of planetary engineering.

Nuclear explosives might help modify the Arctic ice pack which
makes the Arctic region "the refrigerator of the world." According to
this proposal, nuclear explosives would simply blast the ice pack,
greatly roughening its surface and increasing its absorption of solar
radiation. Thus warmed, the ice pack would begin to melt; the appearance of open patches of water would accelerate the process; and
the Arctic Ocean would be open to navigation. In principle, this plan
is equivalent to the much older suggestion that lampblack be spread
over the ice fields from aircraft to increase their absorption of solar
radiation. The Arctic might warm up if either of these two schemes
was successful, but the resultant increase in sea level would certainly
negate any advantages.

It has long been a fond hope of climate modifiers that big artificial
lakes would add enough moisture to the now arid air in America's
Southwest to dampen the climate a bit. James McDonald, a meteo-
rologist, calculated the size lake needed to increase Arizona's summer
rains by 10 percent. The lake, dubbed "Lake Fallacy" by McDonald,
would have to have an area of 20,000 square miles. There is no reason
to suppose that such a lake could be permanently maintained, once
built.

Taking all of these weather and climate control suggestions together, there is not one that appears promising. We have to conclude that despite the immense energy of the atom, we have neither the power nor the knowledge to manipulate weather or climate—yet. Most importantly, we simply do not know enough to predict accurately the consequences of our actions in this field.

Earthquake Control

Intuitively, earthquakes would seem less amenable to human control than the weather. The weather at least is *sometimes* predictable, and it is out in the open where we can apply forces to it. In contrast, earthquakes occur far underground and science has not yet made sufficient progress in forecasting when and where the next quakes will appear. Nevertheless, earthquake modification may be easier than weather or climate modification because of the different types of physical mechanisms involved. As the Earth's crust stretches, compresses, and twists under the influence of internal forces, stresses and strains build up until—suddenly—they are relieved as the crust adjusts itself. Earthquakes are generated during these crustal adjustments. The process can be likened to setting a mousetrap: energy is stored up in the slow displacement of the spring, then the trap is triggered, releasing the energy at once.

Underground nuclear detonations cause alarm because some feel that they might trigger these natural mousetraps. After many nuclear tests, thousands of tiny, highly localized aftershocks are recorded during the next few weeks. The shocks are almost always too small to be felt except by sensitive instruments. The important question here is usually taken to be: Will there be any danger from a planned underground nuclear detonation? That is, may not a large destructive earthquake be stimulated? Taking a more positive approach, perhaps the question should really be: Will there be any danger from natural seismic activity if the accumulated strains are not relieved by underground nuclear explosions?

Strains in the Earth's mantle seem to accumulate until the rock formations snap into new positions, often resulting in large natural earthquakes. The longer the strains build up without relief, the more severe the resultant earthquake—at least evidence points that way. Properly located and timed nuclear explosions might possibly "pull the teeth" of large quakes building along notorious fault zones, such as the San Andreas fault on the California coast. It would be a welcome day for the Atomic Energy Commission when, after public confrontations such as those over the Amchitka tests in Alaska, residents petition it to forestall earthquakes in their area by deliberately firing nuclear charges underground.

All this is not so wild as it appears. Seismology has come a long way since jiggles were first recorded on smoked paper by the first crude seismographs a few decades ago. Under Project Vela, a Department of Defense project with AEC involvement, very large seismograph arrays have been emplaced in the North Central states. In addition to understanding seismic activity better through the Vela Program's attempts to detect Soviet underground nuclear tests, fault zones have been instrumented with lasers, which measure minute relative displacements along the crustal cracks. As our knowledge improves, it is quite reasonable to expect seismologists to identify areas where crustal strains must be relieved to prevent major earthquakes in the future. This would be preventive medicine to be sure, but cracked plaster in the living room is better than a demolished house. A populated area could brace for a known shock, but might panic when a natural (and probably more severe) quake caught people in bed asleep. Despite the logic of such a prescription, human nature is such that some undefinable danger in the uncertain future is usually regarded as being preferable to today's cracked plaster.

Defending the Earth Against Cosmic Projectiles

Some day, the danger of a natural catastrophe may be certain, and humanity may be forced to take drastic action. The ultimate in planetary engineering would be saving the Earth itself from an intruder from outer space. Suppose an asteroid a mile in diameter (not an uncommon denizen in outer space) were to crash into one of our oceans. Equivalent to a 500,000-megaton bomb, it would blast out a crater fifteen miles in diameter on the sea floor and send 100-foot tidal waves racing toward the world's coastal cities. Such astronomical catastrophes have probably occurred before; Hudson Bay and the Carolina Bays are very likely meteor impact craters.

Paul Sandorff, an M.I.T. professor, posed the following problem to some of his students as a term project: suppose that a large asteroid had been detected on a collision course with the Earth. What could technology do to save our planet? The students suggested that a good offense would be the best defense. A good offense in this instance would be a series of Saturn-5-launched, 100-megaton bombs aimed to explode at one side of the on-rushing celestial projectile. Bombs in the 100-megaton class should either pulverize or at least deflect an asteroid one mile in diameter. In the first instance, the many small pieces would be a lot less dangerous to humanity than the original asteroid, because most of the debris would burn up during its fiery plunge through the Earth's atmosphere. In the second case, only a deflection of a fraction of a degree from its original course would be needed if the asteroid were intercepted a million miles from Earth.

The asteroid Icarus is about a mile in diameter. It comes within a few million miles of the earth every nineteen years—a near miss by astronomical reckoning. The last encounter with Icarus was in June 1968. Who knows what minute perturbation, applied as it swings 183 million miles away from the sun, might place it on a collision course for Earth in 1987?

A Few Concluding Thoughts

The contents of this chapter reveal our optimism about the use of nuclear explosives in the service of man. At the same time we recognize that the Plowshare Program (and its ramifications) is a controversial one—so much so that public reaction of an adverse sort may stifle its growth. Contrary to much published comment, society does not blindly accept new technology when clearly informed of its pros and cons. The problem is to present both sides fairly to the public— preferably along with hard evidence rather than innuendos and vague suppositions.

Budget stringencies have now slowed the pace of any permissible nuclear excavation experiments and also suggest that industry will have to make larger financial investments if satisfactory headway is to be made in atomic underground engineering. The Plowshare Program will suffer reverses, but in the long run we believe that its pros will far outweigh the cons and that nuclear explosives will indeed serve mankind in its peaceful pursuits. The need for natural resources—gas, oil, minerals—some already scarce, will weigh heavily in determining the future of Plowshare.

Chapter 7
New Worlds
Above and Below

New Dimensions

According to the mystique of space exploration, the future lies among the stars. Man must escape the planet he is slowly draining of sustenance, and even poisoning. Like all visions, this one can be neither proven nor denied with anything approaching scientific certainty. Our purpose in this chapter is mainly to show *how* the atom figures in our early gropings into interplanetary and interstellar space, but we shall not try to avoid a few semimystical *whys*. As a matter of fact, we see the human race expanding inwardly as well as outwardly. The oceans and the Earth's rocky mantle remain almost untouched. With the application of abundant nuclear energy even these frontiers will open.

Automated Precursors

The possibility that extraterrestrial intelligent life may be mechanical or inorganic rather than organic has not been widely discussed in the scientific and engineering literature but it is a possibility worthy of careful consideration.

Roger A. MacGowan, a proponent of intelligent machines, wrote this in 1962.° In fact, say many who wax enthusiastic about machines, the proper way to explore space is with automata—smart, self-controlled machines that will radio back data about the universe to scientists sitting comfortably in their laboratories. This hardly squares with the science fiction picture of intrepid astronauts battling the extraterrestrial elements and bug-eyed monsters; but it is certainly a possible concept for space exploration. One wonders if other foci of intelligence in the universe—if any—are not taking this approach.

Extraterrestrial machines are energy-hungry machines. Power is needed to communicate with Earth, for propulsion, and for simple survival in the cold space between the sparse planets. In just a decade, we have witnessed the evolution from America's Explorer 1, orbited in 1958 and consuming only a few watts, to OAO 2, the second Orbiting Astronomical Observatory, launched in 1968 and drawing nearly a kilowatt of electricity. Paralleling the increase in power level has been the increasing sophistication of spacecraft computers and greater automation. The parallelism is not accidental because power and sophistication are interdependent.

Solar cells and batteries power most of today's unmanned spacecraft. As long as spacecraft are modest in size and do not penetrate too close to the sun (where they get too hot) or too far past Mars (where sunlight has waned too much), solar cells suffice. Of course, solar cells cannot work on the dark side of a planet or under an opaque atmosphere like that of Venus; and their performance suffers in intense radiation fields, such as those encountered in the radiation belts around the Earth and Jupiter.

It is "lucky" we discovered nuclear energy when we did, for it is essential to the exploration of the outer planets and the nearby stellar systems. Nuclear space power plants work well almost anywhere; they permit us to escape the *cul-de-sac* created by solar cell limitations. On the Earth's surface, coal, oil, and wood fueled early technological development. In space, solar cells gave us a start. Everywhere, the atom will fuel the next stages of development.

Nuclear space power plants extract energy from either radioisotopes or fissionable fuels. The radioisotope power generators com-

° MacGowan, R. A.: "On the Possibilities of the Existence of Extraterrestrial Intelligence," *Advances in Space Science and Technology,* F. I. Ordway, ed. (New York: Academic Press, 1962) vol. 4.

monly convert heat to electricity with thermoelectric elements—thus their abbreviated name RTG, for radioisotope thermoelectric generator. RTGs rarely generate more than a few hundred watts and therefore they find application in the smaller automatic spacecraft. Fission power plants occupy the high end of the power spectrum, generating many kilowatts. They are needed for more ambitious missions.

The unmanned spacecraft of today weigh only a few hundred pounds and consume a few hundreds of watts at the most. But if we discard the constraints imposed by contemporary rockets, we see how the universe lies open to the probing of self-sufficient, semi-intelligent, reactor-powered automata. Launched toward the stars, they will return decades hence from other planetary systems with cargoes of data. If sufficient power is available, these spacecraft could also telemeter back some of their most important findings, but, at interstellar distances, communication would be extremely difficult. Only large nuclear power plants could sustain such long-lived, ambitious machines, particularly during the many years when they are far, far from any source of sunlight or starlight.

Returning to the space missions of today and the near future, the panorama compresses, but is still far beyond what most people would have dreamed possible in 1958. Here we find a critical role for the SNAP * RTGs fueled with the versatile radioisotope plutonium-238, which possesses a half-life of almost ninety years. RTGs first saw service in space in 1961, when SNAP-3A provided 2.7 watts for a Transit navigation satellite. The first space RTG was roughly spherical in shape, with an outside diameter of about five inches. Some later Transit satellites carried 24-watt SNAP-9A RTGs into orbit. The weather satellite, Nimbus 3, launched April 14, 1969, was partially powered by two SNAP-19B, 25-watt RTGs. Nimbus satellites are large weather satellites employed by NASA as test vehicles for new meteorological sensors and equipment. The first attempt to launch Nimbus 3 actually took place almost a year before the successful flight, but the launch rocket's guidance system failed, leading to the destruction of the Nimbus spacecraft and the "splashdown" of the RTGs off the California coast. The SNAP-19Bs were later recovered from the sea floor with their plutonium-238 fuel capsules still intact. Safety engineers had foreseen the possibility of such an accident and designed the fuel cap-

* SNAP = Systems for Nuclear Auxiliary Power, an AEC program.

Cutaway view of the SNAP-19 RTG. The plutonium-238 fuel capsule is located at the center. Heat from the decay of the plutonium flows outward through the thermoelectric elements where part of it is converted into electricity. The unconverted heat is radiated from the RTG body and fins. (*Teledyne Isotopes*)

sules for intact reentry. The Transit and Nimbus satellites can scarcely compare with the interstellar precursors we envisioned in the previous paragraph, although the Nimbus is certainly a highly sophisticated spacecraft—but they are vital steps in our expansion into space.

Another step outward into space will take place in the early 1970s, when spacecraft in NASA's Pioneer series of deep space probes will embark on the long interplanetary trajectories toward ponderous, oblate Jupiter, with its intense radiation belts and wandering Red Spot. Drawing electrical power from four 30-watt RTGs, the Pioneers will relay the first on-the-spot data ever recorded from the vicinity of Jupiter. We wonder today about the radiation belts encompassing this great planet built—apparently—of ices unknown to any terrestrial clime; but the real value of these far-flung expeditions will be in the new and unexpected things they discover about the universe. For example, almost no one conceived of Mars as a heavily cratered planet until Mariner 4 snapped close-up pictures in 1965. In 1975, NASA will carry the exploration of Mars a step further by launching the Viking spacecraft. Once in an orbit around Mars, a landing craft will detach itself from the main part of the spacecraft and descend to the Martian surface. The Viking lander will be powered by two RTGs using plutonium-238 fuel. Viking instruments will carry out meteorological, geological, and biological experiments. In the latter experiments, carbon-14 tracers will be employed to determine whether samples scooped up from the Martian surface harbor any life forms that metabolize carbon like Earth organisms.

Space automata need not be mobile. The instrument packages left behind on the moon's surface by the Apollo astronauts automatically

Artist's concept of the Viking lander resting on the surface of Mars. The two SNAP RTGs are located on top of the spacecraft body. (*Teledyne Isotopes*)

The seismograph installed by Neil Armstrong and Buzz Aldrin and left behind on the moon was kept warm during the long lunar nights by radioisotope heaters. (*NASA*)

record moonquakes, the "solar wind," and other lunar phenomena. Such experiments need power to radio the data they collect back to Earth, as well as heat to help the instruments survive the long, cold lunar nights. RTGs supply both heat and electricity whether the sun is out or not. The first lunar landing mission, Apollo 11, in July 1969, left EASEP behind. EASEP stands for Early Apollo Scientific Experiments Package. EASEP was not powered by an RTG, but two 15-watt (thermal), plutonium-238 heaters kept the package warm. Apollo-12 astronauts Charles Conrad, Jr., and Alan L. Bean set up the Apollo Lunar Surface Experiments Package (ALSEP) during their first foray onto the lunar surface in November 1969. ALSEP drew its power from SNAP-27, a 60-watt RTG fueled with plutonium-238,° which possesses a half-life of almost ninety years. This was the first of a series of RTGs taken to the moon to power scientific equipment.

° The Apollo-12 ALSEP telemetry indicated that its RTG actually generated 73 watts.

One of the Apollo-12 astronauts, Alan Bean, is shown deploying the SNAP-27 RTG on the surface of the moon to provide power for experiments and the transmission of data back to Earth. (*NASA*)

Men Follow Automata

The moon already seems familiar ground, but our lunar knowledge is still really quite limited and more advanced exploration will surely follow the Apollo missions sometime in the future, as will a wide variety of other, more sophisticated space missions. RTGs will not be able to meet the multi-kilowatt electricity demands of such potential missions. Fission reactor power plants seem a likely choice because solar cell power would cease during each two weeks of lunar darkness. In many other potential missions, requiring high power, solar arrays would be ineffective because of the need to operate at long distances from the sun or because of the very large size and great weight of the solar cells.

For a number of years, the AEC has been conducting a space reactor development program to meet the potential demands for electricity in the tens of kilowatts to hundreds of kilowatts power range. This technology is difficult and the lead times from drawing board to flight are long. The principal effort to date has been on the development of a thermal reactor operating on enriched uranium fuel with the slowing of the neutrons accomplished by collisions with hydrogen incorporated in zirconium hydride.

The first milestone was reached in 1965 when the first such "zirconium hydride reactor," a 500-watt experimental system, was flown. This reactor system, designated SNAP-10A, was launched from the Vandenberg Air Force Base, California, in April of that year. While in orbit, this system operated at full power for 43 days before a failure in the satellite's voltage regulator system—not the reactor system —caused a shutdown of the entire satellite. An exact copy of this orbital unit completed over one year of uninterrupted operation on the ground at the Santa Susana, California, test site. This, incidentally, is by far the longest uninterrupted operation of any nuclear reactor in the world to date.

The advanced zirconium hydride reactor systems are designed to produce from 10 to 100 kilowatts of electrical power using static power conversion systems at the lower end of the range and dynamic or rotating conversion systems at the higher end. The reactor core is a small cylinder about the size of a two-gallon gasoline can. A liquid metal called NaK (a mixture of sodium and potassium) is pumped through the core to remove the fission-generated heat. This reactor generates heat at about 1200° F, which drives turbine or thermoelectric conversion systems.

A manned orbiting space station is a good example of a potential mission requiring kilowatts of power. A listing of electrical power requirements for NASA's long-lived manned space station indicates that at least one to one and one-half kilowatts per man will be required in orbit. Initially such orbiting labs will house only a few astronauts for a period of two to three months at a time. For the later semipermanent, very large space stations or bases in operation, supporting perhaps up to 50 spacemen and spacewomen, the zirconium hydride reactor system could be utilized for this application.

Even 100 kilowatts of power will not be sufficient for some of the potential space missions of the 1980s and beyond. For example, deep space manned missions will require from hundreds of kilowatts to a megawatt of power because of the long duration of the missions and the wide variety of energy requirements for such trips, involving both propulsion as well as power. For these requirements the AEC is in the early stages of the development of a nuclear system which it refers to as a thermionic reactor. This system employs hundreds of nuclear fueled diodes grouped to form a reactor core. An individual thermionic diode uses a pellet of uranium fuel to heat tungsten (the element in ordinary light bulbs) to 3000° F, at which temperature it starts emitting copious electrons leading to the flow of electricity.

Following the trails blazed by the unmanned spacecraft, men will follow, briefly at first, and in bulky space suits. Later, perhaps larger spacecraft will bring construction materials for permanent bases (powered, of course, by fission reactors). New automata will be launched from these new bases as machine and man gradually expand outward toward new star systems, using our sun's planets as stepping stones. Each craft and each base will have to make its own environment. The more hostile the outside environment, the more power needed to hold it at bay.

While on the subject of extraterrestrial bases, nuclear excavation must not be bypassed. The pictures we have of the lunar surface and the testimonies of the astronauts tell us that some subtle form of erosion prevails on the moon. The solar wind, micrometeoroids, or some still-unknown force softens the sharp features of fresh craters over the eons. It may be that the bubblelike lunar bases so popular with space artists are not really feasible in the face of this erosive force. Furthermore, men working on the moon will also want to retire to radiation shelters when storms on the sun spew high-energy radiation like a water sprinkler throughout the solar system. An inexpensive lunar base can readily be blasted out of the lunar mantle well below the surface. Indeed, if it were not for the Earth's atmosphere, which is almost opaque to ultraviolet radiation, and the magnetosphere, which deflects most space radiation, we would have to live underground on the Earth, too.

Dandridge M. Cole, in his book *Beyond Tomorrow*, postulates that much of our extraterrestrial colonization will be done underground, where there is relative safety from the solar system "weather." Cole goes further, depositing colonies *inside* hollowed out asteroids, those mile-sized chunks of rock which exist by the thousands throughout the solar system. It would be a concave world inside rather than the convexity we are accustomed to. Before dismissing Cole's dreams out

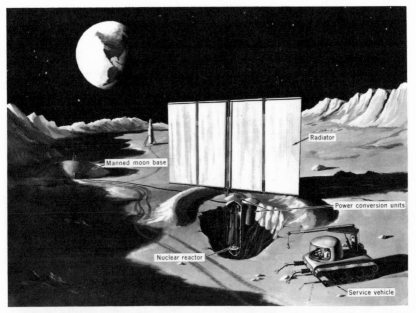

Artist's concept of a nuclear power plant installed in a lunar crater. The vertical panels are waste-heat radiators. A tracked service vehicle with teleoperator arms is shown at the lower right. (*U.S. Atomic Energy Commission*)

of hand, recall that Robert H. Goddard was labeled a crackpot by *The New York Times* in 1919 for even suggesting that an unmanned rocket could be sent to the moon, when, as everyone knew, rockets could not function in a vacuum. Just fifty years later, men walked on the moon.

Proxy Astronauts

Beachcombers say that the waves often come in threes. We terrestrials have already dispatched the first wave of automata out in all directions along the plane of the ecliptic. Men have followed as far as the moon. But, in keeping with the beachcomber's experience, there is—or could be—a wave of hybrid man-machine "creatures" cresting between the waves of true automata and astronauts. Ordinary spacecraft effectively carry man's eyes and ears plus such nonhuman faculties as radiation detectors to distant worlds. Astronauts, comprising the third wave, bring along their brains, their manual dexterity, and their abilities to generalize and solve complex problems that would baffle the largest computers. Men have intuition, too; they can cope with the unexpected. Too bad they are so fragile. If man were more rugged, he might have reached the moon five years earlier. The unique qualities of man and machine can be combined into "man-extension" systems, commonly called "teleoperators." The sensors on unmanned spacecraft substitute for human eyes and ears. The teleoperator goes one vital step further: it transmits man's dexterity across distance and, at the same time, adds the sense of touch or feel to the repertoire of the usual automaton. A man can sit at the terrestrial controls of a teleoperator and, in effect, be transported electronically to the moon (or any other planet) as far as his senses and hands and feet are concerned. A teleoperator extends human capabilities over great distances without putting men on rocketships.

If teleoperators could perfectly project man's senses and manipulatory capabilities across distance into all environments, manned space travel would be superfluous. Teleoperators are not perfect, however, and sending men to the moon is still more efficient than sending machines for such purposes as geological reconnaissance. Actually, teleoperators are in only a rudimentary state of development. Years of improvement await before we can hope to collect selected geological samples from the lunar surface by remote control with the ease the astronauts have displayed on the Apollo flights.

Much of the teleoperator technology that does exist came from AEC radioisotope research and production programs in which man had to manipulate chemical apparatus in the presence of intense radiation fields. The remote manipulators, particularly the electrical types developed at Argonne National Laboratory by Ray Goertz and his associates during the 1950s, will doubtlessly soon be applied to space exploration. One can visualize the great utility to science in having a remotely controlled arm with hand and fingers on the moon. Samples could be collected with ease and all manner of scientific experiments could be consummated. With a good TV link and a sense of feel conveyed to Earth by radio, a geologist could see, feel, weigh, and check the hardnesses of lunar minerals without leaving his terrestrial laboratory. He could not, however, make the taste test that is still favored by many older geologists—at least, not yet.

The distinguishing mark of a teleoperator is the presence of man in the control loop. In contrast, the most modern unmanned satellites and space probes respond to a few score commands from Earth. Except for throwing a few switches, man is out of the control loop most of the time. These spacecraft are therefore very close to being true automata, lacking only artificial intelligence and the complete decision-making function.*

The only machines resembling teleoperators launched into space so far have been the Surveyor surface samplers and, to a lesser degree, the sample retrieval equipment aboard the Soviet moon lander, Luna 16, which returned the sample to Earth. The Surveyors were soft lunar landers, five of which landed successfully on the moon. These spacecraft carried scoop-like devices mounted on the ends of articulated "arms" that could be controlled by radio signals from Earth. By sending coded commands, a terrestrial operator could turn on the motors driving the surface sampler joints for short periods of time. Progress was monitored by television. Touch feedback did not exist, and the surface sampler could hardly be called dexterous. Nevertheless, this simple machine dug trenches, picked up samples, tested rock hardness, and on one occasion dislodged the alpha-scattering experiment when that did not deploy properly. Surveyor experience demonstrated conclusively the great value of even a little dexterity. Similar sampling arms will be installed on the 1975 Viking Martian landers.

The Argonne National Laboratory's electric master-slave manipulators are by far the most sophisticated teleoperators built on Earth to date. With a head-controlled television circuit, an operator can perform fairly deft manipulations across great distances. The television picture moves left when he moves his head left; he feels the solidity of objects handled through touch feedback. One can imagine exploring the moon or Mars from the comfort of Earth with a teleoperator like this—it would be almost like being there in person. This sounds good, but the finite speed of light severely limits this scheme. The delay between the time an operator wiggles his finger on Earth and the time he sees (via TV) the machine on the moon wiggle its mechanical finger amounts to over three seconds. This is extremely disconcerting to an operator. Efficient remote manipulation on Mars will be out of the question unless computers are brought in to help man

---

* Manlike automata are termed "robots." Obviously machines need not be manlike to carry out human instructions effectively.

cope with long time delays between action and perceived reaction. Computers can help in two ways with the time-delay problem. First, they can create a "predictive display" for the operator, using time extrapolations based on the laws of physics and the operator's commands; in other words, they act like time machines, carrying the operator a few seconds or minutes into the future (via synthetic senses) so that he can see the probable consequences of his actions immediately. A second role for computers is in "supervisory control," in which routine tasks are directed by the computer without the need for human assistance. Supervisory control represents a stage one step beyond teleoperators in the direction of complete automata.

The time delay between a satellite in Earth orbit and an operator on the surface below is negligible. The first practical space application of versatile electric manipulators may be on repair and maintenance satellites. Alfred Interian, a GE engineer, described such a satellite in *Astronautics & Aeronautics* in May 1969. The lifetimes of many scientific, weather, and communication satellites could be greatly extended with occasional maintenance. Some of NASA's multimillion-dollar observatory-class satellites have relatively minor problems, such as the boom that did not deploy on the first Orbiting Geophysical Observatory. Replacement of minor electronic parts could revive many spacecraft now defunct for the lack of human-guided maintenance. A maneuverable teleoperator satellite could approach an ailing satellite, adjust its orbit to correspond to the target, and then lock onto it (if it is not spinning too fast) and commence repair and/or maintenance.

The time delay problem would not exist for astronauts exploring the surfaces of hostile planets, such as Venus and Jupiter, from orbital craft. Manipulator-carrying unmanned probes could be dispatched to the surface from orbit to carry out reconnaissance without endangering the astronauts. It would be particularly rewarding to send a probe down through the thick (twenty atmospheres pressure), hot (800° F) atmosphere of Venus to see what lies on the surface of this mysterious planet. As for Jupiter, astronauts may never be able to penetrate safely the deadly radiation belts that ring this giant planet. But an unmanned probe could transport experiments down into the famed Red Spot itself, carry out chemical manipulations, and solve some of this planet's perplexities.

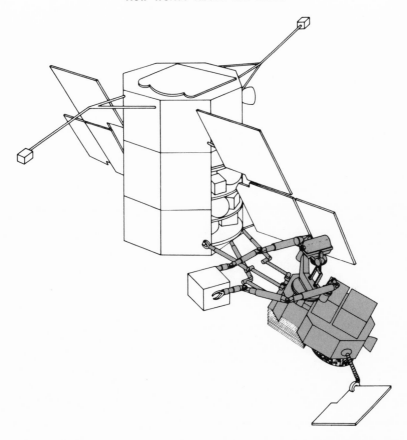

Conceptual drawing of a teleoperator-carrying repair-and-maintenance satellite. The large satellite being serviced is an Orbiting Astronomical Observatory (OAO). (*General Electric Company*)

Interplanetary Shuttle

Hart Crane, in his Whitmanesque, prophetic poem *Cape Hatteras,* wrote these two lines:

> The soul, by naphtha fledged into new reaches,
> Already knows the closer clasp of Mars,—

Crane wrote this three decades before a kerosene-propelled space probe, Mariner 4, intercepted Mars in its orbit around the sun. Because naphtha is a petroleum derivative like kerosene, we can credit Crane with great foresight. We have placed men on the moon through the use of petroleum products and will doubtlessly do much more in space with this fossil fuel. But here again, will not the atom, with its seemingly limitless reservoir of concentrated energy, allow us to expand farther and more quickly into the unexplored light years of space around us? The answer is "yes," but for an unexpected reason.

Part of the value of nuclear power in rocket propulsion lies in the high concentration of energy in nuclear fuel and in the high temperatures of nuclear reactions. A more important factor in rocket propulsion is the amount of thrust obtained from the unrecoverable stream of propellant that roars through the nozzle. The rocket *specific impulse,* measured in *pounds of thrust* for each *pound per second* of propellant flow, is a key criterion of rocket performance. It is a measure of an engine's ability to create thrust while being economical in the use of propellant. The nuclear rocket has a specific impulse nearly twice that of the best hydrogen- or kerosene-burning rockets because of a single not-so-obvious fact: it does not burn anything. The nuclear rocket can use *any* gas that increases specific impulse as its propellant. Low-molecular-weight propellants are best, so nuclear rockets use hydrogen, while chemically fueled rockets are always burdened with the heavy oxidizer atoms needed for combustion, notably oxygen or fluorine. As a result, the nuclear rocket generates about 800 or more pounds of thrust for each pound per second of hydrogen that is heated and expelled from its nozzle. This compares to about 450 for the hydrogen-oxygen chemical rocket. Consequently, for many space missions, payloads are roughly doubled. The more ambitious the mission, the greater the superiority of the nuclear rocket. Here, then, is a step-increase in our capability to explore the solar system.

The promise of fission-heated hydrogen rockets was recognized long before the Soviet Union's Sputnik signaled the beginning of the space age in October of 1957. Chauncey Starr at North American Aviation pioneered the concept in the late 1940s. American projects, it seems, must have code names; and the first important nuclear rocket program was formed and christened Project Rover in 1955. Because of the extremely high hydrogen temperatures required—about 5000° F—Los Alamos scientists concentrated much of their effort on reactors built with graphite fuel elements. Graphite is one of the few materials that increases in strength as temperature rises. Hydrogen gas is pumped past the hot graphite fuel elements and expands through the rocket nozzle to create forward thrust. The nuclear rocket engines being developed today are still based on graphite as the basic structural material.

Initially, nuclear rockets were conceived as direct replacements for the familiar chemical rockets, such as the Saturn moon rockets. "More weight into orbit and into Earth-escape trajectories for the same-sized launch vehicle" was the justification in the late 1950s. However, the hazards of a 10,000-megawatt reactor ascending through the atmosphere with a cargo of newly created and very dangerous radioisotopes vetoed such plans. Chances were against it, but the rocket might go astray and crash. Safety considerations insist that nuclear rocket engines be started in near orbit, where there is negligible danger of their impacting on some populated area. Nuclear rockets, in today's thinking, are strictly upper-stage rockets. Consequently, they are smaller, generating tens of thousands of pounds of thrust rather than the millions of pounds characteristic of the Saturn 5 and other big launch vehicles.

The acronym NERVA stands for Nuclear Engine for Rocket Vehicle Application. Standing some 30 feet high, including its bell-shaped space nozzle, NERVA generates about 75,000 pounds of thrust at a

An experimental nuclear rocket engine being transported by the railroad-mounted Engine Installation Vehicle at the Nuclear Rocket Development Station in Nevada. (*NASA* and *U.S. Atomic Energy Commission*)

specific impulse of 825. Amazingly, the reactor which heats the flowing hydrogen operates at an extremely high power level of 1500 megawatts. The development and useful application of NERVA has been
the principal goal of the joint AEC-NASA Space Nuclear Propulsion
Office since the early 1960s. By funneling resources into an engine of
a specific size, NASA and the AEC have been able to develop hardware that could be ready for space use during the late 1970s.

The technological accomplishments of the NERVA program can be
appreciated better when it is pointed out that this small engine—
about the size of three Volkswagens placed end to end—generates
about as much power as 15,000 "Bugs," more than Hoover Dam. High
temperature capability is another important NERVA achievement.
From a cold start at −420° F, NERVA gas temperatures reach 4500° F
in a few seconds. This temperature is roughly three times those attained in the coolants of modern commercial nuclear power plants.

A thrust of 75,000 pounds seems miniscule in comparison with Saturn 5's liftoff thrust of 7.5 million pounds; but a 75,000-pound-thrust
engine in orbit can accomplish a great deal once the gravitational ties
to Earth are weakened. Of course, a 75,000-pound chemical engine
can do a lot in orbit, too, but only at half of NERVA's specific impulse. The chemical engine is not as "cost effective." In this economic
sense, NERVA has a great advantage. The emphasis in space today is
no longer merely on getting payload into space; rather, it is the economical delivery of payload from Earth to orbit, from orbit to the
moon and back again, and from Earth orbit to other planets. In this
shift of emphasis, we see the whole character of space transportation
changing from the experimental, fingers-crossed launches of the late
1950s to routine, scheduled flights. It is tempting to compare this
transition period with the analogous period in aviation that saw the
heady barnstorming days followed by the first commercial passenger
service. The first commercial aircraft flights never flew at night and
even then took some courage on the part of the passengers. Everything has its beginning, and perhaps this century will yet see excur-
sion trips to the moon.

The nuclear rocket engine, once launched by large chemical rockets, would never return to Earth. Instead, it would shuttle between
different Earth orbits carrying supplies to various space stations. It
would perform the same service between Earth orbit and lunar orbit,
carrying passengers and freight in both directions. Whereas a pound

of payload transported from the Earth's surface to lunar orbit via the Saturn-5/Apollo chemical rocket costs about $6000, the substitution of a reusable nuclear space shuttle would bring the cost down well below $1000 per pound. In other words, a reusable nuclear engine is a cheaper way to move goods and passengers from Earth orbit to lunar orbit. In 1990, a scientist taking a tour of duty at the 500-inch lunar telescope might take the 2:00 P.M. nuclear shuttle from a 270-mile terrestrial orbit to a similar lunar orbit and then down to the lunar surface via chemical rocket.

Furthermore, the solar system with its eight unexplored planets and two-dozen-plus assorted moons still lies before us. With its higher payloads, the nuclear rocket can magnify considerably the results of each foray into the farther reaches of the solar system. Nuclear-propelled automata can, for example, be more than twice the size of those launched by all-chemical rockets, as the following table indicates:

Mission	Chemical plus nuclear	Four-stage, chemical only
Probe to Jupiter	84,000 lbs pay-load	32,000 lbs payload
"Tour" of Jupiter, Mars, and Venus (1978 launch)	52,000 lbs	30,000 lbs
"Far-in" solar probe (to 0.2 Astronomical Units; i.e., 80% of the way to the sun)	40,000 lbs	23,000 lbs

The chemical rocket considered in the table is the Saturn 5, while the middle column shows the payload gained by replacing the S-4B third stage of the Saturn 5 with NERVA. When one realizes that pictures of the Martian surface were transmitted back to Earth (along with other scientific data) by automata weighing only a few hundred pounds, the scientific opportunities available with an 84,000-pound Jupiter probe are enticing. For example, a sophisticated teleoperator could be included to help perform chemical (perhaps biological) analyses, set up experiments, and make repairs that would greatly extend the lifetime of the machine.

The transition between unmanned and manned missions is always marked by the requirement for a large increase in deliverable payload. To illustrate for the case of the moon: the unmanned Surveyor landers weighed only about 620 pounds, while the Apollo-11 astronauts were sustained by the 33,205-pound Lunar Module. Of course, manned missions must also return to Earth; this requirement makes manned missions to the planets with the payloads listed in the table above impossible. However, manned missions to Mars are possible with NERVA technology through the stratagem of assembling larger spacecraft in Earth orbit and injecting them into a Mars trajectory with several NERVA engines strapped together. The first stage of the Saturn 5 rocket really consists of five 1.5 million-pound-thrust rockets operating in parallel, like the team of horses pulling on an old western stagecoach. The ability to gang small engines together and operate them like big ones not only saves money by eliminating the development of a new, much larger engine, but it also improves spacecraft reliability. When one engine on a four-jet aircraft fails, the plane can always land safely; the "engine out" safety factor also applies to multi-engine rockets.

With these facts in mind, the AEC and NASA have put development emphasis on nuclear-powered modules, each with a single NERVA engine. Modules can be strapped together in various combinations for different interplanetary trips. A manned trip to Mars seems feasible by the 1990s, using four nuclear modules strapped together for launch from Earth orbit. This group of four would be detached along the way and returned to Earth orbit for reuse. Braking the spacecraft into Martian orbit and furnishing power for the return flight home would fall to a fifth nuclear module. Descent to the Martian surface and ascent back to the nuclear stage in orbit would be via chemical rockets. A total of 450 days would be required for the manned mission to Mars and return to Earth.

In these days of concern over national priorities and the need for a large-scale manned space program, it is not surprising that the enthusiasm of space and nuclear technologists for the NERVA concept has been dampened somewhat by budgetary pressures. The NERVA engine, by itself, has no application; it must be combined with space vehicles, astronauts, guidance systems, and all the rest of the paraphernalia needed to leave Mother Earth and accomplish a mission in deep space. Only time will tell when or if the NERVA dream can be turned into reality.

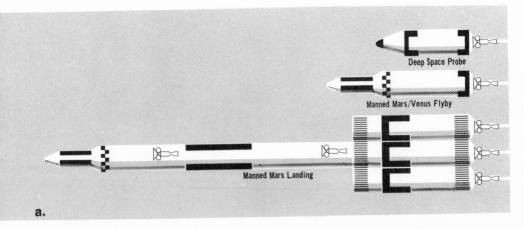

Spacecraft configurations for various missions utilizing NERVA nuclear rocket engines. The ambitious manned landing on Mars would require the clustering and staging of several NERVA engines.

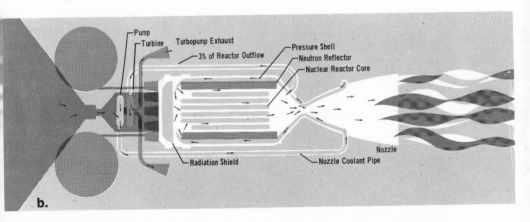

Sectional diagram of the NERVA engine. Hydrogen gas is pumped through the reactor where it is heated close to 4500°F. When the hot gas expands through the nozzle, a forward thrust is generated. (*U.S. Atomic Energy Commission*)

Some Advanced Propulsion Concepts

Nuclear rockets are considered advanced these days, but a century hence our descendants will probably marvel about how we embarked on lengthy space voyages with such crude machinery. We wonder as much about the Vikings and Columbus in their frail craft. There are certainly better ways to span the solar system and probe nearby stellar systems. Once more, abundant energy is the vital ingredient. We can raise the specific impulse of space engines by a factor of ten, even one hundred, if we have enough energy at our disposal. The pessimistic prophecies of multiyear trips to the other planets are tales born of insufficient energy. No law of physics would be broken if a spaceship departed from Earth orbit for Mars at a steady acceleration of one G, a velocity increase of thirty-two feet per second each second, equivalent to gravitational acceleration at the Earth's surface. After just a single day, over 23 million miles would separate the spacecraft from Earth. By the end of the second day, the spacecraft would be almost 100 million miles from Earth. To an energy-rich civilization, the planets would be in the suburbs.

Nuclear rockets of the NERVA type cannot accomplish this; they cannot be made larger and larger, and neither can their performance (i.e., specific impulse) be increased very much. The specific impulse of NERVA-class nuclear engines is limited to 800–1000 by the integrity of the materials of construction at very high temperatures. Nuclear rockets that heat hydrogen through the use of solid fuel elements are limited in physical size and specific impulse. The conventional nuclear reactor just cannot pump enough power into the propellant stream.

Three unconventional approaches lead out of the blind alley:

1. Don't use solid fuel elements.
2. Don't use nuclear reactors for propulsion.
3. Don't use thermal energy to accelerate the propellant.

Solid nuclear fuel elements can be replaced with gaseous fuel to circumvent fuel-element destruction at high temperatures. In a gaseous-core reactor, the hydrogen can by heated by the fissioning uranium atoms to extremely high temperatures—perhaps 10,000° F—far beyond the temperatures achievable by fuel elements because there are no fuel elements. Thus heated, the hydrogen expands through the rocket nozzle to produce thrust. Solid structures in the engine can probably be protected or cooled adequately by cool gas streams. According to studies made over the last decade, specific impulses of about 2000 may be possible with gaseous-core reactors. Some uranium escapes, but perhaps not so much that the engine will be too costly to operate. This concept has not progressed beyond studies and a few experiments dealing with nuclear criticality and gas flow.

By far the boldest concept advanced was embodied in Project Orion, a fitting code name for a rocket ascending under the influence of a succession of hundreds of small nuclear explosions. The Orion concept apparently originated at Los Alamos Scientific Laboratory toward the end of World War II. The idea goes like this: by exploding small bombs (just a kiloton or so) under a disk-shaped spacecraft, perhaps a couple of hundred feet in diameter, the spacecraft will be driven onward by each successive explosion. (Remember the effect of a firecracker under a tin can?) By gas bags and other shock absorbers, the payload atop the disk would be mostly cushioned from the shock waves applied to the bottom "pusher plate." Much of the work on the Orion concept was done at General Dynamics Corporation under a United States Air Force contract. In fact, General Dynamics engineers even launched small simulated spacecraft to low altitudes with a progression of small chemical-explosive charges. The signing of the Nuclear Test Ban Treaty brought an end to Project Orion because more realistic tests were rendered impossible. On paper, at least, hundred-ton payloads could be launched into orbit more cheaply than with chemical rockets.

The third advanced rocket concept is, in principle, not dependent upon nuclear energy at all—the source of energy is irrelevant to the engine; it consumes electricity. Nuclear-electric power plants, however, are generally superior where large quantities of power are needed in space. *Electric rockets* either accelerate ions with electrostatic fields or they accelerate a plasma (a neutral mixture of ions and electrons) with electromagnetic fields. A bewildering variety of electric rockets exists. Several types have already been tested on satellites as part of the SERT (Space Electric Rocket Test) Program. Terrestrial atom smashers demonstrate that ions can be accelerated to near the velocity of light, which infers a possible specific impulse of about 30 million. A rocket with this specific impulse could fly to Mars with scarcely any propellant expenditure, but the electrical power plant it would have to carry along would be colossal. Consequently, most electric rocket designs are compromises: reasonable power plant sizes seem obtainable with specific impulses in the 10,000 range. (Hydrogen would have to be heated to about 700,000° F in a reactor to achieve the velocities attainable with a few thousand volts of electricity.)

As the Space Age dawned, space plans almost as grandiose as the

Jupiter voyage in the film, *2001, A Space Odyssey*, were popular. All of these leviathans of space were imagined as forging paths to the planets with electric rockets and nuclear power. The hopes were premature. Although electric rockets were developed and tested in space with surprising speed, the large nuclear power plants needed to support them have lagged. Sluggish development funds and difficult technical problems are mainly to blame. A megawatt space electric power plant would probably cost a billion dollars to develop and test; national priorities relegate such efforts to studies and technology development rather than full-scale hardware programs. The thermionic reactor mentioned earlier is considered to be the long-range candidate for space missions requiring up to a megawatt in electric power. Many years of work are required before the thermionic reactor will be ready for lift-off.

Planetary Engineering: Phase II

In the preceding chapters we have sketched how the atom can help improve life on a planet stripped of many of its primordial raw materials and, in some places, polluted to the point where life cannot be sustained any longer. On the moon, the situation is also bleak, but for different reasons. The absence of air, water, and readily accessible minerals seems only a little more forbidding than the environmental nadir toward which we seem headed here on Earth. On the moon, at least, we have the opportunity to begin afresh and, we hope, more wisely.

Considering the vast effort it took us to put man on the moon, the effort to inaugurate lunar mining and manufacturing projects seems forbidding indeed. However, it would be foolish to write off the moon because of its present inaccessibility. America was inaccessible to Europe once, too. In trying to solve today's terrestrial problems, we should not turn our eyes inward so much that we fail to see the future prospects for interplanetary commerce.

In his book *The Case for Going to the Moon*, Neil Ruzic devotes two substantial chapters to lunar mining and manufacturing. To get in the proper frame of mind for such ambitious extraterrestrial undertakings, consider the facts that the lunar crust is not too far different from the Earth's and that lunar thermal activity will likewise have segregated many minerals into easily mined deposits. Furthermore, the moon's airless environment will be ideal for vacuum metallurgy. In the cold lunar shadows, cryogenic factories, perhaps using super-conducting magnets, will be technically attractive. The raw materials are there, and so are unusual manufacturing environments. The missing ingredient is energy. We obviously have in mind a lunar Nuplex.

A terrestrial Nuplex is built upon the astute melding of raw materials, people, and energy. Plenty of water and air are available to begin with, although they may be salty or polluted. The moon has plenty of oxygen and water locked in the lunar rocks. For example, terrestrial volcanic rocks, and presumably their lunar counterparts, hold about 1 percent water by weight. Nuclear heat and electricity can free these critical materials for the use of man and the lunar Nuplex. Lunar prospectors should also find iron and nickel in good supply around volcanic areas. Cobalt, chromium, copper, phosphorus, and many other useful minerals also tend to congregate around thermally active areas. The primitive lunar base would, of course, be sustained at first by materials brought from the Earth. But given a nuclear reactor and indigenous lunar materials, the lunar base can pull itself up by its bootstraps, so to speak, and graduate into a lunar Nuplex. It would be a hermetically sealed Nuplex, like those science-fiction bubble cities.

As the lunar Nuplex approaches self-sufficiency, manufacturing for export to Earth will become a possibility. Again, we have to think in the framework of the future, not in terms of the costs of the first flights to the moon. With the development of nuclear-powered Earth-to-moon shuttles, transportation costs back to Earth will probably drop to $100 per pound or thereabouts. Some products of terrestrial manufacture cost far more than that on a weight basis. Extrapolating trends in microelectronics, a pound of circuitry a decade hence may well be worth a hundred or a thousand times the transportation costs back to Earth. By virtue of the moon's vacuum, its cleanliness, and the ready cryogenic environment, it may be cheaper to manufacture some electronic equipment on the moon and then ship it back to Earth. Other potential products include electro-optical devices and pharmaceuticals. Information may also be a valuable lunar product—a product requiring only radio waves for transportation. The lunar vacuum and low temperature make the moon an ideal environment for huge computer complexes operating at cryogenic temperatures and making ample use of superconductivity. Terrestrial computers are already taxed by problems in weather forecasting, urban development, and economics. It would be ironical if some of the Earth's major problems were solved on the moon—in the mathematical sense, that is.

The *Lenin,* the *Savannah,* the *Otto Hahn,* the *Mutsu,* and the *Enrico Fermi*

With this multinational assortment of names, we make the transition from outer space to the depths of the oceans, pausing for a moment at the interface to observe that nuclear power has already transported goods and people on the surfaces of the Earth's watery expanses. The five names are, of course, the names of seagoing atomic vessels built, under construction, or in design. The icebreaker *Lenin* holds the honor of being the first nonmilitary ship to be propelled by the atom. She steamed to her Arctic station in 1959. The American *Savannah* entered commercial service in 1962 and had logged over 450,000 miles by November 1970, still operating on her first reactor core. Originally a combined passenger and cargo carrier, the *Savannah* was later converted to cargo only. Germany's *Otto Hahn* is an ore carrier and in 1970 was in routine commercial service between Germany and Morocco. The Japanese *Mutsu,* now under construction, is a cargo carrier scheduled to go into operation in 1972. The *Enrico Fermi* is in the design stage. It will be built to commercial specifications but will be used to supply the Italian navy.

The *Savannah* received considerable popular attention as she steamed to ports all over the world, first as a goodwill ship and later as part of the United States Merchant Marine. Her sleek lines have graced many a photograph. However, it was never intended that the *Savannah* compete economically with the much larger freighters that ply the seas. She has now been retired from service after opening many ports to nuclear merchant ships and successfully accomplishing her prime mission of demonstrating that nuclear-powered ships are feasible and safe. In time, economically competitive nuclear merchant ships will be built. The monster freighters now being built can employ much larger, more economical reactors than those on the *Savannah*. This, coupled with the rising costs of conventional fuel, will one day make maritime nuclear power more common.

The Atom in Inner Space

Mankind is essentially a two-dimensional society, multiplying prodigiously wherever nature has chanced to mix the proper ingredients for human survival—but so far only on dry land. Humans hardly burrow at all for new habitation; neither do they colonize the three-dimensional sea, which comprises seven-tenths of the surface area of this planet. The sea and the planet's rocky mantle beneath us are certainly more accessible than the moon, planets, and asteroids above. In many ways, the same technologies that have propelled us into space suffice for exploitation of the sea and Earth. Recapitulating the preceding pages on space exploration, the essential ingredients for stepping beyond our cozy natural environment are:

1. Abundant power that is not dependent upon the sun or huge quantities of low-energy-density fossil fuels.
2. Closed-cycle ecologies, transportable and capable of long-term sustenance of man in otherwise hostile environments.
3. Automata to chart the routes for manned vehicles.
4. Teleoperators to aid man in penetrating hostile environments.

Until we can change man himself (see next chapter), expeditions into space and under the sea must carry along microenvironments that fend off the elements. It is strange that in the rush to explore space, the sea and the solid earth beneath our feet have been all but ignored. Does science believe that anything interesting exists "down there"? The answer is, "Yes, but scientific curiosity has not been sufficient to justify the huge costs of such exploration." A beginning has been made, though, because of military interest in this part of the Earth's integument. This interest may lead to carving out habitable niches in what we ordinarily think of as a hostile environment.

The Undersea Frontier

Men penetrated the undersea frontier before unmanned probes and automata, thus reversing the pattern one might expect from the preceding pages. The reason, of course, was the incomparable military advantages of long range and long submersion time gained by nuclear submarines. Now, however, scientists are beginning to use machines to a larger degree as they carry their studies to greater and

greater depths. New research submarines, many of which carry teleoperators, are now rectifying this oversight. Unmanned probes now
relay back information from the deepest submarine trenches. Yet, our
exploration is superficial and our exploitation of the sea, aside from
fisheries, is almost nil. We venture timorously out onto the continental shelves and into the top thousand feet of seawater. Undersea technology seems more primitive than space technology.

Commercial exploitation of the sea will probably always pace scientific exploration. Offshore oil production began in the United States
in 1938 when pumping started at the Creole field off Louisiana.
Offshore oil is worth many billions. But the sea is also bountiful in
other ways. The first of its famous manganese nodules were dredged
up by the British oceanographic research vessel *Challenger* almost a
century ago. Few people know that millions of dollars' worth of dia-
monds are dredged from the sea bottom each year off South Africa.
Tin and iron ores are also mined beneath the sea. The ocean waters
themselves contain various amounts of magnesium, iron, gold, and
many other elements. Perhaps Jules Verne was right when he had
Captain Nemo of the *Nautilus* extract everything his ship needed
from the sea. The sea is truly bountiful, but it is also harsh and unfor-
giving.

Scientists have been probing the sea and its floor with nets,
dredges, and coring devices for many decades. The famed *Challenger*
expedition of 1873–1876, led by Sir Wyville Thomson, is generally
considered the first systematic oceanographic survey. Fifty volumes of
new knowledge evolved from this expedition. In the years that followed, additional surveys were made by various countries, and
oceanography became a solid science. However, researchers were always "fishing in the dark." All information came from samples recovered and from sounding equipment. Not until after World War II
did unmanned automata really begin to be used below the surface.
And only when missile-carrying submarines posed serious threats did
research-oriented submersibles make their appearance. Funds finally
became available for the development of automata and manned vehicles for the *in situ* exploration of the sea. The era of blind bottom
groping from the sea's surface had ended.

The first big technological step toward *in situ* exploration of the
depths came in 1948, when Auguste Piccard's bathyscaphe *FNRS-2*
made a manned descent off Cape Verde. By 1962 the bathyscaphe,

which is the underwater equivalent of a blimp, had taken men down to 31,300 feet in the deep-sea trenches off Japan. It was Jacques Piccard, Auguste Piccard's son, with Lt. Don Walsh, U.S.N., who reached 35,800 feet in the Marianas Trench in 1960 in the bathyscaphe *Trieste*. The *Trieste*, incidentally, was the submersible that retrieved parts of the nuclear submarine *Thresher* in 1963. The bathyscaphe, for all its vertical mobility, cannot stay down for more than a few hours and possesses but limited horizontal mobility. It is a manned vertical hydrospheric probe.

The second quantum jump in our undersea capability came in January 1955, when the nuclear-powered submarine *Nautilus* commenced operations. Freed from dependence upon fuel tankers and submarine tenders, the *Nautilus* and its progeny roam the undersea at will—but not to extreme depth. It was a revolution in horizontal mobility and staying power.

It seems as if a nuclear-powered bathyscaphe would provide us with full three-dimensional mobility as well as immense staying power.

The *NR-1*, a nuclear-powered submersible designed for undersea scientific and rescue missions. It is about 150 feet long and displaces about 400 tons, making it much smaller than most nuclear submarines. The *NR-1* was launched at Groton, Connecticut, on January 25, 1969. (*Electric Boat Division, General Dynamics, Inc.*)

This likely combination of technical breakthroughs has not been consummated as yet, although some studies have been made. As a matter of fact, nuclear submarine propulsion, perhaps the greatest single success story in the practical application of the atom, is just beginning to be applied to undersea research or exploitation. Some auxiliary nuclear-powered submersibles, such as the *NR-1*, a submarine rescue vehicle, have been built, but no bathyscaphes or research submersibles. Two facts block such widespread application of nuclear power: the nuclear power plants are much more expensive than chemical power, and the technology is often in the secret category by virtue of its military significance.

Nuclear power will inevitably send research craft along the axes of the great submarine trenches and the flanks of the mid-ocean ridges. The potential is too great to ignore. Meanwhile, atomic technology is already contributing to undersea research in different ways—through teleoperators.

Almost all of the ocean floor is at least two miles deep. On the shallow continental shelves, divers rarely work below 500 feet. The commercial and military importance of the sea demands that we know how to manipulate machinery and other things in the great depths. Walks in space and walks on the moon are possible, but for man to walk on the sea floor two miles down and manipulate anything is very unlikely in terms of today's technology. Only machines can survive the crushing pressures.

Small manipulator-equipped submersibles are very popular research tools today. Carrying only two or three men, some, like the *Trieste,* can dive to any depth, but most are designed for the continental shelves. The *Beaver, Alvin, Deep Quest, Turtle, Star III,* and *Asherah* are among these new submersibles. Eventually, they may gain the same fame as the lunar equivalents, Apollo's *Columbia* or *Yankee Clipper,* for the world they are exploring is certainly as mysterious as the moon.

The research submersibles, whether searching for marine life or undersea archeological evidence, are much alike. Bright lights illumine the area being searched; the aquanauts view the target through wide ports; and one or two teleoperator arms manipulate rocks, instruments, specimens, and repair tools. The same vehicles also find application around the oil-drilling rigs out on the continental shelves, especially at depths where divers cannot remain for long (over 500

The *Star III* research submersible equipped with teleoperator arms. (*Electric Boat Division, General Dynamics, Inc.*)

feet). Undersea teleoperators, however, are relatively crude by dry-land, hot-cell standards. Most are hydraulically actuated and without force feedback. Therefore, dexterity is low and they are slow and clumsy. Nevertheless, the mere fact of manipulation is as remarkable at 35,000 feet down as it was with the Surveyor-3 surface sampler on the moon in 1967. Eventually, we may see submersibles helping exploit resources of the sea, such as the pavements of manganese nodules that cover some areas of the ocean floor. Wherever a teleoperator can venture, it takes with it many of man's senses as well as his unique manipulatory capabilities.

Teleoperators are also mounted on sea-floor vehicles. RUM (Remote Underwater Manipulator) was a tracked underwater vehicle operated by Scripps Institution of Oceanography. This vehicle was operated (via TV) from shore and was powered by a long cable. Victor C. Anderson, RUM's inventor, has also designed a Benthic Laboratory, which incorporates a one-arm teleoperator for remote maintenance and repair. The Benthic Laboratory bristles with current meters and similar oceanographic instruments. Its lifetime without direct human intervention can be extended months, possibly years, with a simple teleoperator that can replace electronic modules and do limited troubleshooting. This same principle was incorporated in some of the designs for NASA's proposed Automated Biological Laboratory (ABL) for use on Mars and other planets. The technological similarities between inner and outer space are frequent and striking.

Pursuing the parallelism one step further, if moon cities can be built (at least in concept), why not undersea cities? There is no technical reason why we cannot build bubble cities on the ocean floor. The domed walls would be built to keep out seawater rather than the vacuum of outer space. This is a rather simplistic view of the situation. The corrosive seawater environment at 10,000 pounds per square inch pressure is much more forbidding than the relatively benign vacuum of outer space, despite the threats of micrometeoroids, ultraviolet light, and cosmic rays. The sea is heavy-handed in comparison, but abundant energy conquers all environments, or so we would like to think.

The parallelism between space and undersea exploration weakens when one looks at undersea automata. No precise counterparts to the Explorer satellites and Pioneer deep-space probes exist. The reason is that undersea telemetry is at present nigh impossible. Radio waves traverse millions of miles of outer space with ease, but a few feet of ocean water are impervious to them. Communication via sound waves is possible but of limited range ° and somewhat undependable. Therefore, the sea bottom boasts no automata that "broadcast" scientific information back to waiting scientists. True, there are sonic beacons and listening posts with ears cocked for intruding submarines, but these relay their information through cables to nearby islands and continental stations. A few cable-connected research stations have also been installed just offshore. When oceanographers want to learn more about the sea bottom, they usually go there in person.

Yet, radioisotopic power generators, the same RTGs that help power satellites and space probes, have played a role in undersea exploitation. The United States AEC's series of SNAP-7 RTGs power navigational buoys, floating weather stations, and sea-bottom sonic beacons. The first application of nuclear power to offshore petroleum

° The maximum range for voice communication via underwater sound systems is about six miles.

production began operation on June 21, 1965, when SNAP-7F replaced a small diesel power plant on an unmanned oil-and-gas platform out in the Gulf of Mexico. Strontium-90, a radioisotope with a half-life of twenty-eight years, provides heat for the thermoelectric elements, which convert it into sixty watts of electrical power for the navigation lights and foghorn on the platform—a modest but promising beginning.

All of the RTGs in the SNAP-7 and SNAP-21 series had marine applications:

SNAP-7A	10 watts	Powered a navigational buoy
SNAP-7B	60	Powered a fixed navigational light
SNAP-7C	10	In a fixed weather station
SNAP-7D	60	On a floating weather station
SNAP-7E	7.5	Powered an underwater acoustic beacon
SNAP-7F	60	Oil-and-gas platform navigation aids
SNAP-21	10	Undersea pingers and transponders

Other RTGs, some entirely commercial products, operate oceanographic buoys. A 25-watt RTG powered an acoustic transmitter in an underwater sound transmission experiment. The sound source was placed about 2000 feet down and sent signals to receivers several hundred miles away. The SNAP marine applications may not involve inquisitive automata like the planned Pioneer probe to Jupiter, but the analogies are strong between the marine beacons and the RTG-carrying Transit navigation satellites and between the marine weather stations and the Nimbus meteorological satellites. If we could communicate as well beneath the sea as we do in outer space, there would undoubtedly be marine equivalents of scientific satellites.

Of course, an alternative to long-distance communication exists—physical recovery of data stored, say, on magnetic tape. In this category fall small scientific stations placed on the sea bottom for later recovery and tape playback. More intriguing is the possibility of self-propelled undersea probes taught to trace out specific search patterns and then return to base with cargoes of recorded data. Or, the probe could wander for years, rising periodically to view the star fields and compute its position. Sophisticated automata such as this, in lieu of returning to base, might be programmed to surface daily and disgorge their tape recorder memories via radio transmitter to shore stations. Seawater parameters (salinity, temperature, and so on), noise spectrograms, photographs of the sea bottom, and many other data of intense scientific value could be "read out" of the probe's memory as it surfaced on a regular schedule. Naturally, these submersibles would be powered by the atom and equipped with tele-operators.

The cleverer one tries to get with undersea automata, the more power one requires to satisfy increased communication, propulsion, guidance, and control capabilities. RTGs might be able to handle the simpler undersea probes consuming under a kilowatt; but the more sophisticated unmanned machines, especially those that wander widely, would need small reactor power plants producing 10 to 100 kilowatts of electricity for motors and equipment.

When man himself enters the underseas picture—either because he is a computer, observer, and controller par excellence or just because he wants to conquer a new environment—power requirements sky-rocket. We have already mentioned that nuclear power now dominates military submersibles and is under consideration for smaller research craft. The power plants of the latter submersibles would have to generate several megawatts, mostly as shaft power for propulsion. For example, J. Madell of the AEC's Argonne National Laboratory has reported the design of a 7.75 megawatt (thermal) pressurized water reactor that would supply 2000 shaft horsepower to drive a 100-foot oceanographic research submersible. Military submarines are more powerful by more than a factor of ten.

The United States Navy has already sent aquanauts to live on the continental shelves for periods of more than thirty days. In 1970, this was still longer than any astronauts have spent in orbit or on lunar voyages. The marine counterparts of the Gemini and Apollo manned spacecraft are heavy, stationary pressure vessels, such as Sealab I and its progeny. Aquanauts live at ambient pressures within these metal walls; because of this they can exit and reenter freely through a hole in the floor that is open to the sea. Sealab I was stationed in 193 feet of water near Bermuda in 1964. Sealab III, the most ambitious of the series, was designed to support eight aquanauts in over 450 feet of water at ambient pressures. The total power plant capacity was 140 kilowatts—a very high per-capita consumption rate, which was provided by cable from a support ship. Without nuclear power, Sealab-type ventures are far from autonomous. Obviously, the men could not have withstood the pressures prevailing two miles down on the sea floor proper. Nevertheless, the successes of the Sealab and Tektite ventures have encouraged the age-old dream that man might one day live on the sea floor in closed-cycle cities. These "sea colonies" would be much like the lunar colonies that have received the lion's share of attention in the press and science fiction. Sustaining such distant descendants of the Sealabs, we would of course find nuclear power plants. What other kind of power plant does not breathe valuable air and also provides abundant electricity and space heat? The latter is vital since the deep ocean waters are close to freezing regardless of the clime.

Man's first attempt to protect himself from the hostile ocean by nuclear means began in the mid-1960s, when the United States Navy and the AEC started a cooperative effort to develop a diver's swimsuit heated by nuclear energy. Ordinarily, a diver is limited to a few minutes of exposure to the icy seas in many parts of the world. To replace lost body heat, a radioisotope heat source emitting a minimum of penetrating nuclear radiation was used to heat a fluid which circulated through the "veins" of a special swimsuit. Such a suit was built and operated successfully. The isotope used was plutonium-238, the same isotope in the power supply assembled on the moon by the Apollo-12 astronauts. We do not visualize that someday every scuba diver will have his own nuclear heated suit. The radioisotope heat sources cost about $100,000 each at current prices (these will drop somewhat in the future). This fact limits the application of the swimsuit to special situations, such as saving lives in the event of a submarine disaster.

Other applications of the atom are probably closer at hand, for commercial exploitation of the sea bottom may be the greatest stimulant to the opening of this frontier. (Look at offshore oil well drilling technology!) Undersea mining operations, for example, may entail on-site processing of ore and its transportation to surface vehicles or craft submerged nearby. If the deposits are less than 1000 feet down, "water-lift" mining techniques will probably be applicable. Power supplies in the 250- to 1000-electrical-kilowatt range would probably suffice to raise the raw materials to a waiting surface vessel. At depths beyond 1000 feet (as on most of the ocean floor) 10 to 100 megawatts of electricity would be needed to power mechanical conveyors. The atom will probably be the most economical source of power for many deep commercial operations. One can even conceive of sea-floor, megawatt-size power modules supplying several square miles of an undersea mining operation through heavy cables. When the time comes to move on to richer areas, buoyancy tanks would be attached to the power plant, and it could be towed to a new location.

In spite of these undersea achievements and bright prospects for the future, our nation's primary technological momentum does seem to be upward and outward rather than downward.

The Nether Frontier

Our ability to live and operate on the surface of the moon exceeds our ability to live and operate in the abyssal seas. Still closer to home, beneath the surface of the Earth, we confront similar challenging conditions. Although we have built underground factories and crept wonderingly into the great cavities blasted out by nuclear detonations, there is something eerie and forbidding about the nether world. Tolkien's tales sound uncomfortably real to us. Legends of gnomes and cave "spirits" persist all over the world. H. G. Wells in *The Time Machine* wrote of a degenerate race of men who lived in vast catacombs. Even Jules Verne succumbed to the lure of the nether frontier in *A Voyage to the Center of the Earth*. Where science fiction goes, can the atom be far behind?

For all the levity, an inward frontier really exists beneath our feet. Mankind's normal habitat is a thin, essentially two-dimensional layer clinging to hospitable areas on the planet's surface. But not only do space and the undersea realm yield to abundant energy; so does the solid Earth itself. The "bowels of the Earth" is a rather disagreeable phrase for the world of tunnels and caverns carved by nature and man out of the Earth's integument. Coal-blackened miners symbolize the underground life to us, but the cities of the distant future may be carved from clean, honest granite. Perhaps some poet of the future will exclaim over amethyst-lined tunnels or the crystal, nascent waters springing from some underground fountain. Spelunkers are like skin divers: both are beginning to explore beyond our historical, two-dimensional world. Underground cities are just as possible in the technical sense as undersea habitations.

Modern, windowless, air-conditioned buildings closely simulate the underground environment. Waxing philosophical for a moment, a hard, dense cover of rock is a superb protective shield against the polluted ecosphere that society is now brewing. A rock ceiling is also protection against natural catastrophes on an astronomical scale. The Apollo astronauts, for example, found glassy, fused sand at the centers of many lunar craters, indicating (possibly) that an intense flash of solar radiation once seared the moon—and of course, the Earth, too. The Earth's atmosphere would have absorbed much of this pulse of energy, but the ultraviolet light and cosmic rays that penetrated to the surface may have caused considerable biological damage. What happened once could happen again. Finally, large living areas carved out of rock could serve as protection against man-made catastrophes—as man's last retreat in the event of total nuclear war.

The role played by reactor power plants underground is essentially a carbon copy of their undersea assignment. Large quantities of power are essential to a closed-cycle ecology. In particular, the power plant would not consume hard-won oxygen and replace it by noxious fumes. The atom has a great advantage here which has so far not been exploited. Mines, missile silos, and military command sites buried for A-bomb protection all depend upon conventional fossil fuels. They can do this because they still retain air conduits (umbilicals, really) linking them with the atmosphere. These links have to be broken for space and undersea exploration. Long-term, closed-cycle ecologies must depend upon nuclear power for complete independence. Outside of a few studies applying nuclear power to underground military complexes, the field is in its infancy. Perhaps the small nuclear power plant that has been operating since 1962 at McMurdo Sound in Antarctica, furnishing both electricity and desalted water to the American base there, will serve as a prototype for this area of application of nuclear power.

Let us examine the role played by automata in exploring the world beneath our feet. Except for those very simple instrumented probes lowered down oil wells and other drill holes to reconnoiter the geology, nothing exists to compare with scientific satellites or even floating weather stations. There was Project Mole, a fictitious, tongue-in-cheek program conceived by some earth scientists in pique when the space program was getting most of the research and development money. Disintegrating the rock ahead of it, the Mole vehicle was designed to obey the laws of orbital motion, but beneath the Earth's surface!

Teleoperators and nuclear explosives should not be ignored in considering the underground frontier and how we might attack it. The large-scale tunneling and excavation capabilities of Plowshare explosives have already been proven in underground tests. As for teleoperators, why send men into hazardous areas when teleoperators can do the job? Teleoperators could handle mining machinery in high-risk areas, acting (as always) as man's expendable precursors. Gas-filled corridors, tunnels near sources of geothermal power, blasting areas—these are teleoperator environments.

Mining has always been a hazardous occupation, even though the mining industry has instituted many safeguards. An obvious solution is the replacement of men by machines. The application of teleoperators to hazardous mining tasks means that men no longer need be sent into the tunnels and dangerous passageways. The senses of sight, hearing, and touch would be transmitted back to surface operators electronically. Men, in complete safety on the surface, would go through the motions of mining, while their machine replicas far below would reproduce these actions accordingly.

Potentialities and Realities

This has been a chapter for speculation. How can the atom help explore and exploit the new environments opening up in both outer and inner space? Small nuclear RTGs, big fission power plants, teleoperators, and nuclear explosives—these are the atom-based tools that will help us cut the umbilical cord that ties us precariously to the thin layer of air, water, and rock coating this 8000-mile sphere careening through space. Technology gives us the power to become three-dimensional. But this chapter is full of "woulds" and "mights." Three-dimensional life is only a potentiality, not yet a reality, in our time.

Chapter 8

Sustaining
and
Augmenting Man

What about man? The preceding chapters have dealt with remaking cities and continents to human specifications, but suppose that man does not like the way he came from the mold? Firstly, he can do as he has done for millennia: repair and sustain the body when it falters; treat it as another machine susceptible to the ministrations of technology. Or, secondly, he can view the human body as unfinished, something one can mold and perfect despite the natural evolutionary processes we learn about in school. We shall examine both possibilities in this chapter.

More than anything else, the human body is a chemical machine of grand proportions, so we would expect that radioactive tracers would be valuable in diagnosing those malfunctions called disease, and just as some industrial chemical processes are susceptible to modification by nuclear radiation, bodily processes should also respond to radiation therapy.

Nuclear power sources enter the picture, too, in a small way at first, as energy sources for cardiac pacemakers and artificial hearts. However, more and more of our natural organs are being replaced by *ersatz* parts, and we see no technical objection to replacing most of the body by more efficient and reliable machinery—powered by the atom in some cases. Such a man-machine hybrid is often called a *cyborg* or cybernetic organism. Technologically, individuals with artifi-

cial limbs or electronic pacemakers are cyborgs, but cyborgs are just
the beginning.

Medicine has already benefited incalculably from the atom. Far
more lives have been saved and sustained by the atom than were ever
taken at Hiroshima and Nagasaki. The word "atom" should conjure
up visions of modern hospitals in preference to mushroom clouds.

True, it was the atom as a military power that gave birth—really
rebirth, as we will demonstrate—to the field of nuclear medicine. The
nuclear reactor provided an almost limitless supply of radioisotopes.
The few years in the history of man that have passed by since World
War II have witnessed a revolution in medicine. This has been due to
the growth and impact of technology in general, but atomic energy
has played an indispensable part.

Dr. Frederick J. Bonte, Chairman of the Department of Radiology,
University of Texas Southwestern Medical School at Dallas, recently
described the phenomenal rate of growth of nuclear medicine:

> Fifteen years ago in a large and well-known hospital in the
> Southwestern United States, about 100 radioisotope tests were
> made in the course of a year's medical practice conducted there.
> During the year just past [1969], some 8,000 isotope tests, simple
> and complex, were part of the background of good medical care
> in this institution.
>
> Also, 15 years ago, in a renowned Eastern university hospital,
> no radioisotope scan pictures were used in diagnosis. Now, in
> that same hospital, one out of every four patients admitted has
> a radioisotope scan test of some sort.
>
> These are not isolated instances. Fifteen years ago perhaps 500
> American hospitals were served by a like number of physicians
> who were licensed, and therefore adjudged competent, to use ra-
> dioactive material in medical practice. It is estimated that during
> that year they performed 200,000 patient tests. During the past
> year [1969], 2,000 licensed physicians served more than 4,000
> hospitals, and to four million patients they offered almost eight
> million tests, within the framework of Nuclear Medicine.

What are these "tests" and "scans" to which Dr. Bonte refers? How
do they help people? We will trace a little of the history of nuclear
medicine and describe how the all-pervasive atom comes to the res-

cue of man—not just man in general, but each individual—at times of great need.

Atomic Aids to Diagnosis

The potassium-40 in our bodies, and other naturally occurring radioisotopes, make us all radioactive to a slight extent. To the sensitive "eyes" of nuclear particle detectors, we all "glow" with an aura of radioactivity. Each year large numbers of persons make themselves temporarily glow more brightly by imbibing "atomic cocktails." They drink radioisotopes because these labeled atoms help doctors pinpoint places where the body's chemistry has gone awry. The proponents of radioactive tracers in medical diagnosis have gone so far as to claim that they are the greatest aid to diagnosis since the invention of the microscope. Such enthusiasm is shared by many doctors, for, in 1970 alone, about 8 million doses of some thirty different radioisotopes were administered in the United States in diagnostic and therapeutic procedures.

The quantities of radioisotopes administered in diagnosis are very tiny. The goal of the doctor is simply to follow the journeys of specific radioactive chemicals as they pass through the body. When radioisotopes are used in *therapy*, the doses must be much larger than they are in diagnosis because the goal is usually the destruction of abnormal tissues, such as cancer. In therapy, the doctor must make sure that the probable gain to the patient overshadows any damage done to surrounding tissue. With this distinction, let us look briefly at the beginnings of radioactive tracers in medicine.

Georg von Hevesy, the first great practitioner in radioactive tracing, first applied a biological tracer in 1923, when he studied the absorption of lead-212 (called thorium-B then) in plants. Eleven years later, he and his associate swallowed some deuterium to investigate the half-life of water molecules in their own bodies. This was the first recorded use of a *stable* isotope in human biology.

Artificial radioisotopes, as distinct from Hevesy's natural radioisotopes, became part of the medical repertoire in the year 1936, when Joseph G. Hamilton and Robert S. Stone, at the University of California at Berkeley, used cyclotron-produced radiosodium in studies of sodium uptake and excretion. Hamilton's early radioiodine studies also employed iodine-128, which possesses a half-life of only twenty-five minutes. In response to Hamilton's plea for a longer-lived radioisotope of iodine, the senior author and Jack Livingood synthesized iodine-131 in Ernest O. Lawrence's cyclotron. Happily, iodine-131's half-life is eight days, just what the doctor had ordered.

Radioiodine-131, which accounts for more than half of all diagnostic tests employing radioactivity, first became available in quantities sufficient for clinical tests in 1938. By 1939, Hamilton and Soley had demonstrated the selective thyroid uptake of radioiodine in human patients. The predilection of the thyroid gland for iodine had been known for years before Hamilton's request for a better species of radioiodine. Furthermore, the gland itself is close to the skin of the neck, making detection of iodine-131's gamma rays easy. By modern standards, the first efforts at diagnosis of hyperthyroidism, the condition often characterized by thyroid enlargement, with tracers were crude. One simply held a Geiger counter tube directly over the thyroid gland and measured the rate at which the orally administered radioiodine was concentrated by the gland. If the gland was unusually avid for iodine (hyperactive), the patient's metabolism was abnormal and perhaps a tumor existed. Simple enough, but like the microscope, this gave doctors information they could not perceive by ordinary sight and touch.

Doctors got an even better view of the thyroid when the first "scanners" reached the hospitals in the 1950s. Using scintillation detectors rather than Geiger counters, scanners move back and forth across the thyroid area, recording only the concentration of radioiodine immediately below the scanner head. Thick lead collimators prevent the scintillator crystal from seeing other portions of the gland. Thus, a two-dimensional picture of the thyroid gland is built up piece by piece in the same way a television picture on a glass tube in the living room is created. Pictures of the normal thyroid taken in the "light" of iodine-131's gamma rays are butterfly-shaped. Different kinds of thyroid deficiencies produce different gamma-ray pictures. Cancerous portions of a diseased thyroid can be separated from benign tumors in gamma-ray pictures because the latter usually show up brighter.

Scanning is not used only for the detection of thyroid malfunctions. It is sometimes the only means for locating brain tumors and is often used to determine the size and shape of other organs that are difficult to see with ordinary X rays. Again, Dr. Bonte has caught the drama inherent in the physician's desire to diagnose disease quickly and accurately:

> Not all patients come into the radioisotope laboratory by day. A patient may be brought there in the middle of the night, with a history of sudden development of cough and chest pain. The much feared complication of pulmonary embolism, or impaction of a blood clot in the vessels of the lungs, must be considered. The patient is given an intravenous dose of a particulate radio-nuclide tracer which will temporarily lodge within the vessels of his lungs. He will be placed beneath a radioisotope camera, and pictures will be made. A hole which appears in the midst of the normally uniform pattern of lung circulation may betoken a pulmonary embolism, and the proper treatment of this life-threatening situation may begin at once.

In another situation, suppose that cancer of the thyroid is proven. The doctor may not wish to risk an operation. He will then turn to the use of radiation to destroy the cancer in many cases. This approach is described later in this chapter. If bits of the thyroid cancer have broken off and migrated to other parts of the body, the risk of surgery is increased, and it is most important to know if this has occurred. The detection of secondary cancers or *metastases* is possible with the *whole-body scanner,* an instrument developed at the University of California's Donner Laboratory in 1952.

The whole-body scanner should not be confused with the *whole-body counter.* The patient is literally swallowed up by the whole-body counter rather than being scanned systematically by a moving radiation counter. By surrounding the subject on all sides with radiation detectors, radioactivity from all sources and all parts of the body is measured simultaneously. Whole-body counters are particularly valuable in uptake studies, where the amounts of radioactivity resident in the body are very small. For example, in the old Geiger-counter days, 50 to 100 microcuries of radioiodine were usually administered in thyroid uptake analyses. The doctor may not wish to expose an infant or adult to such a large dose of radioactivity. By using a whole-body counter, the dose can be reduced to 0.1 microcurie. Of course, no picture of the thyroid is obtained with a whole-body counter, but the retention of iodine by the body as a whole can

Seven serial scans made with a whole-body scanner were put together to provide this whole-body scan of a patient with thyroid cancer that had spread to a lung. The scans were made 72 hours after the administration of iodine-131. (*Lawrence Radiation Laboratory*)

be determined. Whole-body counters have also been valuable in measuring the uptake of very small amounts of radioisotopes from radioactive fallout. Even though atmospheric tests have been abandoned by the United States and the Soviet Union, scientists are still able to trace the radioisotopes created in weapons tests of the 1950s and early 1960s as they pass through the biosphere from animal to man. Each new Chinese or French atmospheric weapon test can likewise be detected as a surge in the biosphere's burden of radioactivity.

Iodine-131 is more versatile than we have described. It can be used to determine blood volume, cardiac output, plasma volume, liver activity, kidney function, fat metabolism, and brain tumors as well as thyroid disorders. The thyroid patient takes his atomic cocktail. Sometimes human blood serum tagged with radioiodine is injected into a vein. *Rose bengal* is not a wine from India but a chemical dye long used in appraising the effectiveness of the human liver. When rose bengal is injected into a vein, the liver normally removes it from

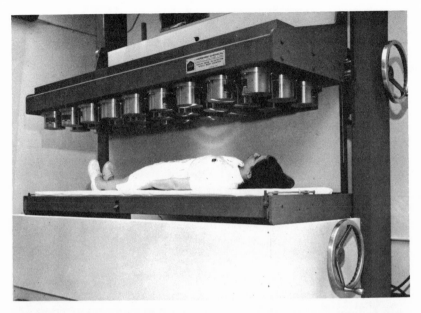

A nurse simulates a patient being scanned by a whole-body counter consisting of 54 sodium-iodide crystal detectors. (27 above, 27 hidden below the patient.) Rather than scanning the patient with a single moving detector, this whole-body scanner creates two 27-element pictures simultaneously. (*Brookhaven National Laboratory*)

the bloodstream and transfers it to the intestines for excretion. The speed with which rose bengal is removed from the bloodstream is thus a measure of liver activity. The speed of removal can be measured chemically, but this diagnostic test can be greatly improved by tagging the rose bengal with iodine-131. Immediately after the administration of the tagged fluid, several detectors begin recording the progress of rose bengal through the body. One detector monitors the liver; another, the small intestine; and the third, the bloodstream at the head or thigh.

Iodine-131 also helps spot brain tumors because of a peculiar property of the brain. There is a barrier between the brain proper and the blood so that albumin in the blood does not invade healthy brain tissue. The brain is unique in this respect. Most substances injected into the bloodstream readily pass into muscles and other tissues. When a tumor is present in the brain, however, this singular barrier is breached and blood albumin can penetrate into the tumorous tissues that have invaded the brain. The role of radioactive tracers is obvious: label a component of the blood and scan the brain for "hot spots" with radiation detectors. Once again iodine-131 is the most popular label. It is chemically attached to human albumin and injected into the patient. Any hot spot pinpointed by gamma ray detectors is likely to be a tumor site.

A radioisotope of unusual interest is technetium-99m, the *m* meaning that the radioisotope is *metastable* or in an energy state higher than normal. As the technetium-99m decays with its characteristic six-hour half-life to become plain "ground-state" technetium-99, the metastable atom's excess energy appears in the form of gamma rays and so-called "conversion electrons." Technetium is one of the four elements lighter than uranium that do not occur naturally. (The others are promethium, francium, and astatine. All are man-made.) Although technetium-99 in the ground state has a half-life greater than 100,000 years, this is still too short for technetium formed at the beginning of the universe to have survived.

The six-hour half-life of technetium-99m is ideal for the diagnosis of disease in the thyroid, the brain, and the liver. Technetium is used in at least 2000 diagnoses *every day* in the United States alone. How can

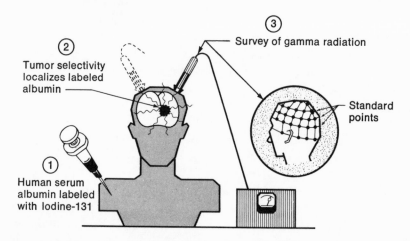

When radioiodine is absorbed by a brain tumor, the tumor's location can be determined accurately with radiation detectors. Older techniques employ the standard points shown above for reference. Today, automatic scanners are common. (*U.S. Atomic Energy Commission*)

it be so popular when it decays away so fast? Most radioisotopes distributed in the United States originate in the reactors at Oak Ridge National Laboratory or some other AEC or commercial installation. By the time a shipment of technetium-99m might arrive at its destination from Oak Ridge, it would be mostly gone. In the parlance of the trade, technetium-99m comes from a radioisotope-giving "cow." The cow consists of a tube filled with an ion exchange resin upon which has been deposited molybdenum-99, a radioisotope which boasts a logistically satisfactory half-life of 2.8 days. The molybdenum-99 decays into technetium-99m, which is "milked" by an appropriate chemical process. Technetium cows are found in any well-stocked medical stable.

Over one hundred radioisotopes have been used in medicine for diagnosis since Hevesy carried out his first tracer experiments in the 1920s. Six of these hundred have been extremely popular: iodine-131, phosphorus-32, gold-198, chromium-51, iron-59, and technetium-99m. All are versatile, like iodine-131, the most popular. Tables 1 and 2 in Appendix I summarize the use of radioisotopes in medical diagnoses during 1966. Table 1 indicates that there were over one-half million non-scanning applications of radioisotopes in 1966, mostly thyroid studies involving iodine-131. Table 2 in the Appendix sums up the scanning applications. Out of the 408,000 applications in 1966, 153,000 used iodine-131 for thyroid scanning. The next two most common applications involved technetium-99m in brain scanning (63,000 cases) and gold-198 in liver scanning (42,000 cases).

We have tried to impress upon the reader that nuclear diagnosis and therapy have unobtrusively become a cornerstone of modern medicine. Unfortunately, statistics do not exist that give the numbers of lives saved or made more bearable through nuclear technology, but the number must run into millions.

Human Activation Analysis. Many radioactive tracers in use today are made in reactors where elements common in body chemistry are exposed to neutrons. Once produced by neutrons the tracers emit the gamma rays that make tracing possible. A rather startling suggestion is that people might be exposed to reactor neutrons directly in order to make them radioactive for purposes of activation analysis. (See Chapter 3.) It all sounds rather deadly. In practice, though, the required neutron doses would amount to only the equivalent of a few ordinary X-ray examinations. As we shall see in the next section on radiation therapy, the body, especially certain parts of it, tolerates considerable radiation, and the medical benefits often outweigh the hazards. Activating the whole human body is feasible, but in practice is not necessary for many studies.

Just what can human activation analysis accomplish? Activation analysis is the best known way to measure the concentrations of very rare "trace" elements in the body without disturbing bodily processes. For example, arsenic exists in the body as a trace element. Arsenic is also a component of cigarette smoke suspected of causing cancer. By far the best way to detect the minute concentrations of arsenic in human beings and tobacco, too, is through activation analysis. In humans, for example, arsenic atoms often end up in the hair, so that hair samples rather than the body proper may be exposed to reactor neutrons for purposes of activation analysis. Samples of blood also suffice for activation analysis of blood trace elements, such as sodium.

Activation analysis with human subjects was pioneered by Keith Boddy of the Scottish Research Reactor Center and W. D. Anderson of Western Infirmary, Glasgow. By bombarding the thyroid for five minutes with a low-intensity neutron beam from a reactor, the *stable* iodine already present in the gland (iodine-127) was activated to produce radioiodine-128, which could be measured. Then, when radioiodine (iodine-131) was administered, the iodine turnover rate in the gland could be ascertained more accurately, knowing what was present in the beginning.

Radiation Therapy

Wilhelm Roentgen discovered X rays in November 1895. Three months later in Chicago, Emil H. Grubbe, in collaboration with R. Ludlam, tried to cure advanced breast cancer with the barely discovered rays. A few months later, J. Daniel at Vanderbilt University described the phenomenon of radiation-induced epilation. He was immediately hailed by the press as the man who had banished forever man's daily chore of shaving. Not an auspicious beginning; and things were about to get worse.

Almost every known form of malignancy, as well as benign afflictions such as acne and warts, were treated indiscriminately with X rays in the early days. Beneath the superficial (but dangerous) enthusiasm for the new, miraculous unseen rays, doctors were discovering that X rays indeed did possess some inexplicable curative powers against cancers. But the stampede was on and the results of early carelessness with X rays are still to be seen in some elderly people today, particularly those with cataracts of the eyes.

Even more exciting than X rays was radium. It was in April 1898 that Marie Curie reported to the French Academy the existence of a new powerful radioactive element in pitchblende. Because radium was known to be the source of "emanations" just as mysterious as Roentgen's X rays, it was assumed that it also must have great curative powers. In 1901, a Paris physician named Danlos borrowed some radium from Pierre Curie and tried it on skin lesions. Pierre Curie himself observed in 1904 that malignant tissues were destroyed more rapidly than healthy tissues when exposed to radium. With this fact, the medicine men could proclaim that radium would cure all human afflictions. Radium chloride was marketed without restrictions as "radiumite" in the 1920s. "Nature's gift to mankind" read the advertisements. People flocked to radioactive springs in the Rockies in the United States—they still have a few customers. How many people today would swallow "Atomic-Nu-Life"? It was on the market at one time.

Amid all this ballyhoo one indisputable fact emerged: radiation does destroy malignant growths preferentially. The technical challenge lies in getting the right amount of radiation to the right spot without endangering the patient or anyone else. Three well-established techniques do this:

1. Radiation from an X-ray machine, from a radioisotopic source, or from a particle accelerator is collimated into a thin beam aimed at the target area. This approach is called *teletherapy*.

2. A radioisotope is confined in a needle, seed, or some other encapsulation and physically inserted into the target area. This treatment is called *brachytherapy*, from the Greek word for "short."

3. A radioisotope is prepared in a chemical form so that the body itself will concentrate it in the desired area, as radioiodine, for example, is concentrated in the thyroid. The technical name for this procedure is *radiopharmaceutical therapy*.

To give the reader a feeling for the popularity of radiotherapy among physicians, we present detailed applications data for 1966 in Appendix I as Tables 3, 4, and 5. They show that teletherapy is by far the most common of the three procedures, with nearly 2 million applications of radioisotope sources alone. Almost everyone knows someone who has had treatments with radioactive cobalt-60, which emits cancer-destroying gamma rays. The hopes of the rash experimenters early in this century were merely premature.

A doctor related the following cobalt-60 case history.[*]

A 75-year-old white male patient, who had been hoarse for one month, was treated unsuccessfully with the usual medications given for a bad cold. Finally, examination of his larynx revealed an ulcerated swelling on the right vocal cord. A biopsy [microscopic examination of a tissue sample] was made, and it was found the swelling was a squamous-cell cancer.

Daily radiation treatment using a cobalt-60 device was started and continued for thirty-one days. This was in September 1959. The cobalt-60 unit is one that can be operated by remote control. It positions radioactive cobalt over a collimator, which determines the size of the radiation beam reaching the patient. The machine may be made to rotate around the patient or can be used at any desired angle or position.

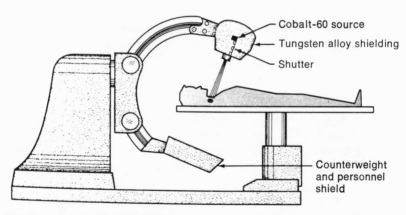

Radiotherapy using a cobalt-60 gamma-ray source. (*U.S. Atomic Energy Commission*)

[*] Quoted from E. W. Phelan, "Radioisotopes in Medicine," AEC Understanding the Atom Series, 1966.

When the treatment series was in progress, the patient's voice was temporarily made worse, but it returned to normal within two months after the treatment ended. The radiation destroyed the cancerous growth, and frequent examinations over six years since have failed to reveal any regrowth.

The treatment spared the patient's vocal cords, and his voice, airway, and food passage were preserved.

In 1903 Alexander Graham Bell first suggested encapsulating radium and inserting it directly into the body. Radium salts or radon gas ("milked" from a radium source) sealed in needles, seeds, or beads have been used in brachytherapy since 1905. Cobalt-60, cesium-137, and gold-198 are reactor-produced and much cheaper than radium these days. Table 4 in Appendix I lists almost 50,000 applications of brachytherapy in 1966. Radium implants in cancer patients are the most common, but strontium-90 is often used for application to the skin or eyes. Radioisotopes are used in the form of wire, ribbon, seeds—whatever shape best suits the condition at hand.

All manner of cancerous growths are treated with these tiny implants. The pituitary gland is a frequent target because its secretions stimulate cell reproduction throughout the body. If one can destroy part of this gland, tumor growth is usually arrested, though not cured. The pituitary, however, cannot be reached surgically without great risk and difficulty. A small, glasslike bead of yttrium-90 oxide, however, can be implanted directly in the gland, often bringing dramatic relief. The beta particles emitted by the yttrium-90 destroy the cancer, but because they have little penetrating power, they do not affect the nearby brain tissue.

A new radioisotope for brachytherapy is californium-252, a radioisotope that is unusual because it is an element that decays by spontaneous fission.[*] The fission-produced neutrons add a new dimension to radiotherapy because neutrons seem to be more efficient in destroying oxygen-deficient cancer cells than X rays and gamma rays. Hitherto, neutrons for therapy had to come directly from reactors or particle accelerators. Now, with californium-252 available, a good neutron source exists outside the reactor or accelerator. Placed in tiny needles, californium-252 is now being evaluated at a number of research

[*] Actually only 3 percent of the decays are fissions; the rest are alpha decays. Half-life = 2.65 years.

centers, such as Brookhaven National Laboratory and the M. D. Anderson Hospital in Houston.

Iodine-131 (in sodium iodide) is typical of the radiopharmaceuticals. In small doses it is an effective tracer, as we have already described. In much larger doses, the radioiodine emanations are powerful enough to alleviate hyperthyroidism and to destroy or at least arrest the growth of thyroid cancers. Table 5 in Appendix I shows over 20,000 such applications of radioactive sodium iodide in 1966. Phosphorus-32 and gold-198 are other radioisotopes important in radiopharmaceuticals.

Before leaving the subject of nuclear medicine, a few overall statistics for the field show its great impact since the discovery of radioactivity. If one includes all therapeutic and diagnostic applications, one finds that radioisotopes are employed in about 8 million treatments annually. More than 2200 physicians in private practice and 4300 hospitals apply nuclear medicine under licenses granted by the AEC or by those states which have agreements with the AEC to issue licenses. Over \$65 million are spent each year for equipment and supplies in this field.

Another strong indicator of the growth of human and humane applications of the atom is its formal recognition by the medical profession. The Society of Nuclear Medicine, whose president in 1971 was Dr. Henry N. Wagner, Jr., of Johns Hopkins Hospital, numbers about 4000 members. The society, together with the American Boards of Internal Medicine, Pathology, and Radiology, are creating an American Board of Nuclear Medicine to establish standards for the training of nuclear physicians. Already more than forty residency training programs in medical schools and teaching hospitals are educating young doctors who have elected to specialize in nuclear medicine. The future is bright for the lifesaving capabilities of the invisible atomic nucleus.

Medical Spin-Off from Centrifuges and Activation Analysis

During the search for ways to separate uranium-235 from uranium-238 for the purpose of weapons manufacture, high-speed centrifuges were developed. The idea is, of course, to separate atoms of different weight after the fashion of the cream-and-milk separators on the farms. Because the difference in the weights of uranium-235 and uranium-238 is so small, the centrifuges had to be highly effective. The so-called Zonal Centrifuge grew out of this work at Oak Ridge National Laboratory. Besides separating isotopes, the ultracentrifuge can produce ultrapure vaccines, separating out the impurity molecules, which have different weights. The Zonal Centrifuge is also now being applied to the isolation of the viruses responsible for hepatitis, polio, rabies, the common cold, animal tumors, and other diseases.

An Oak Ridge scientist adjusts a high-speed liquid zonal centrifuge used to separate viruses and cell components. Highly purified vaccines are produced by this machine. (*Oak Ridge National Laboratory*)

One of the most frustrating of all diseases has been Parkinson's Disease, with its deterioration of the nervous system and, finally and seemingly inevitably, coma and death. Hope for patients with Parkinson's Disease has risen with the development of the drug L-Dopa (L-dihydroxyphenylalanine) by George C. Cotzias and co-workers at the AEC's Brookhaven National Laboratory. Cotzias used the technique of activation analysis (Chapter 3) to trace the reactions of L-Dopa in the body. Treatment with L-Dopa is becoming widespread, allowing a fair proportion of the patients so treated to return to their jobs.

Nuclear-Powered Hearts and Other Devices

Pumping unceasingly, almost 100,000 strokes a day for decades, the human heart is unquestionably one of nature's greatest constructions. Nevertheless, it does falter and fail on occasion. At these times, technology is just beginning to be able to offer assistance. The ailing heart can be stimulated electrically, replaced with another's heart through surgery, or replaced entirely by machinery. The contribution of the atom in cardiac pacemakers and artificial hearts is power—long-term, reliable power.

When the heart is injured or diseased, its control center may not deliver the regular electrical impulses that cause the muscular contractions that result in pumping action. Heart stimulation by artificial means has a long history beginning with simple external mechanical massage by the physician. C. Walton Lillehei, at the University of Minnesota, carried out internal electrical stimulation experiments in 1957. The first successful implantation of an electrical pacemaker was reported by William M. Chardack, at the Veterans Administration Hospital, Buffalo, in 1960. In 1970, about 40,000 persons carried battery-powered pacemakers around with them, and 5000 new pacemakers are "installed" each year. The mercury batteries implanted in the body with the usual pacemaker last only two or three years. Their replacement requires surgery. Further, battery failure may occur without warning. Very small RTGs, generating less than one thousandth of an electrical watt and lasting ten years, would be a lot safer and more convenient.

During 1965, the Atomic Energy Commission began a program with Nuclear Materials and Equipment Corporation (NUMEC) to develop a small plutonium-238-fueled RTG capable of providing 162 millionths of a watt for an implantable pacemaker. NUMEC delivered the first generator in 1967. The alpha particles emitted by the half gram of plutonium-238 metal in the fuel capsule are stopped in the capsule walls, generating heat in the process. The radiation dose to the wearer is no greater than that from a wristwatch with a radium-painted dial. The weight of the 1¼ by 2 by 2½-inch package is only 3.5 ounces. Heat from the decaying plutonium is converted into electricity by a copper-nickel-chromium thermocouple junction rather than the thick lead-telluride thermoelectric elements common in the RTGs producing higher power levels for space and undersea applications. The AEC delivered the NUMEC generator to the National Heart and Lung Institute (NHLI), which had developed the electronic portion of the pacemaker. In May 1968, Andrew G. Morrow and Peter L. Frommer, both at NHLI, implanted the first nuclear-powered pacemaker in a dog named Brunhilde. After a series of tests with Brunhilde and other animals, the pacemaker, if approved, will be ready for human application. In May 1970, two French doctors, Paul Laurens and Armand Piwnica, successfully installed a nuclear pacemaker using plutonium-238 fuel

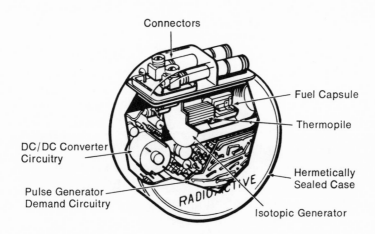

Connectors

DC/DC Converter Circuitry

Pulse Generator Demand Circuitry

Fuel Capsule

Thermopile

Hermetically Sealed Case

Isotopic Generator

RADIOACTIVE

Cutaway view of a French cardiac pacemaker powered by a small radioisotope power generator. The overall diameter is just under three inches. The thermoelectric generator was developed at Alcatel Centre de Recherche Pierre Herreng de Bruyeres-le-Chatel. (*Reprinted with the permission of* Nuclear News)

in a fifty-eight-year-old woman. Hopefully this was the first in a long
line of nuclear pacemakers.

If the heart is past saving, a transplant from another human is possible but not very likely because suitable donors are very rare compared to the large numbers of people with defective, deteriorating
hearts. Besides, the body tends to reject alien tissue, and the scores of
heart transplants have been only temporarily successful. Hundreds of
thousands of people could be helped with a mass-produced mechanical heart that could be tolerated by the body. Long-lived mechanical
pumps made from materials that do not stimulate the body's rejection
mechanism are technically feasible and several types have been developed. One of the major problems is the power source. One cannot
plug a human into the 110-volt A.C. household outlet, or to a storage
battery, for that matter. The wires leading through the skin into the
heart cavity will become an intolerable source of irritation and a
prime route of infection. Like the cardiac pacemaker RTG, the heart
pump power supply should be implanted completely within the body
and work without direct access to the outside environment.

Batteries are out because a heart pump will draw ten to fifteen
watts, depending upon physical activity. As we pointed out in the last
chapter, the only lightweight, reliable power plant that will work for
long periods of time without sunlight employs nuclear fuel. In this,
the heart cavity and the dark side of the moon have something in
common.

The complete heart pump system would consist of a heat-generat-
ing radioisotope driving one of several kinds of engines, which, in
turn, would operate the pump. Tiny steam engines and Stirling engines (a type of high efficiency gas-cycle piston engine) are being
studied in the United States by Aerojet-General, Donald W. Douglas
Laboratories, Thermo Electron Corp., and Westinghouse under the
sponsorship of the AEC and the National Institutes of Health. The
contract for the development of the heat source to go with the various
pumps and engines was awarded to the Hittman Corporation in
April 1969. Plutonium-238 will be the fuel.

The steam engine and Stirling engine convert less than half of the
heat they receive into mechanical power for the pump. The rest of
the heat, fifteen to twenty watts, as well as the heat due to mechanical losses, must be expelled from the body in some way. Conduction
to the surface of the body is out of the question, but the body already

possesses a most elaborate built-in system for rejecting waste heat: the circulatory system. Waste heat can be added to the bloodstream, which then carries it to the body's surface through the capillaries. The patient's body temperature would rise slightly, of course, but that is better than being without a heart. The final element in the heart pump's power supply, then, will be an internal heat exchanger that conveys the waste heat into the bloodstream.

The heart is a powerful organ; in terms of wattage it rivals a fluorescent reading lamp. From an engineering standpoint, it is a marvelous engine and to reproduce it mechanically will require great skill, ingenuity, and imagination. The biological heart in an adult human pumps blood under normal, rather sedentary conditions with a power of about five watts (or about 1/150 of a horsepower). During sleep this may drop to about two watts, as the body's metabolism decreases. An athlete, performing at maximum output, may demand fif-

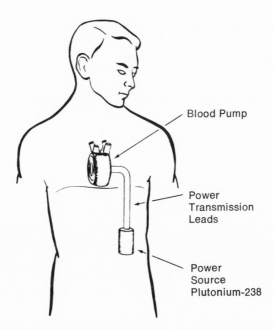

Blood Pump

Power
Transmission
Leads

Power
Source
Plutonium-238

This sketch shows the important components of a nuclear-powered artificial heart. The power source is a small engine which converts the heat from radio-isotope decay into mechanical motion. This energy is conveyed to the artificial blood pump by a hydraulic fluid.

teen watts from his heart. (These rates vary from person to person.) The biological heart is able to accommodate these widely varying loads and yet do so with an efficiency that so far cannot be matched by any artificial device. What is even more remarkable is that the heart and the vascular system automatically adjust to the power demand, without any conscious effort on the part of the owner. Some— but not all—of these attributes will have to be duplicated in an artificial heart.

The radioisotope power supply for an artificial heart ideally should not cause any radiation exposure to its human recipient. This will not be possible in practice, but the radiation levels will be small and acceptable in terms of the extension in life achieved by the artificial heart. Undoubtedly other new complexities will arise in the widespread use of these mechanisms; for example, the small radiation exposure to other persons in close proximity to the bearer of a nuclear-powered artificial heart will have to be taken into account in regulating the use of the devices.

In this context, it should be pointed out that plutonium-238 is now a relatively rare and expensive substance. Because of the limited supply of plutonium-238, a successful nuclear-powered artificial heart today could not save all of the hundreds of thousands who need new hearts. Plutonium-powered artificial hearts of advanced design will need about thirty watts of thermal power. This is equivalent to about sixty grams (two ounces) of plutonium-238. By the turn of this century, twenty-five tons of plutonium-238 could be manufactured. If the artificial heart becomes a reality, this would be sufficient fuel to save many thousands of lives.

The Cyborgs Are Coming

Karel Čapek's play, R.U.R., introduced the world to robots in 1923. Science fiction was never the same again. Robots are automata built in human form; they move like men, too, and R.U.R. suggested that they could also possess the emotions of men and women. In fact, some fictional robots have the audacity to think they are superior to men. Whatever stories reach the science fiction press, the truth of the matter is that we are a long way from building sophisticated robots. Powerful engines and complex computers, yes; but, machines with man's brain and dexterity, no.

Since manlike machines seem unattainable in the near future, let us consider machinelike men! Machine properties include inorganic construction, motors, tirelessness, great strength, stamina, and so forth. The vital human ingredient is the brain. If we leave the brain alone and start replacing the other parts of the body with machinery, we are building *cyborgs* or cybernetic organisms. The term "cyborg" was coined by Manfred Clynes, Rockland State Hospital, Orangeburg, New York. Take the heart out and replace it with a radioisotope-powered pump; replace failing kidneys with mechanical counterparts. Organ by organ, limb by limb, a new kind of man is built. This new man, a cyborg, may be constructed either because the original natural components were failing, or, and this is the controversial part, because someone wants to construct a superman.

Obviously, the first cyborgs are already with us, if anyone with an artificial arm is considered a cyborg, or anyone carrying a cardiac pacemaker in his chest cavity. But our minds leap the decades to the man-machine world foreseen by Arthur C. Clarke in his *Profiles of the Future:*

> I suppose one would call a man in an iron lung a Cyborg, but the concept has far wider applications than this. One day we may be able to enter into temporary unions with any sufficiently sophisticated machines, thus being able not merely to control but to become a spaceship or a submarine or a TV network. This would give far more than purely intellectual satisfaction; the thrill that can be obtained from driving a racing car or flying an airplane may be only a pale ghost of the excitement our great-grandchildren may know, when the individual human consciousness is free to roam at will from machine to machine, through all the reaches of sea and sky and space.

Where does the atom fit in this trend toward a more intimate man-machine symbiosis? In three diverse places:

1. As a compact, long-lived power source.
2. As a source of a teleoperator technology; that is, sophisticated manipulator technology.
3. As a source of radiation to suppress the body's mechanism that rejects foreign matter.

We will not discuss the third area beyond saying that radiation has been used frequently in the case of heart transplants to depress the white blood cell population, thus forestalling the well-known rejection phenomenon. In the case of cyborgs, the objects and machinery introduced into the body are not another's flesh and blood. The introduced materials can be chosen so that rejection and irritation are minimized. Radiation will find little use in cyborg surgery, which is extreme to say the least.

The time may come when we will save only the most unique parts of the body—brain, eyes, ears, and perhaps the hands. Then, it would be a case of grafting a piece of man onto a machine instead of vice versa. Everyone has seen, perhaps against their will, science fiction movies with a "brain-in-a-box." Brains in boxes are inevitably inimical on the screen, but they need not be that way.

In his intriguing book *Cyborg: Evolution of the Superman*, Daniel S. Halacy, Jr., divides cyborgs into two subspecies: the "artificial inner man" and the "artificial outer man." A brain-in-a-box would be both, obviously, but our early cyborgs fit nicely into these two classes —artificial hearts and artificial limbs, for example. Potentially, the atom could provide power to both classes, but the AEC's teleoperator technology is applicable only to the artificial outer man.

Looking first at the artificial inner man, we find that the first artificial "internal" organs were all applied externally; that is, they were so big and cumbersome that the patient had to be plugged into them rather than the opposite. In the first artificial heart-lung machine built by John H. Gibbon in 1934, the patient's blood was pumped into a tank where it was cleansed of carbon dioxide and primed with fresh oxygen. Artificial kidney machines, which are in wide use today, are built the same way. People suffering from kidney diseases spend a night or two a week at the local hospital attached to the artificial kidney through plastic fittings implanted in their arms or legs. Once the accumulated wastes have been extracted from their blood, they can leave the hospital and lead normal lives for several more days. An early goal of cyborg enthusiasts is to follow the successful implantation of artificial hearts and kidneys in animals with long-lived, reliable implants in humans. The first steps toward the development of a portable—and perhaps ultimately implantable—artificial kidney have been taken at Argonne National Laboratory, where a method originally used for winding magnet coils for high energy physics research is being adapted to prepare the multiple layers of plastic sheet that form the artificial membrane for separating metabolic products from the bloodstream.

Cyborgs were far from the AEC's mind in 1948, when Ray Goertz's group at Argonne National Laboratory began designing the first master-slave manipulators. The thought of cyborgs in 1948 was just as ridiculous as, well, heart transplants. The sole AEC objective was working safely with radioactivity through the shielding walls of hot cells. Despite the single purpose of this work, the electrical and electrohydraulic manipulators that evolved have the aura of the future about them. In particular, we see the Hardiman man amplifier under development at General Electric as a harbinger of the cyborg superman of the future.

The Hardiman Project, which is sponsored by the Department of Defense, aims at building a man amplifier that will allow a person to lift 1500 pounds six feet and carry this load twenty-five feet in ten seconds. Because the operator fits inside Hardiman much as a hand fits a glove, Hardiman is termed an "exoskeleton." To operate Hardiman, the operator walks, bends over, and lifts as he would naturally. Pressure controls inside the exoskeleton activate the machine's hydraulic joints and force it to follow the operator's motions—with far superior strength. Hardiman is still in the development stage. Portions of the arms and legs have been built and tested. The cybernetic technology involved derives directly from Handyman, a pair of sophisticated electrohydraulic manipulator arms built by Ralph S. Mosher's group at General Electric in the 1950s for the Aircraft Nuclear Propulsion Program.

Hardiman does not replace any of man's organs or limbs. To see the connection between Hardiman and cyborgs, we must glance at a closely related technical area, that of prosthetics. A prosthesis is an artificial arm, leg, or some other part of the body. Strictly speaking, false teeth are prostheses, but let us dwell on more challenging kinds of machines.

The ancient Egyptians built artificial limbs. In the museums, one can see all manner of mechanical substitutes for parts of the body, everything from wigs to ingeniously contrived, well articulated artificial legs. The object in prosthetics is to make a handicapped person more nearly normal, although down the centuries none of the artificial constructions came very close to the strength and dexterity of the real

Artist's sketch of the Hardiman man-amplifier (a type of teleoperator). (*General Electric Company*)

limb. However, once we begin calling a man with an artificial arm a cyborg, the implication is clear that we eventually expect to outdo nature.

The long-separated fields of prosthetics development and the engineering of dexterous remote manipulators have finally met and are beginning to learn from each other. From master-slave technology come powerful actuators, artificial sensors superior to those of the human, and cybernetic theory, which is that branch of control theory essential to melding man and machine. From prosthetics research come many ingenious mechanical arms, legs, and even fingers, and perhaps the most important of all, electromyographic (EMG) control. The union of these ingredients will help make the external artificial man.

EMG control is based on the fact that whenever a muscle is flexed, distinctive patterns of electrical impulses can be detected by electrodes implanted near the muscle. Properly handled, these impulses can control an artificial limb. For example, the so-called Boston Arm, developed at the Liberty Mutual Rehabilitation Laboratory by Robert W. Mann (an M.I.T. professor) and Allen Cutworth, is controlled by EMG signals from a muscle in the amputee's stump. EMG signals from shoulder muscles can also control artificial limbs. As long as a good muscle site is available, artificial arms and legs theoretically can be controlled with some precision by the wearer of the prosthesis. Still, amputees with prostheses do not have the strength and dexterity of the normal person; the cyborg superman is much farther away.

The ultimate cyborg seems closer when electroneurographic control is considered. Move back a step from EMG, from muscle-originated signals toward the brain itself. Voluntary actions are stimulated by nerve signals from the brain to the target muscle. If these nerve signals could be intercepted and properly interpreted, EMG signals would be unnecessary—harnessed nerve signals form the basis of electroneurographic control. Think about an action and your muscles respond accordingly; think about the same action, and the same signals that stimulated the muscles can be converted into motor currents that drive artificial limbs. Electroneurographic signals exist, although the excised and artificially sustained brain—the brain-in-a-box—does not, at least not yet. But at least we can see how, bit by bit, we might be able to divest ourselves of weak and otherwise unsatisfactory pieces of equipment with which we were endowed. The new equipment could in principle be stronger, less susceptible to the environment, and possibly better suited to survive on the Planet Earth tomorrow.

The Atom in Genetic Research

Content with neither man nor cyborg, the gene tinkerers talk about molding man anew. As the mysteries of the DNA (deoxyribonucleic acid) and RNA (ribonucleic acid) molecules, which apparently carry the genetic code, have unraveled, the new-found knowledge has increased confidence that eventually man can shuffle the units of heredity to his liking. This capability may lead to disastrous consequences, but it seems closer every year. Herman J. Muller summed up the situation nicely: °

> So torrential and unexpected has been the flood of progress of the last ten years in physicochemical genetics, that is, in our understanding of the mechanisms that form the basis of terrestrial life, that it is no wonder that some persons, especially those less familiar with life's elaborations, have been swept off their feet, and have come to think that they have almost arrived at a millennium of biological omniscience and omnipotence. Similarly, some physical scientists of the seventeenth century, dazzled by their new insights, dreamed that all knowledge lay only just beyond their contemporary horizon.

It is fitting to quote Muller's caveat against overoptimism here, for he made the first connection between the atom and genetic research in 1927, when he showed that mutation rates could be enormously increased through exposure to X rays. (For this, he received the Nobel Prize in 1946.) As a matter of fact, radiation comprises a direct means by which atomic technology can aid those who would modify man genetically.

In Chapter 4, we described how the irradiation of seeds by a nuclear reactor or a radioisotope source will generate all manner of "sports" or mutations in the mature plant. Some of the mutations have superior properties according to the standards of the agriculturists and have been consequently adopted by farmers. Whatever the success with plants and fruit flies, genetic improvement by radiation is a shot-gun approach—one never knows what one will get. The experimenter can be sure, however, that the overwhelming majority of the mutants will be defective in some way. Therefore, no one yet countenances the *intentional* use of radiation to improve the higher animals and man. The adjective "intentional" is added because natural radiation may be an important tool in nature's evolution kit. Nature continuously sweeps into oblivion her bad experiments with scant thought of morality, but man prefers not to do this intentionally, although he may be doing it inadvertently as he modifies his chemical and physical environment.

° From his paper "Means and Aims in Human Genetic Betterment," printed in *The Control of Human Heredity and Evolution,* T. M. Sonneborn, ed., Macmillan, New York, 1965.

Some day it may be possible to perform gene surgery and deliberately add or subtract traits and physical characteristics. If this comes to pass, the atom will have played a key role through tracer experiments. In effect, tracers are beacons attached to the immense, complex molecules that carry the genetic code. The important tracers in genetic research are tritium (hydrogen-3), carbon-14, and phosphorus-32. When these radioisotopes are attached to the big molecules, the progress of the molecules can be charted even though they cannot be seen by either optical or electron microscopes.

The substances genetic researchers wish to monitor during the replication of cells are DNA, RNA, and other proteins. We cannot go into the details of modern genetic theory; let us just say that if we can find ways to label these types of molecules uniquely, and thus to follow them, we can more easily unravel the details of cell replication.

The first part of the task is rather easy because DNA possesses four chemical bases, one of which (thymine) is not present in RNA. By labeling the so-called nucleoside of thymine, called thymidine, with carbon-14 or tritium, we can distinguish it from RNA if we can somehow "see" the label. Usually, such tracers are "seen" by radiation detectors, such as scintillator crystals, but they can also be recorded by photographic film because radiation blackens film like visible light. DNA is often detected photographically by the process called autoradiography; that is, DNA labeled with a radioactive isotope takes its own picture when its radiation strikes the film.

To obtain a high resolution picture by autoradiography, two conditions must prevail: (1) the radioactive label must emit particles with such short ranges that the details are not blurred; that is, the lines in the drawing must be narrow and sharp (tritium is excellent); and (2) the cells being examined must remain in close contact with the photographic emulsion. A popular method for achieving these conditions is called *dipcoating autoradiography*. A glass slide carrying the cells that have been exposed to labeled thymidine is dipped into melted photographic emulsion, picking up a thin layer as it is retrieved. After giving the tritium enough time to expose the film, the whole emulsion-coated slide is developed and fixed as if it were an ordinary photograph. All that remains of the emulsion is a thin layer of gelatin containing black grains of silver where tritium beta particles (electrons) exposed the film. Specific chemical stains are used to bring out cell details, with the black dots of silver highlighting those cells actively synthesizing DNA at the time the exposure took place.

RNA can be labeled and detected by autoradiography just like DNA. As thymidine is unique to DNA, uridine is unique to RNA and can be labeled specifically with tritium. It is also possible to separate chemically the DNA and RNA that were in the cells and then measure the radioactivity of the DNA or RNA by a scintillation counter. The rate at which either of these biological molecules was taken up by the cells can be computed from this kind of information.

Given extended "sight" by radioisotope tracers, electron microscopes, and the other paraphernalia of modern science, the cell will yield information about its innermost workings, as the atom and the nucleus did long ago; but as Muller implies, victory is never complete. Nevertheless, Muller never rules out gene tinkering, although he believes that men of the near future will more likely be upgraded by the wise selection of parents. He says further:

> It must be admitted that in both these schemes considerable flights of fancy are involved, at which many safe and sane scientists may laugh. Nevertheless, their laughs would in due time be drowned in the water that keeps flowing under the bridge.

The history of science, both theoretical and applied, assures us that our best tools of medicine, our knowledge of the cell, and our abilities to unite man and machine may ultimately shape new species we cannot conceive today.

PART III
THE ATOM
AND
SOCIETY

Chapter 9
The Atom
as a
Moving Force
in Society

Whatever the atom does is done through people. People run the machines, and people also make the decisions about what the machines will make. Going back in history, people discovered the atom, nuclear radiation, the atomic bomb, power reactors, all the ingredients of the atomic world through their own volition. Some historical milestones, such as Becquerel's discovery of radioactivity, were fortuitous; but, by and large, we have nuclear electricity, nuclear submarines, nuclear weapons, and nuclear medicine because people planned things thus, built facilities thus, and built nuclear technology thus. We often overlook two great facts of life:

1. Technology is not an inhuman monolith; it is people, education, management, discovery, ideas, personal commitment, even passion in many instances. It is not all whirling gears and flashing computer lights.
2. Technology is not a juggernaut; being a human construction it can be torn down, augmented, and modified at will. The question of today is: "Whose will?"

The subject of this chapter is fact number one. As for the crucial question, "Whose will?" we must assume that the solutions to today's pressing social problems must involve plentiful food and water for all,

clean air, clean cities, abundant power, and the wherewithal to investigate further the mysteries of human existence. Whether we like it or
not, the lives of nearly four billion people depend completely upon
maintenance of the present technological foundation or *infrastructure*.
In fact, as populations rise and the wastes of the technical society increase, even more technology—wisely applied technology—is needed
to keep the Earth livable. That idyllic pastoral world that so many
long for became impossible when the world population climbed over
a few hundred million.

Building the Technological Infrastructure

The technological infrastructure is that interplay of people, facilities,
technical know-how, and management expertise that enables a nation
to choose its course—or even more than one course at a time. A
higher standard of living, military goals, or exploration of the universe may predominate in one nation's plans for the future, while a
different selection or emphasis might be typical of another nation's
outlook. Manifestly, the atom is only one stone in this foundation—
we prefer to think it a cornerstone. Communications, transportation,
automation, and education make up a few other parts of the edifice.

We could go back to the taming of fire and the discovery of nuclear
fission, but the atomic portion of the foundation—the part we are
concerned with here—only began to exert its muscle during World
War II, when the Manhattan Project was created to build the atomic
bomb. Since then, the nuclear industry has grown (with considerable
federal cultivation) into one of the more important industries in the
United States.

The immense power of nuclear weapons has required government
control of nuclear fuels and their production facilities during most of
the atomic age. Government licensing and regulatory groups have
played a correspondingly larger role in the development of the nuclear industry than they have in the railroad and communication industries, for example. Some in industry say that the government has
played too strong a role. There is no question that governments all
over the world have chosen nuclear technology for special cultivation
through government funding and close regulation. From the beginning, this very interest has led to the careful, planned development of
nuclear power and other nuclear processes. This early government in-

volvement has ensured an extraordinary regard for the health and safety of the general populace, a regard which is unique in the history of the scientific revolution. This interest also testifies to the value of nuclear technology in the overall technological strength and makeup of a country. Governments view nuclear technology much as they view space technology:

—As a stimulant and catalyst to the economy.
—As a source of advanced and highly marketable technology.
—As a mechanism to upgrade the skills of their people.
—As a source of international prestige.

One might well argue that the above represent only secondary effects compared to the military value of nuclear weapons and the tremendous atomic payoff in terms of tools for building a better world for four billion people. Whatever the justifications, the nuclear portion of the technological foundation of all countries is largely government-owned or government-dependent. Only in the late 1960s, when private ownership of nuclear fuel became possible and when nuclear power plant orders increased substantially, has the nuclear industry in the United States begun to become less dependent upon the federal government. We will discuss the role of the nuclear industry in the economic structure of the United States later in this chapter.

Putting men on the moon and developing nuclear power require commitments of men and money too great for private industry. Before 1940, the small-scale nuclear research that existed was the province of university laboratories. The first government appropriation for nuclear energy purposes was made in February 1940 when the Army and the Navy transferred $6000 to the National Bureau of Standards for the purchase of research materials; that is, materials for potential use in the nuclear fission chain reaction. Albert Einstein's letter to President Roosevelt urging investigation of nuclear fission was then several months old. But the wheels of government machinery were beginning to turn. By 1942, it was clear that nuclear weapons might indeed be feasible and that the Nazis were apparently doing something about harnessing nuclear fission, too. On June 18, 1942, Colonel J. C. Marshall of the Army Corps of Engineers was charged with forming a new district to build the foundation necessary to the development and manufacture of nuclear weapons. The famed Manhattan Engineer District was created on August 13, 1942. For public purposes it was called the DSM Project (for Development of Substitute Materials). General Leslie R. Groves took charge on September 17, 1942, and did a brilliant job of leading the project to successful completion. By the time the responsibilities of the Manhattan District were transferred to the new Atomic Energy Commission on January 1, 1947, Manhattan District appropriations had been about $2.2 billion—a large sum for research and development in those days. This money not only built the bomb but also produced a huge complex of facilities from the Atlantic to the Pacific, thousands of men trained in a mysterious new technology, and a great fund of knowledge about the nucleus and new materials. This was the first great federal investment in technology. It was greater than the investment in radar, the Tennessee Valley Authority, or any other portion of the technological base up to that time.

Many of our most brilliant scientists and engineers, both American and foreign-born, worked themselves to almost complete physical and mental exhaustion to bring the project to a successful conclusion. The names are too many to list, but fleeting through the memories of those eventful years are such distinguished figures as Vannevar Bush and James B. Conant, who furnished critical policy guidance from their vantage point in the nation's capital; J. R. Oppenheimer, the first director of "Site Y," known later as Los Alamos Scientific Laboratory; A. H. Compton, head of the Metallurgical Project for plutonium production; H. C. Urey, then of Columbia University, and E. O. Lawrence, of the University of California, both leaders in developing methods of enriching uranium; and E. Fermi, whose guiding genius permeated the entire Metallurgical Project.

The British Commonwealth made crucial contributions at the very beginning when the effort was still centered in England. When this work was moved to Canada, it was headed by J. D. Cockcroft (later Sir John).

Basic to the postwar development of nuclear-electric power was the Manhattan District gaseous diffusion process for uranium enrichment. In this process, uranium isotopes in gaseous form are separated by virtue of their different rates of diffusion through the fine pores of a barrier material. The process was based on the key wartime contributions of J. R. Dunning, A. O. Nier, E. T. Booth, and A. V. Grosse. The postwar availability of these gaseous diffusion plants was paramount in the American decision to concentrate on water-cooled power reactors fueled with enriched uranium. Together, the diffusion plants and the consequent choice of enriched uranium commercial power reactors ensured the leadership now enjoyed by the United States nuclear power industry.

Since those fateful days in the early 1940s when the leadership of the federal government decided to "get into the nuclear business," federal investment (including AEC expenditures for military applications) has amounted to almost $50 billion—more than in the Apollo lunar landing program. Private industry has also invested tens of billions in mining, fuel-processing, waste-disposal, and electric-power facilities. What do we have to show for this cumulative investment made over the past thirty years? °

The preceding eight chapters have described briefly the nonmilitary products bought with the $50 billion: millions of lives prolonged and made more comfortable, electric power, tracers in agriculture and industry, and a great potential for coping with the demands for high standards of life made by the new millions added to the planet's population each year. These are the direct, tangible results of nuclear technology, not the foundation itself.

Federally owned nuclear facilities are worth about $10 billion. Major installations are located in nineteen states, as summarized in Table 13. This table also indicates that most AEC facilities are operated by private contractors. While the AEC itself employs only about 7000, its contractors employ about 100,000. Another 50,000 people work in nuclear facilities built and owned by private industry. Collectively, then, part of the nuclear technological foundation consists of a $10 billion plant plus about 160,000 trained personnel—an important national resource. The facilities are predominantly oriented toward the production and processing of fuel for weapons and nuclear power plants; but the people represent a large, highly trained cadre that can be applied to almost any technological venture in the national interest.

Naturally, we have stressed here the evolution of the American atomic energy program. For all practical purposes, the same events have been re-created in other advanced countries. The Soviet Union began its program during World War II under the leadership of I. V.

° Note that the total thirty-year federal investment in the atom is roughly matched today in a *single year* by American society-oriented programs, such as the health, welfare, and education programs. The annual United States military budget is also much larger than the cumulative nuclear investment.

Table 13
Principal AEC Facilities *

State	Name and Operating Contractor(s)
California	*E. O. Lawrence Radiation Laboratory*, Berkeley and Livermore. University of California.
	Stanford Linear Accelerator Center, Stanford University.
	Sandia-Livermore Laboratory, Livermore. Sandia Corp.
	Laboratory of Nuclear Medicine and Radiation, Los Angeles. University of California.
Colorado	*Rocky Flats Plant*, Rocky Flats. Dow Chemical Co.
	Uranium Ore and Concentrate Servicing Center, Grand Junction. Lucius Pitkin, Inc.
Idaho	*National Reactor Testing Station*, Idaho Falls. Several contractors.
Illinois	*Argonne National Laboratory*, Argonne, University of Chicago and Argonne Universities Association.
	National Accelerator Laboratory, Batavia. Universities Research Association.
Iowa	*Ames Laboratory*, Ames. Iowa State University of Science and Technology.
	Burlington Plant, Burlington. Mason & Hanger-Silas Mason Co., Inc.
Kentucky	*Paducah Gaseous Diffusion Plant*, Paducah. Union Carbide Corp., Nuclear Division.
Maryland	*AEC Headquarters*, Germantown. †
Massachusetts	*Cambridge Electron Accelerator*, Cambridge. M.I.T. and Harvard University.
Missouri	*Kansas City Plant*, Kansas City. The Bendix Corp.
Nevada	*Nevada Test Site*, Mercury. Several contractors.
	Nuclear Rocket Development Station, Jackass Flats. Several contractors.
New Jersey	*Princeton Plasma Physics Laboratory*, Princeton. Princeton University.
New Mexico	*Los Alamos Scientific Laboratory*, Los Alamos. University of California.
	Sandia Laboratory, Albuquerque. Sandia Corp.
New York	*Brookhaven National Laboratory*, Upton. Associated Universities, Inc.

State	Name and Operating Contractor(s)
	Knolls Atomic Power Laboratory, Schenectady. General Electric Co.
	University of Rochester Atomic Energy Project, Rochester. University of Rochester.
Ohio	*Feed Materials Production Center*, Fernald. National Lead Company of Ohio.
	Mound Laboratory, Miamisburg. Monsanto Research Corp.
	Portsmouth Gaseous Diffusion Plant, Piketon. Goodyear Atomic Corp.
Pennsylvania	*Bettis Atomic Power Laboratory*, Pittsburgh. Westinghouse Electric Co.
South Carolina	*Savannah River facilities*, Aiken. Savannah River Laboratory and the Savannah River Plant, both operated by E. I. du Pont de Nemours & Co.
Tennessee	*Oak Ridge facilities*, Oak Ridge National Laboratory, Oak Ridge Gaseous Diffusion Plant, and the Y-12 Plant, all operated by Union Carbide Corp., Nuclear Division.
	Agricultural Research Laboratory, Oak Ridge. University of Tennessee.
Texas	*Pantex Plant*, Amarillo. Mason & Hanger-Silas Mason Co., Inc.
Washington	*Hanford facility*, Richland. Many contractors.

° Those with facility investments exceeding $25 million.

† The principal headquarters building is at Germantown near Washington, D.C. The regulatory staff is located at Bethesda, Md., and additional offices are maintained in Washington.

Kurchatov, a nuclear physicist perhaps much like Fermi, although comparatively little is known about him in the Western World. Britain (and its Commonwealth associates) and France quickly established their own independent atomic programs, followed years later by China. Each of these countries first used the military power of the atom as the incentive for the investment of capital and manpower. Later, other countries demonstrated that this was not a mandatory membership fee for entering the nuclear age. Today, approximately seventy nations of the world have formal Atomic Energy Commissions or their organizational counterparts.

Impact of Atomic Technology Centers

Every new nuclear power plant is an economic stimulus to the area
where it is built. The small town of Vernon, in the southeast corner of
Vermont, typifies the effects of the injection of money, facilities,
know-how, and trained people into a rural area. Prior to the building
of a 540-megawatt, $120-million nuclear power plant, Vernon was pri-
marily an agricultural community. Like any new industry, the Vernon
nuclear power plant, built by Vermont Yankee, has created new jobs,
reduced taxes, and helped build new roads and new schools. But
from the moment the plant was proposed, the emotional and intellectual climate of the town and the area around it changed. First, there
was great controversy concerning the nuclear safety of the plant and
the potential thermal pollution of the Connecticut River. There were
full-page newspaper ads against the plant, town meetings, and conferences between officials of Vermont, New Hampshire, and Massachusetts. People began to wonder about their environment for the first
time. Originally, the waste heat of the power plant was to be dumped
directly into the Connecticut River, but the controversy forced reeval-
uation and the installation of cooling towers to reduce thermal pollution to acceptable levels. The completed power plant required highly
trained people to operate it. Townspeople were trained in new skills
and new people were brought in from outside. The point here is that
Vernon and the area surrounding it received far more than a new industry. In addition to new jobs and tax monies, Vernon was made part
of the technological revolution and part and parcel of the great debate surrounding it. Further, more industry will doubtlessly be attracted by the power center, just as it is by a new highway or university.

Economically at least, the Vernon plant is an asset to the whole
region. Many people, however, no longer consider favorable economic impact to be automatically "good." Another part of the beauti-
ful Vermont countryside will soon see light industries popping up,
with motels, drive-ins, and the other baggage of "developing" areas.
We all mourn the loss of pastoral America, but where are the new
millions born each year to go? These people must have jobs, housing,
and power, and technology's role is to see that they have them. With
nuclear and space technology, government control can help direct the
distribution of technological largess to meet national goals—social,

military, or otherwise. National goals, in fact, include raising the standard of living and the skills of workers in rural areas. Infusions of technology are key instruments in this endeavor.

It is no accident that the National Aeronautics and Space Administration (NASA) built major new facilities in Texas, Alabama, and Mississippi. Space centers are sometimes deliberately used to inject new life into technologically starved regions. The AEC has not had the geographical freedom that NASA has had because nuclear facilities cannot be built just anywhere. The major fuel production and processing plants, such as the Hanford Works, Savannah River Plant, and Oak Ridge facilities, are located well away from population centers. However, the Argonne and Brookhaven National Laboratories and the Lawrence Radiation Laboratory are near large cities where they draw upon the talents of nearby universities. Los Alamos Scientific Laboratory, which was built in secrecy in the mountains of New Mexico during World War II, is still to a large extent an island to itself, but to a diminishing extent as the Los Alamos Meson Physics Facility and other projects have increased the proportion of non-weapons research. For its large force of scientists, engineers, and technicians, Los Alamos built its own schools, theaters, and other cultural activities. Most AEC facilities listed in the preceding table augment and stimulate communities already existing in their vicinities as well as neighboring colleges and universities.

A large nuclear installation—or any facility employing advanced technology—usually extends its effects statewide. The National Accelerator Laboratory (NAL) at Batavia, Illinois, is a case in point. Responding to physicists' desire for a 200-Bev (billion electron volt) proton accelerator to probe deeper into the microstructure of matter, Congress authorized the AEC to go ahead with the $250 million project. The history of this project is unique and interesting. In April 1965, the AEC began the long difficult task of selecting the most appropriate site for the big new facility. Almost every state in the Union vied for the accelerator. Various groups prepared proposals extolling the sites within their states' boundaries almost as if they were industrial firms competing for a contract. Government facilities have always been sought after for their economic benefits, but never was the competition so candid and spirited. The wide interest in a facility conducting research on such esoteric objects as "mesons" and "quarks" is a tribute to the stimulating value of advanced technology.

Aerial view of the 200-Bev accelerator at Batavia, Illinois. The average diameter of the main ring is 6560 feet. (*National Accelerator Laboratory*)

To finish the NAL story, over 200 sites were considered by the AEC, and a selection committee headed by Emanuel Piore was appointed by the National Academy of Sciences. Seven sites were recommended by Piore's committee; the AEC chose the one near Batavia, Illinois, on December 16, 1966. But this was not to be the end of the tale. As soon as the selection of the Batavia site was announced, the AEC was criticized for choosing an area of the country where there did not seem to be adequate housing available for people of all races, creeds, colors, and national origins. In other words, the area selected did not seem concerned enough about civil rights. The 200 Bev protons seemed incidental to the whole issue, but because of the accelerator fifteen communities within commuting radius of the proposed laboratory strengthened or adopted for the first time open occupancy laws. Indeed, the National Accelerator Laboratory has been more than a catalyst. It has promoted equal employment opportunity through cooperation with local unions and through the establishment of preapprenticeship and other training programs. This facet of nuclear technology is more subtle than providing more jobs, but it is just as important.

By the very nature of its programs, the AEC must work hand in hand with American universities. The table presented earlier, listing

Table 14

Facilities Involved in University-AEC Laboratory Cooperative Program

AEC Facility	Universities Utilizing the Facility °
Ames Laboratory, Ames, Iowa	Independent cooperative programs
Argonne National Laboratory, Argonne, Illinois	Argonne Universities Association Associated Colleges of the Chicago Area Associated Colleges of the Midwest Central States Universities, Inc.
Brookhaven National Laboratory, Upton, New York	Associated Universities, Inc.
Lawrence Radiation Laboratory, Berkeley and Livermore, California	Associated Western Universities
Los Alamos Scientific Laboratory, Los Alamos, New Mexico	Associated Western Universities
National Accelerator Laboratory, Batavia, Illinois	Universities Research Association
National Reactor Testing Station, Idaho Falls, Idaho	Associated Western Universities
Oak Ridge National Laboratory, Oak Ridge, Tennessee	Oak Ridge Associated Universities (ORAU)
Pacific Northwest Laboratory, Richland, Washington	Northwest College and University Association for Science
Rochester Atomic Energy Project, Rochester, New York	Independent cooperative programs
Sandia Laboratory, Albuquerque, New Mexico	Associated Western Universities
Savannah River Laboratory, Aiken, South Carolina	Savannah River Nuclear Education Committee of ORAU and some independent institutions
Stanford Linear Accelerator, Palo Alto, California	Associated Western Universities

° See Appendix II for lists of members of university associations.

major AEC facilities, demonstrates how closely knit the atom and education are. Brookhaven National Laboratory on Long Island, for example, is operated by a group of major universities; so is the new National Accelerator Laboratory. In fact, the AEC has a very broad, country-wide program of cooperation with colleges and universities through its national laboratories, as can be seen in Table 14. In addition, the AEC directly supports specialized laboratories on university campuses (Table 15) and almost a thousand smaller research projects at higher institutions of learning. If no university exists at a large AEC

Table 15
On-Campus Laboratories Supported by the AEC

Laboratory	University
Ames Laboratory	Iowa State University of Science and Technology
Lawrence Radiation Laboratory	University of California at Berkeley
Atomic Energy Project	University of Rochester Medical School
Argonne Cancer Research Hospital	University of Chicago Medical School
Radiological Laboratory	University of California Medical Center
Laboratory of Nuclear Medicine and Radiation Biology	University of California at Los Angeles School of Medicine
Cambridge Electron Accelerator	Harvard University and M.I.T.
Computer and Applied Mathematics Center	New York University
Radiation Laboratory	Notre Dame University
Princeton Plasma Physics Laboratory	Princeton University
Stanford Linear Accelerator Center	Stanford University

site, before long extension courses are established to meet the demands of employees for more knowledge. Local communities receive opportunities never offered before.

When the eminent British physicist Lord Bowden delivered his Graham Clark Lecture in 1967, he said:

> The Great American West was conquered in the laboratories of the Land Grant Colleges, and their graduates tamed the continent. We owe two-thirds of all the food which is grown today in the United States to new crops and new techniques which they developed and to students whom they taught. They studied the Mechanic Arts, their ideas and their graduates transformed American industry, they transformed the very nature of the university, and they helped to create the world as we know it.

We do not believe that it is overly presumptuous to state that universities allied with advanced technology centers become "centers of excellence," performing the same tasks the land grant colleges did a century earlier. The technology is different and instead of taming the land the goals are the molding of a better society, the assurance of the necessities of life for everyone, and the exploration of the universe.

Future Roles of United States AEC Laboratories

During World War II, some of the most talented scientists in the country assembled at Los Alamos, the Radiation Laboratory at Berkeley, the University of Chicago, Columbia University, Oak Ridge, and other Manhattan District facilities to carry out man's greatest technical effort up to that day. Together, the scientists, industry, and the military project officers built the bomb. With that success, all looked forward to applying their new-found skills in large-scale research and development to a peaceful world. This was a constant theme for discussion during the birth of the United States AEC. In particular, how would the Manhattan District laboratories be organized for peaceful pursuits? The scientists had been dominant in the development of the bomb, and it was expected that many would elect to remain in the Manhattan District laboratories to help direct postwar efforts. With this in mind, the AEC was established primarily as a management organization, with important scientific work being done under contract to the national laboratories. The objectives of the national laboratories would include:

1. Pure science, which was expected to flourish in the unique environment carried over from the war and carefully fostered in the new laboratories.
2. Applied science, which would translate the discoveries of pure science into socially useful technology.

The national laboratories that were created—Argonne, Brookhaven, Oak Ridge, and so on—were remarkably successful in attaining these objectives. The structure of matter was probed, breeder reactors were built; so was the H-bomb. The 1955 Atoms for Peace Conference in Geneva focused world attention upon America's development of the "peaceful atom." However, as the field of nuclear technology matured, more and more top scientists began drifting back to the universities, industry developed the capability to translate nuclear science into salable products in the marketplace, and, finally, space technology and other exciting new fields began to compete with the atom for attention. Bernard I. Spinrad, a senior physicist at Argonne, described the situation at the national laboratories in 1966 in this way in the *Bulletin of the Atomic Scientists:*

> . . . considering that the laboratories have both grown in size and in number, comparison of the two decades indicates that the laboratories are doing now a smaller fraction of the exciting and significant work of the country, or even of the Commission.

Today, the national laboratories rarely display the same enthusiasm that charged the air in the heady days following the end of the war. Alvin Weinberg, Director of Oak Ridge National Laboratory, has one answer for this new situation: "redeployment." Oak Ridge has moved energetically into the area of large-scale sociological problems. In Chapter 4, some of the Oak Ridge work on water desalting was described. Oak Ridge scientists also work with the National Institutes of Health on the influence of environmental pollutants upon the incidence of cancer. Because of the great concern over the connection between radiation and cancer, the biology division has always been the largest single division at Oak Ridge. It was only natural to turn this talent to the problem of chemical pollutants in the air and water. The Oak Ridge zonal centrifuge work also illustrates once more the non-nuclear value of national laboratory talents and hardware.

Almost all of the national laboratories possess strong potentials for solving society-oriented problems. These resources have barely been tapped for nonnuclear purposes. Without question the chance to work on some of the major problems afflicting society would help recapture some of that excitement and sense of commitment that prevailed immediately after World War II. That this is realized by the laboratories themselves is evidenced by their efforts at diversification. Argonne is performing ecological studies for the nearby city of Chicago and has contracts with the federal air pollution and water quality agencies. It also does medically oriented research for the National Institutes of Health. Brookhaven has financial support from the National Science Foundation for its studies of air pollution, and the National Institutes of Health funds some of this country's most outstanding medical research at this laboratory. At the Lawrence Radiation Laboratory and at the Los Alamos Scientific Laboratory similar efforts are underway in the environmental field. However, such redeployment is not without peril; Weinberg says redeployment must be gradual, in such a way that new activities are natural extensions of the old ones.

A way to make the AEC research laboratories more effective in these changing times has been proposed by AEC Commissioner Wilfrid E. Johnson. He would like to see the separate organizational structures of the labs fused into a single national organization—the AEC Research Institute, perhaps—which would oversee the balance and direction of the research effort. Employees at one laboratory of the Institute would have exactly the same employment rights of any other laboratory, so that project efforts could be changed, old projects abandoned, and new ones started without layoffs and new hiring. To accomplish this radical change without making the employees civil servants (they are now employees of the contractors who operate the labs) would require a great deal of administrative foresight and skill. Other arrangements would have to be made for the weapons laboratories and the specialized engineering labs that perform little or no basic research.

If we assume that astutely applied technology is part of the answer to our environmental and social maladies, the national laboratories of the AEC (also NASA and other federal agencies) are basic national resources that should be used. As John Platt writes in an article in *Science*:

> In the past, we have had science for intellectual pleasure, and science for the control of nature. We have had science for war. But today, the whole human experiment may hang on the question of how fast we now press the development of science for survival.

Management of Large-scale Systems

Technical expertise and facilities can never be the whole answer to social problems. Given a national commitment of resources, one still needs a profound skill in orchestrating scientists (the physical, biological, and social varieties), the engineers, the facilities they employ, and the political environment. Because any large-scale societal project will rarely be supported unanimously, political skill may often be more important than technical skill. The AEC, NASA, and the other federal agencies created to perform specific tasks (the "mission-oriented" agencies) have never been free of controversy. The political lessons learned in working with conservation groups—during the Plowshare program, for example—should be applicable to the political intricacies associated with cleaning up the environment. (Everyone wants a clean environment, but most do not yet realize what they will have to give up to get it.) Perhaps politics seem trivial compared to the magnitudes of the problems facing us, but it takes people to get things done, and people, whether on the right or left, are political animals.

We are just learning how to manage really big projects—projects so large that hundreds of thousands of people and billions of dollars in facilities are involved. Good management means making good decisions. Decisions must lead toward the objective desired with the least expenditure of resources. The computer, of course, has given managers better data faster; but the management of truly large-scale systems entails more than just information gathering and summarizing. The task of depolluting the environment, for example, involves thousands of interacting physical variables, to say nothing of social and political factors. Faced with such complexity, it is tempting to fall back on experience and employ rule-of-thumb methods. A conservationist might say: *No* more chemicals or hot water may be discharged into our streams. This is an understandable "gut" reaction after seeing what has been done to some of our rivers and lakes. Even if such a dictum were tempered with a list of maximum allowable concentrations of chemicals (as is already the case for radioisotopes) and minimum allowable temperature increases, the ramifications would be far-reaching in terms of cost increases in the affected products. Watchdog personnel would have to monitor the rivers. "Gut" or rule-of-thumb decisions may sometimes be correct because man's intuition somehow cuts through morasses of data to find good solutions in many circumstances. Nevertheless, the bad decisions of history were generally made because of bad data or bad analysis of data. When playing with the environment, world peace, and delicate social problems, management errors cannot be tolerated. New techniques for managing large-scale systems must be created.

The Manhattan District gave the United States its first real taste of managing a large-scale technological effort. Another management success involving the atom was the development of the Polaris nuclear submarine weapons system under the direction of Admiral William F. Raborn, Jr. Here again was a big, complex job that had to be done quickly. It was during the Polaris program that the PERT (Program Evaluation Review Technique) management technique was born. This computer-based management approach has spawned several other management systems that enable the decision maker to see critical problems and their interrelationships clearly amid the glut of information presented to him. The Apollo moon rocket effort has also contributed new management techniques. It has been, in fact, these big technical programs that have revitalized management thinking. Industry has been quick to adopt PERT and its descendants, but the impact upon large-scale, societal systems, such as big city government, has been slight so far. PERT and the computer are not cure-alls, rather they are part of the technological foundation that enables us to encompass the complexities of modern society. There is no going back to the subsistence farm and there is no returning to the old style of management.

Nuclear Economics on a Grand Scale

When the United States Congress was drafting the Atomic Energy Act at the end of World War II, it was very concerned about how to bring the peaceful applications of nuclear energy into the country's traditional free enterprise system. In the act, Congress established the policy of encouraging the uses of nuclear energy in ways that strengthened free competition in private enterprise.

Under this policy and those of successive administrations, the AEC has consistently encouraged private, competitive participation in the nation's nuclear program. It has also discontinued work in government facilities as competence in private industry increased. As a result, there has been a smooth transition from complete government dominance to a vigorous, competitive private industrial complex.

The Power Reactor Business. The United States private investment in the construction of nuclear power plants was approximately $4 billion in 1970. Looking ahead in terms of 1970 dollars; by the end of 1980 this investment should total over $50 billion, rising at an annual rate of about $7 billion. Cumulative expenditures for nuclear fuel will reach roughly $15 billion by the end of 1980, and the annual figure will be about $2.5 billion. Thus the total cumulative expenditures by 1980 will likely be around $65 billion, increasing about $10 billion annually.

If the nuclear power growth projections presented in Table 2 hold true to the year 2000, almost $400 billion will have been invested in nuclear power plants by that time, increasing then at the rate of $60 billion annually. Again these figures might be increased by 25 to 30 percent to include expenditures for nuclear fuel.

The Uranium Enrichment Business. Uranium enrichment is the only portion of the fuel cycle—from mine to power generation—that is not in the hands of private industry in the United States today. The huge gaseous diffusion plants in the United States are now owned by the federal government, which enriches uranium as a service to nuclear power plant owners all over the world—an operation known as *toll enrichment.* It is inevitable, though, that private industry will become involved in the enrichment part of the fuel cycle. Only with industry's participation can the great expansion in the required production be achieved. AEC Commissioners Wilfrid E. Johnson and Clarence E. Larson have been prominent in encouraging a stronger industry role.

Affecting the transfer to private industry has been the development of enrichment processes that strongly compete with the diffusion process. So far, the most promising of these late competitors is the gas centrifugation process, in which the heavier uranium-238 atoms are spun out and separated from the uranium-235 atoms. Tens of thousands of centrifuges would be required to enrich enough uranium for power reactors being built. As of 1970, no full-scale gas centrifuging plant with appreciable capacity had been built. One drawback has been the reluctance of the nuclear powers to divulge enrichment technology of any kind. The reason, of course, is that enrichment technology is important in the manufacture of nuclear weapons. Pressures to pull back the curtain obscuring enrichment methods are increasing. Although it is unlikely that enrichment technology will be made freely available to everyone, it is probable that a limited number of industrial concerns in the United States and friendly nations will participate in the expansion of this technology.

Uranium enrichment is big business and will require large investments. Already, $100 million is spent annually to enrich uranium for American power reactors. By 1980, this figure should rise to $600 million, with a cumulative total of $3 billion, exclusive of fuel services to foreign countries. As plutonium begins to take the place of enriched uranium as the primary nuclear fuel (see later discussion in this section), the demand for enriched uranium will tend to level off—probably in the early 1990s—but by 2000 the cumulative United States investment in enriching services should reach $30 billion, with an annual rate of $1.5 billion. These figures are consistent with those given earlier for the growth of the reactor industry. Present enrichment plants cannot begin to meet these projected demands. Large new plants, using either gaseous diffusion or centrifugation or some other process, will have to be added, probably at the rate of one each year or two in the early 1980s. Satisfying the world's desire for electric power is obviously big business.

The Radioisotopes and Allied Products Business. The AEC has successfully encouraged the growth of a viable private radioisotopes industry. At present, about 100 United States firms produce radioisotopes, sealed sources, radiopharmaceuticals, and related equipment for medicine, science, and industry. The market for radioisotopes has been growing at an annual rate of 10–12 percent; radiopharmaceuticals enjoy a 25 percent annual growth rate.

Sales for 1970 in the United States are estimated at $10 million for basic radioisotopes, about $12 million for radiochemicals, $32 million for radiopharmaceuticals, and $5 million for sealed radiation sources. The total is about $60 million in annual sales. In addition, devices and equipment in which radioisotopes are employed have an estimated market of slightly over $50 million annually. The overall annual commerce in radioisotopes in the United States was about $300 million in 1970, including auxiliary materials, radiation processing, and related services.

The Plutonium and Heavy Isotopes Business. Both water-cooled reactors and fast breeders produce plutonium in quantities that will rise rapidly with time. This is an inescapable consequence of the growth of nuclear power. As this plutonium is returned to power reactors in the form of new fuel, substituting in part for enriched uranium, it will be subject to the law of supply and demand. In this context, we present some long-range forecasts.

By 1980, the annual rate of plutonium recovery from spent reactor fuels will be about 20,000 kilograms (20 tons), with a value of $150 million. The cumulative production for the 1970–1980 decade (consistent with previous power projections) will be more than 80,000 kilograms. By 1990, water-cooled reactors might be producing 60,000 kilograms of plutonium per year for refueling purposes. By this time, its value on a weight basis will be even higher because it will be possible to use plutonium in the more efficient fast breeder reactors; we estimate this value to be about half a billion dollars for the production that year. The value of *all* the plutonium in existence that year could be something like $6 billion. For comparison, United States investor-owned electric utilities spent $3 billion in 1969 for all fuel needs.

By the year 2000, the United States water reactors will have been partially displaced by fast breeder reactors and will produce only about 35,000 kilograms of plutonium per year. Plutonium will be in short supply as the need to start up new fast breeders sharpens, even though breeders at that time will be producing more plutonium than they are consuming, the excess being about 80,000 kilograms per year. The combined value of the plutonium from both kinds of reactors may then have reached $1.5 billion per year and the value of all the plutonium in existence in the year 2000 could approximate $18 billion.

A little arithmetic will show that plutonium will be fluctuating in value from about $7 per gram to about $15 per gram in the last three decades of the twentieth century. In equivalent English units, this is a range of $220 to $470 per troy ounce, compared to the present value of gold at roughly $35 a troy ounce. Even more striking, since the value of the *world's annual production* of gold in 1968 was $1.4 billion and can be expected to remain fairly constant, we can foresee that the value of annual plutonium production in the United States alone will exceed the value of the world's annual gold production around the year 2000. Some have surmised that plutonium could even replace gold as the international monetary standard—at least it has real intrinsic value.

Should high-temperature gas-cooled reactors, molten-salt reactors, and light water breeder reactors come into their own toward the end of this century, they will probably operate on the fuel cycle which converts natural thorium to uranium-233. Uranium-233 would then supplement plutonium production. However, it is difficult to forecast the degrees of success these various types of reactors will enjoy.

Several other valuable materials can be produced in power reactors and as by-products of nuclear fuel reprocessing. Plutonium-238 and curium-244, for example, are in demand for heat sources. Californium-252 can also be manufactured as part of the power economy; it has medical and industrial applications. If nuclear power expands as our illustrations suggest, it is possible that by the year 2000 we could be producing tens of thousands of kilograms of plutonium-238, thousands of kilograms of curium-244, and kilograms of californium-252 as valuable by-products of commercial power generation. The aggregate value of such by-products would be approximately the same as that of plutonium fuel (plutonium-239) suggested above, thus doubling the total value of the synthetic heavy isotope business.

The increasing magnitude of commerce in radioisotopes requires that more and more shipments of these potentially dangerous materials be made along the nation's transportation arteries. From one standpoint, this means increased business for the transportation industry; but we also must recognize that hazards to the public are also increased. The AEC and industry are already planning ahead to meet these expanded requirements while still maintaining the present excellent safety record.

The Atom and Automation

Computers and automatic machines have displaced men and women from many menial jobs—and generally for the better because the humans have generally moved to more challenging tasks. Machines can in effect upgrade employees by forcing them upward out of the mindless office and factory jobs that are still too common today. Not that all menial jobs have now been taken over by automata; the industrial revolution will not be complete until all men are upgraded into endeavors requiring man's unique capabilities of innovation, artistry, synthesis, and morality. Automation, therefore, comprises one of the strongest social forces unleashed by technology.

The mind's eye sees automation in terms of factories filled with shiny, marvelously contrived machinery turning out endless streams of canned soup, light bulbs, and automobile engines. The atom seems remote—and probably not involved at all amid the whirrings and oscillations. Such vision is based on superficialities. Just as the atom is present in almost every hospital, so it is ubiquitous in automated factories.

The history of the industrial atom is, in microcosm, the story of how government-supported science can be applied to increase the productivity of industry and thereby expand the national economy. Fifty percent of the 500 largest manufacturing concerns in the United States employ radioisotopes in one way or another. About 4500 other firms are licensed by the AEC or by the individual states to use radioisotopes. Innumerable other companies and laboratories use small quantities of radioisotopes which do not require the issuance of licenses. Virtually every type of industry benefits from these silent atomic servants; the following industries are, for example, important users:

Metals	Electrical
Chemicals	Transportation
Plastics	Pharmaceutical
Paper	Petroleum-refining
Rubber	Stone, clay, and glass
Food	Tobacco
Textile	Mining

Radioactivity is an ideal source of radiant energy for automated processes. Radioactivity is 100 percent reliable; it cannot be turned on or off or affected by the machine environment around it. Radioisotopes are used in many important ways in industry: for example, in (1) gauging (measurement); (2) radiography; (3) activation analysis; (4) tracing; (5) radiation processing; and (6) luminescent signs. Some examples of these applications will demonstrate the pervasiveness of the industrial atom.

Gauging. The atom, it seems, is a measurer *extraordinaire*. The thickness of paper, sheet metal, virtually any material can be measured, accurately and automatically, by radioisotope gauges. By placing a radioactive source on one side of sheet metal as it moves along a processing line, changes in thickness can be detected as variations in radiation intensity monitored by an instrument on the other side. Conversely, if the thickness of the material is known accurately, the same sort of gauge can measure the density of the product. Radioisotopic weight gauges are also in common use. In the cement industry, for example, the mechanical weighing of cement clinkers is very difficult in the dusty, poorly lit processing tunnels. A radioisotope gauge weighs the clinker within ±1 percent. Anything from potatoes to coal in railroad cars can be weighed automatically and with precision. Level gauges are also prevalent in industry; when a fluid pouring into a bottle or other container interrupts a beam of gamma rays, a signal from a gamma-ray detector immediately shuts off the valve while another container is automatically moved into place. Radioisotope gauges are in essence the "eyes" of these automata. Many of the people who used to perform these tasks are now putting their talents to better uses. (See illustration on following page.)

Radiography. Rather than employ large X-ray machines to examine big castings and other similar products, industry makes use of the penetrating gamma rays emitted by radioisotopes. The radioisotope sources are portable and more versatile. For example, in checking the welds on heavy-walled steel pressure vessels for cracks and voids, it is a lot easier to move in a gamma-ray source than a high-voltage X-ray machine. The same conditions hold for ship construction and other large fabrication tasks. Even for smaller pieces of equipment, such as railroad car wheels, gamma radiography is cheaper and more convenient than X rays.

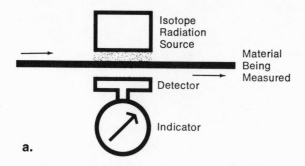

a.

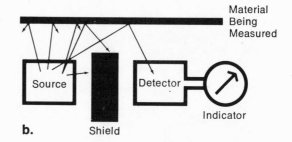

b.

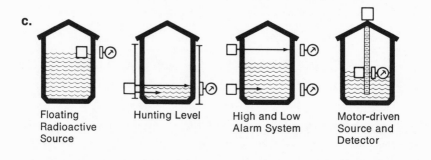

c.

| Floating Radioactive Source | Hunting Level | High and Low Alarm System | Motor-driven Source and Detector |

(a) In this radioisotope thickness gauge, the fraction of radiation absorbed by the moving material is a measure of its thickness. (b) The backscatter gauge measures the thicknesses of thin coatings applied to materials. The thicker the coating, the more radiation is scattered back into the radiation detector. (c) Many of the different types of radioisotope level gauges depend upon the absorption of some of the radiation by the fluid being measured. The gauge on the far left is, of course, an exception to this generalization. (*U.S. Atomic Energy Commission*)

Activation Analysis. Activation analysis is so widely used in all aspects of technology that it has permeated many portions of this book. In industry, activation analysis checks chemical products for impurities, which may be present in such minute quantities that conventional chemical analysis is impractical. Automated systems which reject faulty or "nonspec" material with the help of activation analysis equipment are now being generally accepted by industry. In one instance, the continuous sorting of coal according to its sulfur content appears promising. Such quality control of coal is essential in meeting the tight new restrictions on sulfur content of coal burned in urban areas. The same approach is being studied for sorting metal ores; that is, picking out the richer rocks from the gangue. In another case, activation analysis controls the addition of small quantities of metallic salts in the paper-making process.

Tracing. Radioactive tracers are not used so much for routine industrial operations as they are for solving unusual problems or for optimizing a process before it is scaled up to mass production. The technique is usually quite simple in principle: a small amount of radioactivity is added to a much larger quantity of nonradioactive material, such as a petrochemical in a gasoline refinery. The unparalleled sensitivity of radiation detection instruments allows engineers to follow (trace) the "tagged" material through the entire process. A few examples follow.

Radioactive tracers are especially good for measuring the wear of machine parts. If a piston ring of an internal combustion engine is made radioactive by inserting it in a nuclear reactor for a short time, tiny bits of it that wear off during engine operation can be measured. In practice, piston ring wear at different engine speeds, with different lubricant types, etc., is determined by taking samples of the lubricating oil and measuring their radioactive contents. Wear rates far too small to be seen as dimensional changes can be measured quickly and accurately by tracers. Similarly, the cylinder walls of diesel engines and the gears of earth-moving machinery are studied for wear with tracers. Gasket leakage can also be detected by adding a tracer to the engine coolant and measuring the speed with which it appears in the lubricating oil.

One of the earliest applications of tracers—and one still used—was in locating obstructions in underground (or otherwise inaccessible) pipes. The standard technique is to propel an encapsulated radioactive source down the pipe by mechanical or pneumatic means until the obstruction stops it. A survey party then travels along the path of the pipe above ground until they detect an increase of radiation with a portable meter. In one case, a pipeline twenty-five miles long, which had not been used for many years, was found to be plugged when an attempt was made to put it back in service. A small capsule of radioactive gold was sent along the pipe until it stopped. The above ground crew located the obstruction to within twenty-five feet and removed it. The expensive alternative was the abandonment and replacement of the pipeline.

Today, almost every industrial laboratory of any size applies tracers in myriad ways, ranging from microchemical analysis to tests in complete refineries and chemical plants. The huge catalytic cracking chambers in oil refineries, which convert heavy oils into the lighter fractions needed for transportation and the manufacture of many chemicals and plastics, are designed with the help of tracers. The catalyst, which causes the rapid breakdown of the heavy oil, is a suspension of solid particles. To determine how these particles circulate in the "cat cracker" and to see if they break down by abrasion or chemical reaction, some of them have radioactive material incorporated, or they are made radioactive by exposure in a reactor. Because the radioactively tagged particles follow the same paths as the much more abundant nonradioactive particles, engineers can deduce overall flow patterns by merely watching the tracers with radiation detectors. Radioactive gas can be used in a similar way. The upshot is that gas-flow patterns can be improved, leading to increased process efficiency. Considering the fact that billions of dollars' worth of chemicals flow through chambers of this type each year, tracer experiments leading to even small increases in efficiency pay for themselves many, many times over.

Some other typical industrial applications include the determination of the efficacy of detergents in washing clothes, the measurement of the flow rates of large volumes of material (sewage, for example), and the measurement of the rate of abrasion of paints on highway surfaces.

Radiation Processing. Processing with nuclear radiation is an infant but growing industry. In this application, radiation substitutes for the catalysts in accelerating chemical reactions. Radiation, for example, stimulates cross linking in plastics and thus makes them harder and more durable at high temperatures. Radiation-induced polymerization occurs when radiation causes molecules to link up in the desired fashion. Polyethylene is a well-known plastic that can be made by this method. Pilot plants are already manufacturing plastics with radiation catalysis. The new plants using radiation are more easily controlled and produce a purer material than the conventional plants.

Composites are materials in which two or more other materials are combined to create a new substance with better properties than any of its constituents. To illustrate, combinations of wood and plastic are already on the market as flooring. Because of the plastic content, the flooring can be made almost any color, and it is more durable than wood alone. This flooring is manufactured with the help of radiation. Another new, very interesting composite material combines plastic and concrete. The radiation-treated plastic-concrete composite is much stronger and more resistant to corrosion than concrete alone.

Luminescent Signs. The venerable radium-dial wristwatch illustrates how radioactivity can stimulate light emission from a phosphor. The large-scale availability of man-made radioisotopes, particularly tritium and krypton-85, has made it possible to apply this phenomenon to many other industrial uses. All large passenger aircraft are now required to have exit markers that will glow in the dark without the use of electrical power in the event of an accident. The Arizona Highway Department employs the same technique in lighting highway signs that are not located near a supply of electricity. This application of radioisotopes is just beginning to be exploited for practical purposes.

A New Isotope, New Uses. Although the atom now plays an important part in industry as an adjunct to or replacement of man, it is destined to do even more. We will take as an example the single radioisotope, californium-252, already shown to hold much promise in medicine (see Chapter 8). Known only since 1952 and considered a laboratory curiosity until a few short years ago, californium-252 is now being produced in quantity at the AEC's Savannah River Plant. By "quantity" here we mean a gram or more per year. But with this isotope a gram goes a long way. Californium-252 is unique among the 2000 or so known isotopes. It is the only one that emits neutrons spontaneously at a copious rate (more than 2 trillion per second for each gram) and yet decays slowly enough so that it can be produced, distributed, and used effectively.

The neutrons given off by californium-252 have the same properties as those generated by the fission of uranium or plutonium in a reactor. With californium-252, however, one can take the neutrons to the place of application in a convenient and inexpensive manner. It is this portability factor that makes californium-252 useful and reliable in industry as well as in medicine. Indeed, we have called californium-252 sources "hip-pocket reactors"—figuratively, not literally.

Dozens of United States industrial and government organizations now have been supplied with californium-252 sources (the largest sources are a few thousandths of a gram and many are only a few millionths of a gram) so that uses of this peculiar isotope can be developed and exploited. Although only a beginning has been made so far, the uses seem to fall into two categories: neutron activation analysis and neutron radiography.

We have explained the technique of neutron activation analysis at some length previously, so it remains to be demonstrated how the use of portable neutron sources is advantageous. Take the minerals industry. Typically, a geologist goes into a wilderness area to try to find a new deposit of a valuable ore—gold, copper, titanium, nickel, and a host of others. He looks for promising surface indications and takes a few samples for laboratory analysis. He then returns to his home base to await the result of the analysis. This sequence may have to be repeated several times. With a portable neutron source and suitable electronics equipment, the nuclear geologist can analyze the surface of the ground without even picking up a sample—and the analysis will show the concentrations of essentially all metals and not just the one the conventional geologist might have sought. If this sounds a little farfetched, it isn't. The technique has been demonstrated already by the United States Geological Survey.

In similar fashion, in field or factory installation, portable neutron activation equipment can be used for quality control of chemical products, for sorting out high sulfur coal from the more desirable low sulfur coal, for locating oil bearing strata in well logging operations, for beneficiation of iron ore—and the list goes on and on.

What about neutron radiography? Unlike X rays and gamma rays, neutrons are absorbed more by certain low density materials (such as water, plastics, and organic compounds) than by metals and other dense materials. Thus a "neutrograph" of an animal (or a human) will show soft tissue variations with great clarity, but bone is hardly distinguishable. An ordinary X-ray picture of a rifle cartridge will clearly define the metallic components but cannot show the explosive well; the neutron radiograph shows the powder in stark contrast. Industrial radiography with X rays and gamma rays is now commonplace, and a whole new vista will be exposed when portable neutron radiography becomes available.

Atomic Forensics

Forensic science has come to mean the science of crime detection and resolution. Since crime is without question a significant force in modern society, the part the atom plays in tracking down criminals is pertinent to this chapter. Crime has been estimated to be a $40-billion "industry," with most of the cash flow originating in the poverty-stricken segments of society. We cannot claim that atomic science gets at the roots of crime, but it is a powerful weapon in modern criminology.

Activation analysis is the atomic sleuth. Activation analysis enables police to detect and identify minute bits of matter, such as a residue of gunpowder on a suspect's hand, and thus connect crime and criminal. Activation analysis is such a powerful weapon that international conferences on forensic activation analysis are held regularly. Sherlock Holmes was able to connect crime and criminal by observing a singular deposit of soil on a suspect's shoes or pants cuffs; the modern detective uses activation analysis to prove incontrovertibly that the suspect carried away a bit of the soil from the scene of the crime.

The *forte* of activation analysis is, of course, the measurement of extremely small concentrations of specific elements in a sample. By bombarding the sample with neutrons in a reactor and then observing the gamma rays and nuclear particles released, samples from the criminal and the scene of the crime can be matched with precision sufficient enough to constitute legal evidence in many criminal cases. It is in these forensic applications that activation analysis has been described as taking "atomic fingerprints."

Analysis of gunpowder residues is one of the most common forensic

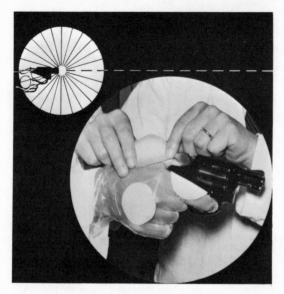

Minute quantities of gunpowder residue from muzzle flashback can often be detected on a suspect's hand by activation analysis. Frequently, the powder manufacturer can be pinpointed. (*U.S. Atomic Energy Commission.*)

applications of activation analysis. The powder in the cartridges fired during a crime can frequently be linked positively to the "flashback" residues on suspects' hands. However, the classical stories of activation analysis involve long-dead Napoleon and King Eric XIV of Sweden. Both monarchs died under suspect circumstances. In Napoleon's case, as previously mentioned, activation analysis of his hair revealed unusual concentrations of arsenic. Perhaps he was systematically poisoned to death. When King Eric's body was exhumed recently, rumors four centuries old that he had been poisoned were confirmed by activation analysis. Even time cannot conceal crimes from probing neutrons.

Looking ahead a few decades, one can see how teleoperators might also come to the aid of hard-pressed law officers. Imagine a heavily armed criminal barricaded in a house surrounded by police. Rather than risk a shootout, the officer in charge sends in a teleoperator, an armored, dexterous machine (not necessarily manlike in appearance) which, controlled by a policeman located in a safe place, can do almost anything a flesh-and-blood policeman can do. With its superhuman strength, the teleoperator forces the door, finds the criminal with its television eyes, and subdues him with a tranquilizer gun or gas.

Chapter 10

The International Atom

On December 2, 1942, the birth of the nuclear age was announced by the following message, "The Italian navigator has landed in the New World." This was the secret code that told waiting scientists that Enrico Fermi and his co-workers had succeeded in releasing and controlling the energy in the nucleus of the atom. This newly released energy was first applied for military purposes, but the United States has sought to give meaning to the prophecy of Isaiah—"And they shall beat their swords into plowshares, and their spears into pruning hooks." This will be the theme of this chapter.

Scientists from many countries over a period of fifty years had contributed to Fermi's remarkable achievement. A listing of some of the transuranium elements: curium, einsteinium, fermium, hahnium, lawrencium, mendelevium, nobelium, rutherfordium (or kurchatovium), reads like a roster of the towering figures of international science. United States initiative, however, set in motion the program for international cooperation known as the Atoms for Peace program. History may well view this initiative, which resulted in unprecedented dissemination of scientific and technical information across national boundaries, as one of the greatest contributions the United States has made for the promotion of peace and the betterment of mankind.

In December 1953, President Eisenhower made his historic address, "Atomic Power for Peace," before the General Assembly of the United

Nations, proposing an Atoms for Peace program and the establishment of an international agency to promote peaceful applications of nuclear energy. On that occasion the President said:

> I would be prepared to submit to the Congress of the United States, and with every expectation of approval, any . . . plan [for international cooperation] that would:
>
> First—encourage worldwide investigation into the most effective peacetime uses of fissionable material, and with the certainty that they [all countries] had all the material needed for the conduct of all experiments that were appropriate;
>
> Second—begin to diminish the potential destructive power of the world's atomic stockpiles;
>
> Third—allow all peoples of all nations to see that, in this enlightened age, the great powers of the Earth, both of the East and of the West, are interested in human aspirations first, rather than in building up the armaments of war;
>
> Fourth—open up a new channel for peaceful discussion, and initiate at least a new approach to the many difficult problems that must be solved in both private and public conversations, if the world is to shake off the inertia imposed by fear, and is to make positive progress toward peace.
>
> Against the dark background of the atomic bomb, the United States does not wish merely to present strength, but also the desire and the hope for peace.

The response to President Eisenhower's proposals was overwhelming, and discussions were begun that ultimately culminated in establishment of the International Atomic Energy Agency (IAEA) in 1957.

Meanwhile, in order to pursue the United States Atoms for Peace program, President Eisenhower submitted recommendations in February 1954 for amending the Atomic Energy Act of 1946, which had severely limited American cooperation with other nations in development of peaceful uses of the atom. At that time, the President noted that the recommended revisions of the Atomic Energy Act would enable "American atomic energy development, public and private, to play a full and effective part in leading mankind into a new era of progress and peace." The Congress shared the administration's views and enacted the Atomic Energy Act of 1954, which authorized broad domestic and international programs of peaceful nuclear development.

The following month, the ninth session of the United Nations General Assembly was convened. In line with the President's 1953 proposals, the United States submitted a resolution for the development of an international cooperative program in the nuclear energy field.

In November 1954 the United States ambassador to the United Nations, Henry Cabot Lodge, gave a report to the United Nations on United States efforts to establish a reactor training school, to provide courses in safety and other constructive applications of the atom, and to offer technical information and "Atoms for Peace" libraries to other countries. Ambassador Lodge noted that there were already ten such libraries available. That number has since grown manyfold, as more and more countries have moved into the Atomic Age.

Cooperative Arrangements

The new Atomic Energy Act was signed by President Eisenhower on August 30 of 1954.

Early in 1955 the Department of State and the Atomic Energy Commission began negotiating bilateral agreements for cooperation under the new Act. By the end of 1955, some twenty-five such agreements had been negotiated. These agreements were designed to increase the pace of peaceful nuclear energy activities throughout the world; to provide an opportunity or a vehicle for making available assistance to other countries; and to speed peaceful nuclear applications in cooperative countries in order to strengthen these countries economically and technologically.

The first agreement was negotiated with Turkey—an indication that the program was intended from the start not solely for technologically advanced countries, but particularly for nations that saw in science and technology one means to better the lot of their peoples. At one time, these agreements were in effect with more than forty individual countries. With the development of the International Atomic Energy Agency and the European Atomic Energy Community (Euratom), however, some of America's bilateral partners agreed to let the bilateral agreements expire and to obtain the benefits available under the agreements through these two international organizations. At the end of 1970, the United States had agreements for cooperation with twenty-eight nations and with the IAEA and Euratom. Under the terms of these agreements, the United States has supplied her partners with nuclear research tools, including research reactors, and nuclear fuels for both research and power reactors and, of course, information on various peaceful applications.

These agreements are divided into two types: (1) the research agreements are so named because they provide for the supply of nuclear materials—especially enriched uranium—for research reactors. The amount of nuclear material supplied under these agreements is rather limited, and the term of the agreement is usually only five to ten years; (2) the power agreements authorize a broad exchange of unclassified technical information on power reactor technology and the application of nuclear energy to peaceful uses. These agreements, of duration up to thirty years, also provide for the sale of thousands of kilograms of uranium-235 fuel and smaller amounts of plutonium for reactors constructed overseas.

Both of these types of agreements include unique safeguards provisions against the diversion of this fissionable material to military uses. The importance of these safeguards is so great that we will later discuss this subject in some detail.

East-West Cooperation. In a world often torn by East-West strife, it is somewhat surprising to find that the Soviet Union and the United States have had a formal agreement for cooperation on the peaceful uses of atomic energy since 1959. Informal exchanges of technical information occurred even before that. The basic agreement was signed in Moscow by Ambassador L. E. Thompson for the United States and G. A. Zhukov for the Soviet Union. The agreement has been implemented by a series of "Memoranda for Cooperation" signed by the respective heads of the two nations' atomic energy programs. The first such memorandum was signed in Washington, in November 1959, by AEC Chairman J. A. McCone and Professor V. S. Emelyanov, Chairman of the Soviet Main Administration for Atomic Energy. In May 1963, a new Memorandum for Cooperation was signed in Moscow by the senior author and A. M. Petrosyants, the Chairman of the U.S.S.R. State Committee for the Utilization of Atomic Energy. The memorandum was renewed again in July of 1968, in Moscow, by former AEC Commissioner G. F. Tape and I. D. Morokhov, and once again, in February 1970, by the senior author and Chairman Petrosyants. The agreements provide for the exchange of technical reports and for visits by expert delegations in such fields as nuclear reactor technology, plasma physics, controlled thermonuclear reactions, high energy physics, radiation chemistry, biology and medicine, and radioactive waste disposal. Long-term assignments of scientists from one country to the other country's laboratories are also arranged under these memoranda.

A hopeful sign for the extension of this kind of cooperation may be found in the agreement between Romania and the United States. This document was signed in 1968 by Professor Horia Hulebei, Chairman of the Romanian Committee for Nuclear Energy, and the senior author. It was renewed in 1970. One final illustration: an informal agreement for the exchange of technical information exists between the AEC and the Central Research Institute for Physics in Budapest.

World Nuclear Power

Despite the great importance of the many existing applications of nuclear technology, such as radioisotope tracers, the most significant form of cooperation under the Atoms for Peace program has involved nuclear power. The technology which the United States has made available, and perhaps even more importantly the nuclear fuel and reactor construction materials it has supplied, have been important ingredients in the nuclear power programs of most other nations. In addition to the political importance of this form of cooperation, it has tremendous economic impact both for the United States as a supplier of the nuclear fuel and for cooperating countries, which can thereby reap the economic benefits of reliable, low-cost nuclear power. Nuclear power has proven particularly attractive to those countries which lack adequate low-cost domestic fuel resources. This includes some of the most heavily industrialized countries in the world, such as Japan and Italy.

During his tenure as Chairman of the Atomic Energy Commission, the senior author has visited some sixty countries, including virtually all those countries in the world (some twenty-five) now actively applying nuclear power. Each of these countries realizes that its rapidly growing demand for electrical power requires the efficient use of all energy sources available in the world today—nuclear, fossil fuel, and hydro power. By 1980, it is estimated that some 100 million kilowatts of electrical generating capacity—two thirds of the expected United States total—will be in use abroad, employing light water reactors of the type pioneered by the United States and fueled largely with enriched uranium provided by the United States. This enriched uranium is made available through the provision of enriching services on the very same terms and conditions, including prices, which apply to domestic customers of the AEC. By 1980, overseas sales of these enriching services are expected to total more than $2 billion and to reach an annual rate of $400 million. These earnings make an impressive contribution toward balancing the flow of hard currency into and out of the United States. At the same time, foreign customers receive an important service at reasonable prices.

By the time the year 2000 rolls around, the world will have to build the equivalent of five thousand 1000-megawatt power plants to meet global needs, according to present estimates. To attain this growth, new plants must be placed on the line at a rate equivalent to one 1000-megawatt plant per day at that time. Considering that only 350 such plants would be needed to supply the entire United States today, these figures for A.D. 2000 are almost incomprehensible. All energy resources of our small planet will be needed to satiate our appetite for power. Nuclear power, of necessity, will be a predominant factor.

World interest in and world desire for nuclear power bring certain problems. To create the framework for a sober discussion of how the benefits of nuclear electric power generation might be provided to other nations without upsetting fragile political balances, we would like to introduce several aspects of international nuclear power that go beyond the statistics and projections given in Chapter 2 and above.

First, it must be recognized that nuclear power may be economical in one corner of the world but not in another. The United States produces conventional electric power very cheaply by world standards. Nuclear power must sell for about the same price as fossil-fuel power to be competitive. To meet this price competition, nuclear power must be supplied in large blocks—800 megawatts and above. In a small, developing country, a nuclear power plant generating 800 megawatts could easily supply one-half the country's entire demand, but the country would not have the facilities to distribute the power from such a large, central source. In addition, it could not afford to have such a large fraction of its power capacity tied up in a single unit. After all, nuclear power plants do have to be shut down occasionally for maintenance and refueling. What the small countries need, then, if they do not have adequate indigenous supplies of fossil fuels, are modern, efficient, nuclear power plants in the power range from 200 to 500 megawatts. Unfortunately, the development of small nuclear power plants that are also economically competitive has proven to be a difficult task. Continuing efforts to determine the economic feasibility of these smaller reactors represents an important phase of the nuclear power scenario.

Many of the more advanced countries are exporters of power reactors. Each exporting country offers some feature or collection of features designed to make its reactor system attractive to others. Each country has had some successes in the marketplace. The Canadian reactors, for example, utilize natural uranium. These are attractive for countries possessing their own uranium resources. However, the capital costs of Canadian reactors are high because they require heavy water (at $30 per pound) for moderating and cooling. This may be counterbalanced by lower fuel-cycle costs. In a similar fashion, Great Britain's gas-cooled nuclear power plants, which use natural or slightly enriched uranium, have high capital costs combined with the promise of low fuel costs. Several countries besides the United States and the Soviet Union offer the water-cooled, enriched-uranium nuclear power plants, but with various distinctive features. Germany, Sweden, Belgium, and Japan are notable examples.

When enriched uranium reactors are offered, the exporting country must be in a position to supply, directly or indirectly, the fuel needed throughout the useful life of the reactor. Aside from the Soviet Union and the countries in its sphere of economic influence, the other countries of the world—exporters and buyers alike—have had to rely on the United States and its massive uranium enrichment facilities. This does not mean that all of the original uranium ore comes from the United States. Natural uranium may be shipped to the United States, be enriched there, and then returned to the country of origin. Still, these other countries have begun to worry about their nearly complete dependence upon the United States. Consequently, many discussions and the first concrete actions have transpired in an effort to reduce this dependency. Three European countries (Great Britain, the Netherlands, and Germany) have banded together to develop and build gas centrifuge plants (described briefly in Chapter 9) to manufacture the slightly enriched uranium needed for power plants. (They have agreed that the uranium thus produced will not be used for military purposes.) It is unlikely that the plants when completed will be able to produce initially a fuel competitive in price with the enriched uranium from the larger United States enrichment plants. However, the independence achieved is thought to be worth the cost.

By 1971 a number of additional sources of uranium enriching services were evolving to augment that in the United States. The Soviet Union has entered into contracts to furnish uranium toll enriching services utilizing her gaseous diffusion plants to Finland and France and has offered such services to additional interested countries. England plans to use its diffusion plant to supply the slightly enriched uranium required in its nuclear power program. A number of countries including France, Canada, Japan, Australia have expressed interest in participating in multinational diffusion plants and some of these and others are also interested in constructing or participating in gas centrifuge plants. South Africa has announced plans to build a uranium enrichment plant of its own unique design.

The United States Government has considered making gaseous diffusion technology available to friendly countries under some appropriate system of secrecy and safeguards which would protect the information and ensure that the material produced would be used only for peaceful purposes. Even the secrecy wraps may be discarded sometime in the future.

Financing the construction of large nuclear power plants in foreign countries does not seem to be a major problem—at least for the time being. Although these plants cost around $100 million to build, they are heavy revenue producers when completed and thus good investments. Countries building reactors for export usually offer some kind of financing arrangement if it is needed by the recipient country, often through a government-controlled bank. Under certain conditions, the Export-Import Bank of the United States, for example, will extend credits for the construction of nuclear power plants and for the supply of nuclear fuel. In the fiscal year ending June 30, 1970, "Ex-Im" authorized loans in these categories totaling over $170 million for four different countries. Credits extended to the Kansai Electric Power Company, in Japan, have added up to $170 million over the past several years. In the 1971–1975 time period, the bank anticipates that it will make loans of over $1.5 billion for nuclear projects.

These figures do not represent the complete financing of the power plants; typically the bank provides only about 40 percent.

International Exchange of Technology

Conferences provide an important way to transfer technical information to other countries. As in other fields of science, many different types of conferences are employed, depending on the scope of the subject matter and the interest in it. However, the Atoms for Peace program pioneered a particularly significant form of international conference. Early in 1954, AEC Chairman Lewis L. Strauss announced:

> I am privileged to state that it is the President's intention to . . . convene an international conference of scientists at a later date this year. This conference, which it is hoped will be largely attended and will include the outstanding men in their professions from all over the world, will be devoted to the exploration of the benign and peaceful uses of atomic energy. It will be the first time that any such body has been convoked, and its purpose, also in the words of the President, will be "to hasten the day when the fear of the atom will begin to disappear from the minds of people and the governments of the East and of the West."

As a result of the United States proposal, the General Assembly approved the convening of the first United Nations International Conference on the Peaceful Uses of Atomic Energy at Geneva in August 1955. Subsequent international conferences of this type were scheduled for 1958, 1964 and 1971.

The First Conference was successful beyond all expectations. It was, at that time, the largest meeting ever convened under the auspices of the United Nations. Thirty-eight nations were represented. Over 1000 papers were submitted, and over 2700 participants attended. It was a dramatic conference, wide in scope, and a significant step in opening many international doors previously closed to the new technology of the atom. The United States, the United Kingdom, and the U.S.S.R.—the nations which had already developed military application of nuclear energy—as well as virtually every other nation which had made progress in the peaceful use of nuclear energy—

contributed to the success of this conference by their exhibits and papers.

In his assessment of the First Conference, Professor Walter G. Whitman (the conference secretary general) said:

> To laymen everywhere, the knowledge that the world's scientific elite was exchanging information and ideas about nuclear energy, with the purpose of developing its potential benefits to mankind, was most heartening. Here was long-delayed evidence of international cooperation.

The Second Conference in 1958 was even more dramatic and wider in scope than the first. Forty-six nations and six international organizations were represented. Over 2000 papers were submitted, and over 6000 participants attended. This conference helped break down further some of the formidable barriers to the open exchange of nuclear technology between nations.

The Third International Conference was held in 1964. It came at a major turning point in the development of peaceful uses of nuclear energy, especially the generation of nuclear power. The pessimism generated by early technical and economic problems was being dispelled by the successful operation of prototype nuclear power plants in several countries, and the offer on a commercial basis of large nuclear plants which could compete with conventionally fueled power plants in many parts of the world.

The Fourth Geneva Conference was scheduled under the auspices of the United Nations in cooperation with the International Atomic Energy Agency (IAEA) for September 1971.

Multilateral Cooperation

The IAEA. One of the most important developments in the Atoms for Peace program was the establishment of the International Atomic Energy Agency (IAEA). The IAEA, a member of the UN family, helps all nations benefit from the peaceful atom. It allows them to share scientific and technical knowledge and nuclear materials under international agreements and safeguards. It also serves as a world forum on nuclear knowledge, and it operates its own radioisotope laboratory and institute of theoretical physics. First proposed by President Eisenhower in 1953, the agency was established in 1957 with headquarters in Vienna and, in 1970, had 103 members. The IAEA has enjoyed the distinguished and effective leadership of Sigvard Eklund, a Swedish national, who has served as its director general since 1961. The director general has at his disposal a ten-member Science Advisory Committee made up of top-level scientific representatives from Brazil, Canada, Czechoslovakia, France, India, Japan, the Soviet Union, the United Arab Republic, the United Kingdom, and the United States. Internationally known nuclear pioneers who have served in this capacity include Bertrand Goldschmidt of France, W. B. Lewis of Canada, I. I. Rabi of the United States, V. I. Spitsyn of the Soviet Union, and Sir William Penney of the United Kingdom.

The United States, like several other countries, maintains a permanent mission with formal diplomatic status in Vienna to ensure continuous liaison with the director general and IAEA staff. This mission was headed for many years by Ambassador Henry D. Smyth, a former AEC commissioner and the author of the famous "Smyth Report" published shortly after World War II. It was the "Smyth Report" that first disclosed the nature of the Manhattan Project. Upon Smyth's retirement in 1970, his place was taken by T. Keith Glennan, former administrator of NASA and also a former AEC commissioner.

Over the years, the IAEA has been involved in many areas, ranging from radiation applications in medicine, industry, and agriculture to the promotion of the effective use of research reactors, desalting studies, and establishment of international standards in the transport of irradiated materials. In the field of nuclear power, the agency has sponsored the exchange of information on a global basis. Conferences organized by the agency have been devoted to such important subjects as the siting of nuclear power plants, the safety and environmental problems related to nuclear power, the comparison of nuclear power costs, and the use of plutonium as a reactor fuel. The development of smaller nuclear power plants for the less advanced countries is now a subject under active discussion in the IAEA, with Director General Eklund being a strong supporter of this approach.

The United States has strongly supported the IAEA through financial contributions, the provision of fellowships, experts, equipment grants, technical information, special nuclear materials, and assistance in developing a safeguards inspection system. In the senior author's visits to many of the countries receiving IAEA assistance, he has been able to observe firsthand the effective programs and the benefits resulting from the agency's efforts.

Euratom. Since the establishment of the United Nations, United States foreign policy has stressed the importance of multilateral cooperation to world peace and effective development efforts. The Atoms for Peace program has made use of the multilateral approach in several interesting and important ways. Following the signing of the U.S.-Euratom Agreement for Cooperation in 1958, the U.S.-Euratom Joint Power Reactor Program and the U.S.-Euratom Joint Research and Development Program were initiated. The technical purpose of the joint reactor program was to foster the building of large-scale power plants using reactors that had already been developed in the United States. Such a cooperative program would also serve to strengthen Euratom, one of the important institutions designed to further the goal of European integration, as well as to advance Europe technologically and economically. The three pioneer reactors built under the program in Europe were a key factor in the establishment of the vigorous European nuclear power industry.

The United States and Euratom also have arrangements to exchange information on fast breeder reactor programs and in certain other fields. American supplies of special nuclear materials have been made available for both commercial power programs and research projects through Euratom under lease, sale, and toll enrichment arrangements.

IANEC and ENEA. Other major multilateral organizations have established special organs to promote cooperation among their members in the peaceful uses of nuclear energy. For example, the Organization of American States formed the Inter-American Nuclear Energy Commission (IANEC), which provides a forum for consultation and a mechanism for allocating technical assistance available to the Organization of American States in the field of nuclear energy. The Organization for Economic Cooperation and Development (OECD), formerly the Organization for European Economic Cooperation, created the European Nuclear Energy Agency (ENEA). This body has served as the focal point for the successful formation of a number of important cooperative projects in Europe, including the Dragon High Temperature Gas-cooled Reactor Program in England, the Halden (Norway) Boiling Heavy-Water Reactor Project, the European Chemical Reprocessing Company at Mol, Belgium (Eurochemic), and the Paris Convention on Third Party Liability for Nuclear Damage.

The United States is a member of IANEC, and, since the creation of the OECD, has been an associate member of the ENEA, and has cooperated closely with both organizations in making the benefits of the peaceful uses of nuclear energy available to their members.

Origin of Safeguards

Since the beginning of the Atoms for Peace program, as we have seen, the United States has realized that the success of the program was dependent upon reasonable guarantees that the nuclear technology and nuclear material to be shared with other countries would not be diverted to any military purpose. Guarantees were needed so that none of this material or assistance would ever create a threat to international security. Thus the United States introduced the concept of *safeguards* to implement such guarantees in its bilateral agreements. It was also recognized that a multilateral control system would be more efficient and objective than bilateral safeguards, and that it could contribute to the evolution of a broader system of arms limitation. Pending the establishment of such a system, the United States insisted that the other governments involved provide assurances that nuclear equipment, materials, and their products supplied by the United States would be used only for peaceful purposes. The agreement also gave the United States the right of actual on-site inspection so that it could assure itself that this provision was being carried out. The United States has always considered these bilateral safeguards arrangements as a prelude to a more comprehensive international system which would be needed as additional nations embarked on nuclear programs.

The Atomic Energy Act of 1954 does not require that inspections or any other type of safeguards have to be applied to nuclear materials as a precondition to export. The act stipulates only that the country concerned must guarantee that any assistance given by the United States under the agreement will not be used for nuclear weapons or any other military purposes.

The decision to apply safeguards, to permit verification by the

United States that the sovereign guarantee given by the recipient government was being fulfilled, was a policy decision taken by the United States Executive Branch in 1955, after very extensive consideration. It is a policy that has since been strongly supported by the Kennedy, Johnson, and Nixon administrations and by every Congress since its inception.

Safeguards provide a warning light that a diversion has occurred or may be occurring. They normally include three basic elements: (1) the maintenance and review of records concerning the utilization of nuclear materials; (2) the performance of on-site inspections; and (3) technical security measures to prevent the loss or theft of materials in the course of processing, use, storage, or transit. Safeguards are not foolproof nor should there develop a complacent feeling that they are all that is needed to prevent or detect nuclear weapons programs being carried on secretly in violation of treaty commitments. They have to be supplemented by political and legal restraints. They represent, however, the best political and technical method yet devised to meet the problem of verifying compliance with the solemn pledges of additional countries not to develop nuclear weapons.

The growth of an international system has been gradual but has kept pace with developments in the nuclear field. As a first step, reactors with outputs smaller than 100 thermal megawatts—mostly research, training and test reactors—were covered by the IAEA's safeguards system. Then the system was broadened to include those reactors larger than 100 thermal megawatts, and finally was expanded to include conversion, fabrication, and reprocessing plants.

To further the acceptance of the agency safeguards system, the United States has strongly encouraged its bilateral partners to accept agency safeguards procedures on material and equipment supplied by the United States under the bilateral agreements. This has been done through the negotiation of trilateral safeguards agreements among the United States, the IAEA, and the other country involved. Twenty such agreements were in effect in 1970.

To assist the development of IAEA safeguards, the United States, in 1962, voluntarily placed four of its civilian prototype power and research reactors under the agency system, both as a means of testing the system, and also to give the IAEA safeguards staff experience in conducting safeguard inspections. The agreement was renewed in 1964 and extended to include a 175-megawatt power reactor with the cooperation of the owner, the Yankee Atomic Electric Company, of Rowe, Massachusetts. In 1966, it was announced that a plant for the chemical processing of irradiated fuel owned by the Nuclear Fuel Services, Inc., at West Valley, New York, was being made available to the IAEA to provide additional training for agency safeguards inspectors. The agreement expired on July 31, 1970. Cooperation with the agency will continue in the form of agreed-upon experiments aiding the IAEA in developing safeguards techniques and training its personnel. In December 1967, on the twenty-fifth anniversary of the first controlled nuclear chain reaction, President Lyndon B. Johnson announced that when safeguards were applied under the terms of the treaty to prevent proliferation of nuclear weapons, ". . . the United States will permit the IAEA to apply its safeguards to all nuclear activities in the United States—excluding only those with direct national security significance. . . ." President Nixon reaffirmed this offer of the United States to put these activities under IAEA safeguards control.

On Keeping the Peaceful Atom Peaceful

The United States has taken many steps to share with other countries its accomplishments in the peaceful uses of nuclear energy. The Atoms for Peace program was unprecedented in that it made available to the world great quantities of information dealing with *applied* science and technology—distinct from *basic* scientific information, which is generally available on an international basis. This technological and applied scientific information had been acquired at considerable effort and expense and, most importantly, had great potential value for providing material benefits to mankind. It had enormous economic value for other countries which could put the information to use directly or which could use it as a foundation for further development and improvement of their technology, even at times in competition with American industry. That result was foreseen and it has, in fact, occurred, without serious adverse consequences for United States industry. The United States nuclear power industry competes successfully with British, French, German, Swedish, and other power reactor manufacturers. The international trade in nuclear power reactors, nuclear fuel, and associated services and materials has already become very important, and significant growth is foreseen in the need for energy throughout the world in the years ahead.

Another possible result was foreseen by those who formulated and carried out the Atoms for Peace program. They knew very well that nuclear reactors for the commercial production of heat or electrical energy unavoidably produce fissionable material such as plutonium, which could be used in the manufacture of nuclear weapons. They knew also that a country which increases its capabilities to develop and exploit the peaceful uses of nuclear energy, by training people or by building up its scientific and industrial capacity, inescapably increases its *potential* ability to develop and manufacture nuclear weapons.

The planners of the United States Atoms for Peace program were thus confronted by a dilemma: how to provide the enormous benefits of peaceful nuclear science and technology to mankind without, at the same time, increasing the risks of mankind's destruction from the misuse of the skills, the techniques, and the materials inherent in that effort.

We have already mentioned some of the steps deliberately taken in designing and implementing the Atoms for Peace program in order to resolve that classical dilemma. The important point to note here is that the people involved in that program in the United States Executive Branch and Congress were not immobilized by that dilemma. They could not and did not accept the easy conclusion that the risk of harmful consequences was so overwhelming that mankind had to be denied the benefits. Rather, they carefully and deliberately evaluated the alternatives and charted new courses in international relations to reduce the risks to manageable levels to achieve the high purposes of the Atoms for Peace program.

Early Attempts at International Control

Let us take a more detailed look at the development of United States policy in dealing with this fateful dilemma. The future development and control of nuclear energy was a popular topic of conversation among informed people—primarily physical scientists—long before the end of World War II. As early as the summer of 1944, many scientists had begun to visualize definite postwar problems, goals, and possibilities. An important focal point of this thinking was the Metallurgical Laboratory at the University of Chicago. During this time, the urgent concern of one group of nuclear scientists over the proposed use of nuclear weapons led them to sign the Franck Report, a memorandum urging that the power of a nuclear bomb be demonstrated to possibly forestall its actual destructive use over Japan. It is fruitless to debate what might have happened had this petition been followed.

The first governmental declaration concerning nuclear arms control took place on November 15, 1945, in Washington, at a time when the United States alone possessed nuclear weapons. On that date President Truman, Prime Minister Attlee, and Prime Minister King issued a joint declaration stressing the willingness of the United States, the United Kingdom, and Canada, the three nations which had cooperated in developing the bomb, to join with other nations in sharing, on a reciprocal basis, information on nuclear energy for peaceful purposes. The declaration recommended the creation of a special UN commission to prepare recommendations on the international control of nuclear energy.

Since 1946, American foreign policy concerning international control of the atom and cooperation for its peaceful development has gone through two distinct phases. The first phase lasted from 1946 to 1954. It was a period of embargo and secrecy, when the United States vigorously sought to preserve its monopoly of nuclear weapons on the one hand and to establish the most comprehensive form of international control over the atom on the other. During this comparatively brief and disillusioning period, two significant actions were taken in America. In the first instance the United States passed the Atomic Energy Act of 1946; its provisions not only created a civilian agency—the AEC—but also placed an embargo on the export of nuclear information and materials from the United States. On the international front, during that period, the United States offered some sweeping and important proposals concerning international control of nuclear energy, which were presented to the United Nations by Bernard Baruch, the United States representative to the UN Atomic Energy Commission. These proposals, developed by a committee of United States scientific, legal, and industrial leaders under Mr. Baruch's direction, and in consultation with the United States Congress, were based on the Acheson-Lilienthal Report on International Control of Atomic Energy. This report, published on March 28, 1946, provided the basis for informed public discussion in this country of some of the military implications of the nuclear age.

The Baruch plan, as the United States proposals came to be known, contemplated establishment of an International Atomic Development Authority. Its functions would include:

1. Control or ownership of all nuclear energy activities potentially dangerous to world security.
2. Control, inspection, and licensing of all other nuclear activities.
3. Fostering of the beneficial uses of nuclear energy.
4. Research and development activities intended to put the Authority in the forefront of scientific and technical knowledge of nuclear energy, thus enabling it to comprehend, and therefore detect, any misuse of the atom.
5. Power to control nuclear raw materials and primary nuclear production plants.

Operation of the system would have been by an international civil service, and "immediate, swift, and sure punishment" was to be provided for violators.

Under the Baruch plan, the manufacture of nuclear weapons would have ceased. All existing weapons, then held only by the United States, would have been destroyed as weapons, and the useful nuclear material transferred to the international agency for peaceful purposes.

This United States offer, if accepted and universally adhered to, would have meant the removal of the threat of nuclear weapons at the outset. It would have allowed the nations of the world to enter the nuclear age in a joint and peaceful endeavor. However, the Soviet Union was not ready to take so gigantic a step which might restrict its freedom of action at that time. It is understandable, in retrospect, that this nation should want to achieve a situation of parity with the United States before making serious moves toward arms limitations.

By 1953 it became apparent that the United States no longer had a monopoly of nuclear technology either for military or for peaceful purposes. Several countries had developed substantial nuclear programs of their own. The U.S.S.R. had developed nuclear weapons. No significant progress was being made in the field of arms control or disarmament through negotiations at the United Nations. Progress was being made by scientists, however, in exploring the peaceful uses of nuclear energy. The reports of such constructive uses stimulated a growing interest in the benign uses of the atom both in the United States and abroad.

These considerations led to the second, or *Atoms for Peace,* phase in American thinking—a new and significant phase which has continued without interruption to the present day. It began with the statement by President Eisenhower that the American people should be informed in frank and realistic terms of the then ominous consequences of the impasse on nuclear arms control. At the same time, however, it was thought desirable to place some new and constructive proposals before the world with the aim of converting the dismal climate to one of hope. It was hoped that in the process a new channel of communication could be developed between the United States and the Soviet Union.

Atoms for Peace and International Safeguards

Throughout the history of the Atoms for Peace program, there have been critics who have felt that this program itself has added to the problem of proliferation of nuclear weapons by enhancing the nuclear capabilities of many countries throughout the world. Many of these critics have felt that the United States could, somehow, hold back the hands of time, not cooperate with other countries, and thus forestall the spread of nuclear weapons. These critics have overlooked the fact that it is impossible to keep science under lock and key for an extended period of time.

Before 1950, the United States realized that it no longer had a monopoly of nuclear technology or of the brainpower to exploit this new field. The United States recognized its responsibility to perfect and share these promising benefits with other nations rather than to stand aloof in a posture of splendid isolation. Besides, the United States realized that shortly some other country or countries would be willing to provide nuclear materials and technology to these nations without firm assurances that such assistance would be used solely for peaceful purposes.

The Atoms for Peace program has been one of the United States principal tools in the fight against the proliferation of nuclear weapons. It has been an important vehicle in helping orient the efforts of other countries to peaceful rather than military uses of nuclear energy.

As a result of steady progress through the years, the IAEA now has in operation an effective safeguards system that is suitable for application to a wide variety of peaceful nuclear activities. Moreover, a growing, albeit still incomplete acceptance of the principle of IAEA safeguards has developed among most nations of the world.

As a result of these developments, international safeguards are now actively being applied on a daily basis. Several thousands of inspections have actually been conducted, on either a bilateral, regional, or international basis. In the overwhelming proportion of cases, these safeguards have worked smoothly and to the satisfaction of the parties directly concerned. Thus, they have done more than simply serve their immediate purpose of assuring that particular peaceful nuclear activities are not being used for military purposes. They have demonstrated that the techniques of international inspection are feasible and effective and that they need not be considered an unacceptable invasion of national sovereignty. They have also stimulated the development of an institutional framework and a cadre of properly trained people to be used in implementing any broader agreements in the future. In so doing they have created much of the foundation upon which the Treaty on the Non-Proliferation of Nuclear Weapons has been structured. Safeguards have significance far beyond the field of nuclear energy—for, in applying them, we have seen sovereignty and the old concepts of national prerogatives give way to the broader interests of international cooperation and security. Perhaps man, if he is ever able to live in a secure world founded on the rule of law, may look upon these activities, in retrospect, as fundamental, pioneering steps on the road to true international stability.

But we must remember that the success of the United States in introducing the concept of international safeguards was due in large part to the leadership of American science and industry in the peaceful uses of nuclear energy. What United States science and industry had to offer in this glamorous and exciting technological field was so substantial in both quality and quantity that other countries were willing to give up a measure of their national sovereignty to take advantage of the benefits of United States cooperation. As other countries become less dependent upon American science and industry in the peaceful uses of nuclear energy and undertake programs on their own, or in cooperation with other nations, a growing proportion of the world's ostensibly peaceful nuclear energy programs could be conducted without commitments that they will remain exclusively peaceful. That sobering thought leads us to the discussion of the general subject of proliferation.

Possibility of Nuclear Proliferation

For many years it has been apparent to a number of observers that the progress that was being achieved in the application of safeguards to international cooperation in peaceful uses of nuclear energy would not, in itself, be sufficient to prevent the spread of nuclear weapons. It was recognized that the possession of their own nuclear weapons might appear attractive to some countries for a variety of reasons. In the years since the early United States monopoly, the United Kingdom, Union of Soviet Socialist Republics, France, and China have developed and tested nuclear weapons. Moreover, the general capability of additional nations to acquire their own weapons has been growing steadily.

There are a number of reasons why some nations might feel that acquiring their own nuclear arms would be to their advantage. Nuclear weapons, unfortunately, have been identified in the minds of many observers with the achievement of big power status. Some countries may believe that they must acquire a nuclear arsenal for prestige purposes. Some countries might feel that their security or their political influence might be enhanced by acquiring nuclear weapons. Some countries might seek short-term military or strategic advantage over a neighboring state. Others might be concerned about future threats to their national security should they share a common boundary with an aggressive, expansionist, or hostile nation possessing nuclear arms. Other states might be interested in nuclear weapons as insurance to preserve their traditional neutrality.

These possible reasons why a country might choose to acquire nuclear weapons are mentioned to underscore the point that the risk of proliferation, while real, is based on a variety of political circumstances, and should not be the subject of sweeping generalities. The fortunate thing is that most countries have recognized that seeking their own nuclear weapons would eventually diminish rather than enhance their security. They have recognized that their acquisition of nuclear weapons would inevitably be followed by similar acquisitions by their supposed adversaries, with the result being no improvement in their own military posture, plus a terrible new risk to their stability and survival.

The risk of proliferation remains great. We must face the reality, therefore, that quite a few nations, not now members of the "nuclear club," have the technical capability to join the club if they so desire; the task is to keep this from happening.

There are several ways in which a non-nuclear-weapon state could acquire nuclear weapons. First, it could try to steal them from a nuclear power. This is not a serious threat, however, since it is a reasonable assumption that the nuclear-weapon powers keep their nuclear weapons under the closest control. A non-nuclear-weapon state, however, might try to persuade a nuclear-weapon nation to provide it with weapons. Fortunately, to date, the nuclear-weapon states have shown no inclination to share their capabilities with other nations, although in the absence of a binding international agreement they would be free to do so.

A non-nuclear-weapon state could also produce nuclear weapons on its own. The special fissionable material required might be obtained either on a covert or overt basis, using the fissionable materials produced through its peaceful nuclear power program, or by producing the material in facilities built especially for that purpose. Nations desiring to acquire nuclear weapons are most likely to follow one of these roads.

The basic chemical element required for a nuclear weapons program, uranium, is widely distributed in the earth's crust and is readily available to most countries. If a nation really wishes to achieve an independent weapons capability, it could build plants for separating uranium-235 from natural uranium or acquire reactors to transmute uranium-238 to plutonium-239, or both. Either highly enriched uranium-235 or plutonium-239 is suitable for a nuclear weapon. Nuclear-weapon countries produce both highly enriched uranium and plutonium for weapons purposes. It is not necessary, however, for a country to follow *both* routes to acquire a nuclear arsenal, although the use of both routes gives weapons designers a greater flexibility. If a country had limited resources, it might elect to choose only one path to a nuclear arsenal. At the present state of technology, most countries would probably favor the route to plutonium production. However, the situation could change if uranium enrichment technology became more widely known and more readily available.

At present the major portion, if not all, of the enriched uranium in the world is produced by the gaseous diffusion process, a process requiring highly specialized technology. Diffusion plants are costly to build because they must be large in order to produce the product at an acceptable unit cost, and they consume large amounts of electricity. Because of these characteristics, gaseous diffusion plants have been constructed only by nuclear-weapon countries. Moreover, these countries have until now treated the technological information of the gaseous diffusion process with virtually the same secrecy that is applied to weapons information itself. The large sizes and large electric power consumptions of such plants seem to make clandestine construction and operation unlikely.

Since enriched uranium is in demand for use as fuel in nuclear power reactors, non-nuclear-weapon countries have shown considerable interest in other uranium enrichment processes.

As indicated earlier in this chapter, the most important of these processes uses the gas centrifuge. The centrifuge, of course, is a very old machine, but to apply its principles to effect a separation of uranium isotopes—and to do so economically—presents a number of very difficult engineering problems. If the process is developed, however, and the technology made public, it could open the way to the construction of such plants in non-nuclear-weapon countries—either to supply fuel for nuclear reactors or material for nuclear weapons or both. Although the unit cost of product might be higher than that of the gaseous diffusion process, the centrifuge plant can be built in small sizes and is a relatively small consumer of electric power. Because of this, the centrifuge process might be attractive to other countries interested in meeting at least a part of their requirements for nuclear fuel from domestic sources. However, the process is equally capable of producing highly enriched uranium for weapons purposes, and, unfortunately, lends itself to the establishment of small, clandestine plants having no external indications of their existence.

There are, of course, other methods for the enrichment of uranium besides gaseous diffusion or gas centrifuging. None are well-developed, but interest is great and the future will surely see the practical realization of other processes.

If a country should decide to try to acquire nuclear weapons, it is more likely to choose the plutonium route. As a first step it would probably refine natural uranium to a pure oxide or metal. (Enriched uranium can, of course, be used but this involves the additional complexity of acquiring a source of this material.) The processes here are relatively well known and the equipment is not particularly difficult to make. The natural uranium would then be placed in a reactor. This, also, would be relatively simple since the technology of nuclear reactors is now widely known. The reactor could be designed solely for manufacturing plutonium or producing both plutonium and usable energy in the form of heat or electricity. The first alternative is simpler, but since it does not produce electric power to help compensate for the cost of building and operating the reactor, it produces plutonium at much higher cost. After production of the plutonium in the reactor, the nation would also need a plant to process the irradiated fuel elements discharged from the reactor in order to separate the plutonium into a pure chemical form. Here, too, the technology is generally well known, and a nation with a reasonably competent chemical industry could develop the capacity.

Following production of the highly enriched uranium or pure plutonium, the country concerned would then obviously need a plant to fabricate the material into suitable forms and shapes in accordance with the specifications of its weapon designers. It would also need a capability in the field of electronics and access to chemical explosive components. Ideally, the country also would need to have a capability to test its weapons (although this would not be absolutely necessary) as well as a capability to manufacture delivery systems.

A word of explanation on the relationship of the Limited Test Ban Treaty to the question of proliferation of nuclear weapons is in order here. The work done by the United States in the drafting and ratification of this important treaty, was one of the outstanding accomplishments of the Kennedy administration. The Limited Test Ban Treaty has been ratified by more than 100 nations, including several that unfortunately may not become parties to the Non-Proliferation Treaty (NPT). The Limited Test Ban Treaty does not, of course, in and of itself prevent a non-nuclear-weapon nation from developing its own nuclear weapons. However, it does require such a nation, if it is a party to the treaty, to carry out underground any tests which it considers necessary. While underground testing is not likely to be beyond the capability of a nation able to develop its own nuclear weapon, it is more difficult than atmospheric testing. Finally, underground testing could constitute a less dramatic demonstration that a particular nation had in fact achieved a workable nuclear explosive than would atmospheric testing. Of course, whether a particular country would feel that any test was necessary is a question that would depend upon its specific objectives and the degree of confidence in its scientists and engineers. One can imagine situations where a country would prefer to conceal the fact that it had acquired a nuclear weapon until it had some number on hand and therefore would forego testing, at least for a time. Thus, the Limited Test Ban Treaty has some effect in discouraging proliferation, although it is by no means a complete legal or scientific bar.

A number of countries—Argentina, Australia, Belgium, Brazil, Canada, Czechoslovakia, the Federal Republic of Germany, India, Israel, Italy, Japan, the Netherlands, and Sweden are examples—probably have the capability to manufacture nuclear weapons and systems for their delivery within a few years following a national decision to do so. Many additional states could achieve this in longer periods of time. (This is not to imply that the countries mentioned have the desire or plans to build nuclear weapons—fortunately, it appears that in general they do not.)

Concern over proliferation has been intensified by the growth of nuclear power, which will increase enormously the amount of plutonium in existence in the world. As previously indicated, plutonium will be produced in nuclear power reactors as a by-product of the generation of electricity. It will also be the primary nuclear fuel in the breeder reactor economy of the future. As a result of the anticipated growth of nuclear power, it is estimated that roughly 150,000 kilograms of plutonium will have accumulated in the world by 1980. Much of this material will be produced in, and hence be available to, nations that do not now manufacture nuclear weapons. By that time, the plutonium production rates will be sufficient for the production of thousands of weapons each year. The critical problem is to make sure that this plutonium ends up in breeder reactors rather than bombs. Here lies the importance of the Non-Proliferation Treaty.

Non-Proliferation Treaty (NPT)

Many difficulties were encountered in the drafting of the NPT, and at times the outcome was unclear. An agreed-upon text was finally developed and commended by a vote of 95 to 4 by a special session of the United Nations General Assembly in June 1968. To enter into force, the treaty required the deposit of instruments of ratification by the three depository governments—the United States, the United Kingdom, and the Soviet Union—and at least forty additional nations. This requirement was met and the treaty entered into force on March 5, 1970, during special ceremonies in Washington, Moscow, and London, at which the instruments of ratification of the United States, the United Kingdom, and the Soviet Union were deposited along with those of a number of non-nuclear states. By the end of 1970, sixty-two countries had ratified and become parties to the Non-Proliferation Treaty. Thirty-six additional countries had signed but had not yet ratified the treaty.

The IAEA will play an important role here since it was assigned the responsibility under Article III of the NPT of assuring through safeguards inspection that nuclear materials for peaceful purposes are not diverted to the manufacture of nuclear weapons or other nuclear explosive devices. Under the terms of Article IV of the NPT, provision is made for cooperation in the peaceful uses of nuclear energy. Article V contains a commitment for all parties to take measures to assure that the benefits of the applications of peaceful uses of nuclear explosions will be made available to non-nuclear parties on a nondiscriminatory basis.

There have, of course, been criticisms expressed about the Non-Proliferation Treaty. Some nations, having unique problems of national security, have questioned whether they should give up their option to make nuclear weapons even though they have no intention of embarking on any such weapons programs in the near future. Others have refused to sign because a rival state has not yet done so. Some feel the need for a more specific guarantee that a nuclear power will come to their aid in the event they are threatened with nuclear attack. It is to be hoped, however, that these nations will come to appreciate that their true security and the security of the world at large will be better served if they became parties to the Non-Proliferation Treaty.

A problem that proved especially difficult in the negotiations was the incorporation of safeguards provisions (Article III of the treaty). This manifested itself in several significant ways. Some industrialized non-nuclear-weapon countries expressed apprehensions as to whether they would be placed at a disadvantage commercially if their nuclear programs were placed under IAEA safeguards while comparable programs in the nuclear-weapon states were not. For example, they have expressed a concern, which has been widely publicized, that their industrial secrets might be discovered and compromised by the IAEA inspectors, putting them at a disadvantage in comparison with the nuclear-weapon states. The United States has sought to assuage these worries, which it believes are groundless. It has sought to remind these states that the IAEA inspectors are placed under the strictest injunctions not to reveal to unauthorized parties the information they obtain. More importantly, the United States has stressed that the information an inspector normally requires is no more sensitive than that available in the public domain. Furthermore, under IAEA procedures, no country is required to accept a particular inspector; rather, the procedures require that the host country agree on each inspector.

The United States has also stressed that it is not asking nations to accept safeguards that it is not prepared to accept itself. In fact, a large private power reactor in the United States, at the Yankee Atomic Power Station, Rowe, Massachusetts, has been subject to inspection by the IAEA since 1964. To date the Yankee experience has disclosed no unduly burdensome problems in the form of interference with operations, increased costs, or disclosure of confidential company information. On December 2, 1967, the twenty-fifth anniversary of the nuclear age, President Johnson announced that the United States would be prepared to permit the IAEA to apply its safeguards to all of the nuclear activities in the United States, excepting only those that have a direct national security significance, when such safeguards are applied under the NPT. The United Kingdom has made a similar offer.

The most serious safeguards problem that materialized in the NPT negotiations concerned the safeguard relationships of the IAEA to the European Atomic Energy Community. The text of the treaty is worded in such a way as to leave the details of this relationship open for negotiation between the IAEA and the Euratom countries, within certain principles intended to guide the negotiation of all the safeguards agreements called for by the treaty. Those principles specify that each such agreement must enable the IAEA to carry out its responsibility of providing assurance to all parties to the treaty that diversion is not taking place, but that, in doing so, the IAEA should make appropriate use of existing records and safeguards. These guidelines should permit an arrangement between the IAEA and Euratom which will be consistent with the objectives of both the Non-Proliferation Treaty and the Treaty of Rome, which established Euratom. Such an arrangement should be acceptable to all parties. The solution will require flexibility and imagination on both sides.

Beyond the Non-Proliferation Treaty

While the Non-Proliferation Treaty represents an extremely important element in making available to all mankind the benefits of the peaceful uses of nuclear energy, while reducing the risks of additional countries acquiring nuclear weapons, that treaty will not solve all the problems presented by the existence of nuclear weapons. That treaty will not, in any way, directly affect the existing arsenals in the five countries which already have their own nuclear weapons. Nor will that treaty affect the continued development and production of nuclear weapons in those five countries. As difficult as it has been for some non-nuclear-weapon countries to foreswear the acquisition of nuclear weapons by becoming a party to the Non-Proliferation Treaty, it will be much more difficult to limit the extension of nuclear weapons in the hands of those nations which already have them. But it will be even more difficult to achieve such limitations unless a line is drawn by the Non-Proliferation Treaty so that the scope and dimension of the problem can be defined. Conversely, of course, the stability of the Non-Proliferation Treaty can be threatened—and it may not even achieve the broadest possible adherence—unless the nuclear-weapon powers show some progress in limiting their own nuclear weapons. Article VI of the treaty, in fact, requires that the parties concerned pursue in good faith negotiations relating to the cessation of the nuclear arms race.

In order for progress to be made in reaching agreement among the nuclear-weapon countries concerning limitations of their weapons, there must be a mutuality of interest. In fact, each party must conclude that the elements of such an agreement will at least not result in a decrease of its national security. Where a situation exists, as between the United States and the Soviet Union, in which mutual deterrence is provided by their respective nuclear weapons, the construction of such an agreement is likely to be complex and difficult and time-consuming. In late 1969, the United States and the Soviet

Union began the first substantive discussions called the Strategic Arms Limitation Talks or SALT. The first round of talks was held in Helsinki, followed by the next three rounds alternating between Vienna and Helsinki. The chief United States negotiator during the talks was Gerard Smith. His Soviet counterpart was Vladimir Semyonov. The SALT discussions have proceeded on a businesslike basis, with both sides interested in defining areas of possible agreement. President Nixon expressed his hopes for the success of the SALT talks in a message to Ambassador Smith at the opening of the first round in Helsinki on November 17, 1969:

> Today . . . you will begin what all of your fellow citizens in the United States and, I believe, all people throughout the world, profoundly hope will be a sustained effort not only to limit the build-up of strategic forces but to reverse it.

Just as in the case of dealing with the dilemma of fostering Atoms for Peace while minimizing the risk of nuclear proliferation, imaginative and novel solutions will be necessary to arrive at mutually satisfactory arrangements for agreed limitations on nuclear weapons in the hands of the nuclear-weapon countries. This will require statesmanship of the highest order to arrive at a solution acceptable not only to the negotiators of each side but to the citizens of the nations they represent and to the entire world.

After SALT

If we look ahead beyond the conclusion of an initial agreement at the Strategic Arms Limitation Talks, we find many important tasks remaining in the field of arms control and disarmament. Important as agreement on strategic arms limitations would be, such an initial agreement, as generally visualized, would be confined to restricting in some degree the number and type of strategic nuclear delivery systems possessed by the United States and the Soviet Union. The long-term, announced goal of the United States in the field of disarmament is not merely strategic arms limitation, but effectively controlled general and complete disarmament, or "GCD" in the shorthand of the disarmament experts.

Before that seemingly distant goal is reached, many intermediate steps of disarmament can be visualized. Most or perhaps all of these will have to be taken as stepping-stones on the way to general and complete disarmament. In the nuclear field, for example, one of the proposed further steps is the cessation of production of fissionable material for military purposes. Such a step would complement and go beyond strategic arms limitation in that it would place a ceiling on all nuclear weapons—both tactical and strategic—by limiting the amount of material available for their manufacture.

Beyond this, we can visualize the reduction, possibly on a phased or stepwise basis, of existing stocks of nuclear weapons, with their material being dedicated to peaceful purposes. Both of these steps have been the subject of serious proposals by the United States.

Another step which has been the subject of negotiations in the past is that of a Comprehensive Test Ban Treaty, one which would go beyond the present Limited Test Ban Treaty and prohibit all nuclear weapons tests, including those underground. This, too, is an established objective of the United States which is given explicit recognition in the Limited Test Ban Treaty.

Each of these steps depends upon the growing confidence which can be created only by the faithful observance of previous disarmament agreements. Each, in turn, will add its own increment of confidence so that someday the long-awaited goal of general and complete disarmament may become a reality.

Chapter 11

New Understanding

The Atomic Age owes its birth to human curiosity. Niels Bohr, Enrico Fermi, Otto Hahn, Lise Meitner, all of those who inaugurated the Atomic Age sought clues to the nature of the universe rather than electrical power plants or nuclear rockets. In this sense, it is fitting we conclude with a discussion of how the atom helps the scientists and humanists of today in their own searchings.

Probing the Structure of Matter

Two of the most popular areas of nuclear research strive in opposite directions. In particle physics, the objective is to tear atoms down into their subatomic parts and then, in as simple and elegant a way as possible, describe the mesons, antiparticles, neutrinos, and other particles that result. The denizens of the subatomic jungle, however, now number several score, making pedigrees hard to draw up. And not satisfied with the elements nature has provided, many other researchers are attempting to synthesize ever larger atoms, thus extending the Periodic Table well beyond uranium. One group breaks matter down, the other builds it up.

In his 1968 Nobel lecture, Luis W. Alvarez recounts how, when he received his B.S. degree in 1932, two particles, the proton and electron, were thought to be the only truly fundamental particles. Later in 1932, however, the number of fundamental particles was doubled when James Chadwick discovered the neutron and Carl D. Anderson photographed unmistakable positron tracks. The roster of fundamental particles has been increasing ever since. Spurts of discovery occur every time a new, more powerful particle accelerator is placed in operation and every time theorists with fresh insight predict the properties of undiscovered particles. To say that there is an end in sight to the parade of new particles would be as short-sighted as the view of those who were content with the electron and proton in 1932. So many new particles are being discovered nowadays that describing them and their manifold interactions is out of the question here. Instead we ask what general problems the particle physicist faces, and mention some tools the AEC is developing to help him bring some order to the field.

An experimental particle physicist cannot sit back and wait for nature to display subatomic particles for him; he must stimulate nature with huge machines—the so-called atom smashers, the particle accelerators. He must watch the subatomic debris created when one particle hits another. By studying the ranges of these particles and the secondary reactions, he can get some idea of particle energy, mass, and electric charge as well as more obscure properties such as spin and "strangeness."

Over the past three decades, theorists have tried again and again to relate the many fundamental particles one to the other. Highlights in this work include the overthrow of the theory of conservation of parity ° by the Chinese-American physicists, T. D. Lee and C. N. Yang, and Murray Gell-Mann's formulation of the Eightfold Way to impose order on the fundamental particles. Even the language used in these deliberations is strange. Gell-Mann's Eightfold Way draws its name from Buddhism. The "hot" topic of the moment in particle physics is the hunting of the quark, an elusive particle with a name taken from Joyce's *Finnegans Wake*. Particle theorists always assume that the microscopic universe is not only comprehensible by man but that it can also be described in simple, elegant terms. Scientists feel intuitively that all of the fundamental particles must be related somehow. A scientist would not be a scientist unless he followed his curiosity into these seemingly impractical frontier lands. Practical

° Parity is a scientific term expressing the (expected) symmetry of nuclear particles and reactions.

people, however, can take comfort in the fact that today's "useless" knowledge is often the foundation of tomorrow's technology. Who can say that a better understanding of subatomic forces will not lead to an antigravity machine or some construction never dreamed of in the science fiction magazines. But the pure scientist needs no such rationalizations.

The physics of the very small depends upon very big machines for elucidation. Allied with the big accelerators in the pursuit of the fundamental particles are big particle detectors. The former create the particles, the latter permit us to see their effects. In fact, the availability of new accelerators and new detectors is critical to progress in the field. Just as a big new telescope allows astronomers to see farther into the seas of stars around us, big new accelerators expand our horizons in the microscopic realm. An important role of the AEC is to help build these new machines.

Particle accelerators have a long history. The first important machine was the electrostatic proton accelerator built by John D. Cockcroft and Ernest T. S. Walton in 1932 in Lord Rutherford's laboratory. Robert Van de Graaff, in the United States, followed with a different type of electrostatic machine. In 1930, Ernest O. Lawrence built his first cyclotron at the University of California at Berkeley. The first cyclotron was only four inches in diameter; but larger ones came quickly, including the famous 184-inch synchrocyclotron at Berkeley's Radiation Laboratory. Lawrence's laboratory, with its unparalleled group of accelerators, became a focal point for nuclear physics and chemistry, producing no less than seven Nobel Prize winners.

The machines have been getting bigger and bigger; the bigger they are, the greater the energies of the particles they propel into target nuclei. Table 16 illustrates the trend toward higher and higher energies. Radioactivity rarely produces particles with more than 6-Mev (million electron volts) energy; thus we expect man-made forces in the billion-electron-volt range completely to overwhelm the internal forces holding the nucleus together. The resulting debris is, of course, rich in the subatomic pieces prized by modern physicists.

The trend toward larger machines has been modified to some extent by another approach to the extraordinarily difficult study of the forces and strange particles that scientists are finding in the nucleus. Instead of hitting a stationary target with proton or electron "bullets,"

Table 16
Intermediate and High-Energy Particle Accelerators Throughout the World

Location	Name	Energy (Mev)	Particles accelerated	Year of startup	Funding [1]
OPERATING					
Lawrence Radiation Laboratory, Berkeley, Calif.	184-inch Synchrocyclotron	740 460 910 1140	Proton Deuteron Alpha Helium-3	1946	(State of Calif.),° AEC
Joint Institute for Nuclear Research, Dubna, U.S.S.R.	Synchrocyclotron	680	Proton	1949	U.S.S.R.
Columbia University, New York, N.Y.	Synchrocyclotron	400	Proton	1950	(ONR),° NSF
Stanford University, Palo Alto, Calif.	Mark III Linac	1200	Electron	1950	(ONR),° NSF
University of Chicago, Chicago, Ill.	Synchrocyclotron	450	Proton	1951	(ONR),° NSF
University of Birmingham, Birmingham, England	Proton Synchrotron	1000 650	Proton Deuteron	1953	DSIR
Lawrence Radiation Laboratory, Berkeley, Calif.	Bevatron	6200	Proton	1954	AEC
CERN, Geneva, Switzerland	CERN Synchrocyclotron	600	Proton	1957	CERN
Joint Institute for Nuclear Research, Dubna, U.S.S.R.	Proton Synchrophazotron	10,000	Proton Deuteron	1957 1971	U.S.S.R.
Saclay Nuclear Research Center, Saclay, France	Saturne (Proton Synchrotron)	3000 2400	Proton Deuteron	1958	CEA-France
Laboratori Nazionali di Frascati, Frascati, Italy	Electron Synchrotron	1100	Electron	1959	CNEN
Faculté des Sciences Orsay, Orsay, France	Linear Accelerator	2100	Electron	1959	NEM
CERN, Geneva, Switzerland	CERN Proton Synchrotron	28,000	Proton	1959	CERN
University of Lund, Lund, Sweden	LUSY (Electron Synchrotron)	1200	Electron	1960	Swedish Atomic Research Council

Institution, Location	Accelerator	Energy	Particle	Year	Sponsor
Brookhaven National Laboratory, Upton, N.Y.	Alternating Gradient Synchrotron	33,000	Proton	1960	AEC
University of Tokyo, Tokyo, Japan	Electron Synchrotron	1300	Electron	1961	Japanese Government
Institute of Theoretical and Experimental Physics, Moscow, U.S.S.R.	Proton Synchrotron	7200	Proton	1961	U.S.S.R.
Harvard University, Cambridge, Mass.	Cambridge Electron Accelerator	6000 / 6000	Electron / Positron	1962	AEC
	CEA Bypass Project [2]	3500	Electron/Positron [3]	1971	
Princeton University, Princeton, N.J.	Princeton-Pennsylvania Accelerator	3000 / 2300	Proton / Deuteron	1963 / 1970	AEC
Rutherford Laboratory, Chelton, Berkshire, England	NIMROD	7000	Proton	1963	SRC
Argonne National Laboratory, Argonne, Ill.	Zero-Gradient Synchrotron	12,700	Proton	1963	AEC
Polytechnical Institute, Tomsk, Siberia, U.S.S.R.	Syrius (Electron Synchrotron)	1300	Electron	1964	U.S.S.R.
Physical Technical Institute, Kharkov, U.S.S.R.	Electron Linear Accelerator	1800	Electron	1964	U.S.S.R.
Deutsches Elektron-Synchrotron, Hamburg, Germany	DESY (Electron-Synchrotron)	7500	Electron	1964	FRG, City of Hamburg, VW Foundation
Laboratoire de l'Accélérateur Linéaire, Orsay, France	ACO (storage ring)	540	Electron/Positron [3,6]	1965	NEM
Space Radiation Effects Laboratory, Newport News, Va.	Synchrocyclotron	600	Proton	1965	NASA
Daresbury Nuclear Physics Laboratory, Daresbury, England	NINA (Electron Synchrotron)	5000	Electron	1966	SRC
Stanford University, Palo Alto, Calif.	Stanford Linear Accelerator (SLAC)	22,000 / 15,000	Electron / Positron	1966 / 1969	AEC
	SPEAR (SLAC Storage Rings)	3000	Electron/Positron [3]	1972	
Institute of Nuclear Physics, Novosibirsk, Siberia, U.S.S.R.	VEPP II	750	Electron/Positron [3,6]	1967	U.S.S.R.

Table 16 (Continued)

Location	Name	Energy (Mev)	Particles accelerated	Year of startup	Funding [1]
Radiotechnical Institute, Moscow, U.S.S.R.	Model Cybernetic Accelerator	960	Proton	1967	U.S.S.R.
Physikalisches Institut, Bonn, Germany	Bonn Electron Synchrotron	2300	Electron	1967	FRG, State of Nordheim-Westfalen
Institute of Physics, Yerevan, Armenian S.S.R., U.S.S.R.	Yerevan Electron Synchrotron	6100	Electron	1967	U.S.S.R.
Cornell University, Ithaca, N.Y.	10-Bev Synchrotron	10,000	Electron	1967	NSF
Institute of High Energy Physics, Serpukhov, U.S.S.R.	Alternating Gradient Synchrotron	76,000	Proton	1967	U.S.S.R.
Saclay Nuclear Research Center, Saclay, France	Electron Linac	640	Electron	1968	CEA-France
Laboratori Nazionali di Frascati, Frascati, Italy	ADONE (Storage Ring)	1200	Electron/Positron [3,6]	1969	CNEN/CNR
Institute of Nuclear Physics, Novosibirsk, Siberia, U.S.S.R.	VEPP III	3500	Electron/Positron [3,6]	1971	U.S.S.R.
CERN, Geneva, Switzerland	CERN Intersecting Storage Rings	28,000	Proton/Antiproton [3]	1971	CERN members
PLANNED OR UNDER CONSTRUCTION					
M.I.T., Cambridge, Mass.	High Duty Factor Electron Linac	400	Electron	1971	AEC
Institute of Nuclear Physics, Novosibirsk, Siberia, U.S.S.R.	VEPP II' [4]	700	Electron/Positron [3,6]	1971	U.S.S.R.
National Accelerator Laboratory, Batavia, Ill.	200-Bev Synchrotron	200,000	Proton	1971	AEC

Laboratory	Machine	Energy (MeV)	Particle	Year	Funding[1]
Los Alamos Scientific Laboratory, Los Alamos, N.Mex.	Los Alamos Meson Physics Facility (LAMPF)[5]	800	Proton	1972	AEC
Institute of Nuclear Physics, Novosibirsk, Siberia, U.S.S.R.	VEPP IV[5]	10,000	Electron/Positron[3,6]	1972	U.S.S.R.
	VAP[5]	23,500	Proton/Antiproton[3,6]		
TRIUMF, Vancouver, B.C., Canada	TRIUMF	500	H-ion	1973	AECB, TRIUMF Universities
Swiss Institute for Nuclear Research, Villigen, Switzerland	Isochronous Ring Cyclotron	585	Proton	1973	Swiss Government
Stanford University, Palo Alto, Calif.	Superconducting Mark III Electron Linac	2000	Electron	1973	ONR,° NSF
DESY, Hamburg, Germany	DORIS (DESY Storage Ring)	3500	Electron/Positron[3,6]	1974	FRG
National Institute for High Energy Physics, Tsukuba, Japan	Synchrotron	8000	Proton	1974?	Ministry of Education, Japanese Government
CERN, Geneva, Switzerland	300-Bev Proton Synchrotron	300,000	Proton	1975	CERN members

[1] Funding abbreviations
AECB = Atomic Energy Control Board (Canada)
CEA-France = Commissariat a l'Energie Atomique
CERN = Conseil Européen pour la Recherche Nucléaire
CNEN = Comitato Nazionale per l'Energia Nucleare (Italy)
CNR = Consiglio Nazionale delle Ricerche (Italy)
DSIR = Department of Scientific and Industrial Research (Great Britain)
FRG = Federal Republic of Germany
NASA = National Aeronautics and Space Administration
NEM = Ministere Education National (French)
NSF = National Science Foundation
ONR = Office of Naval Research

SRC = Science Research Council (Great Britain)
TRIUMF = Universities of Alberta, British Columbia, Simon Fraser, and Victoria
° = Construction funds. ( ) = Not currently providing funds.
[2] The CEA Bypass project utilizes the main accelerator ring itself as the storage ring to provide a colliding beam capability.
[3] Colliding beam capability.
[4] VEPP II is a new ring attached to VEPP II which will have a higher luminosity.
[5] Alternate uses for the same machine. In addition, plans have been discussed for electron-proton colliding beams, and positron-antiproton colliding beams using this machine.
[6] Also usable as a conventional accelerator.

the new approach is to hit bullets with bullets flying in opposite directions. These bullets may be the same particles or particles of opposite charge, such as electrons against electrons or electrons against positrons (positive electrons); protons against protons, or protons against antiprotons (negative protons). Several different schemes are being pursued, but they all have the same basic geometry: particles of one kind are accelerated to high speeds and injected into a circular ring where they are "stored" by letting them fly around in circles like stacked aircraft over a fogbound airport. Meanwhile, the same thing is done with the other particles in another ring (or even the same ring), except these particles travel in the opposite direction. When enough particles have been accumulated in both rings, the two beams are brought together on a collision course. Table 16 also shows where some of these devices are operating or being built.

As accelerators have increased in size, so have the detectors needed to discern and identify the elusive, short-lived subatomic fragments. The familiar cloud chamber was used to identify new particles in the era of the 1940s. Nowadays, it has been largely superseded by the spark chamber and the bubble chamber, which also permit the experimenter to see the trails left by the passage of particles. When one searches for rare particles one must search a lot of territory to find a significant number. Thus chambers, which once rested on table tops, now occupy a large portion of the room.

Accelerator installations like the Stanford Linear Accelerator and the new 200-Bev machine being built at the National Accelerator Laboratory involve the expenditure of hundreds of millions of dollars. They are important elements in our technological foundation as well as scientific tools.

Synthetic Elements

Switching now from atom smashing to atom building, we look at the synthesis of elements beyond uranium. Although the transuranium elements probably existed in the universe at the beginning of time, they disappeared through radioactive decay and had to be re-created through the ingenuity of man. They have been produced through nuclear transmutation. Table 17 lists the transuranium elements known in 1970 with their discoverers.

Gallery of the two-mile-long Stanford Linear Accelerator. (*Stanford Linear Accelerator Center*)

Plutonium is by far the most important of the transuranium elements in view of its use as the explosive ingredient in nuclear weapons and its almost unlimited peaceful potential as a nuclear fuel for the production of electric power. The crucial experiments in the discovery and demonstration of the great value of plutonium were performed during the late winter and spring of 1940–41. The key experiment in its discovery was conducted in a little chemistry laboratory, Room 307, Gilman Hall, on the campus of the University of California, Berkeley, on February 23, 1941. The demonstration of the value of plutonium as an energy source took place on the same campus on March 28, 1941, when it was first shown that its important isotope (plutonium-239) can be split—that is, it undergoes fission—when bombarded with "thermal" or slow neutrons.

A large number of the transuranium isotopes were discovered at Lawrence Radiation Laboratory, in Berkeley. In their production by nuclear transmutation, lighter elements were converted into heavier ones by increasing the number of protons in the nucleus. Up through

Table 17
The Transuranium Elements

Atomic Number	Element	Symbol	Discoverers and Date of Discovery	First Isolation in Weighable Amount
93	neptunium	Np	E. M. McMillan and P. H. Abelson 1940	L. B. Magnusson and T. J. La-Chapelle 1944
94	plutonium plutonium-239	Pu	G. T. Seaborg, E. M. McMillan, J. W. Kennedy and A. C. Wahl 1940–41 J. W. Kennedy, G. T. Seaborg, E. Segrè and A. C. Wahl 1941	B. B. Cunningham and L. B. Werner 1942
95	americium	Am	G. T. Seaborg, R. A. James, L. O. Morgan and A. Ghiorso 1944–45	B. B. Cunningham 1945
96	curium	Cm	G. T. Seaborg, R. A. James and A. Ghiorso 1944	L. B. Werner and I. Perlman 1947
97	berkelium	Bk	S. G. Thompson, A. Ghiorso and G. T. Seaborg 1949	S. G. Thompson and B. B. Cunningham 1958
98	californium	Cf	S. G. Thompson, K. Street, Jr., A. Ghiorso and G. T. Seaborg 1950	B. B. Cunningham and S. G. Thompson 1958
99	einsteinium	Es	A. Ghiorso, S. G. Thompson, G. H. Higgins, G. T. Seaborg, M. H. Studier, P. R. Fields, S. M. Fried, H. Diamond, J. F. Mech, G. L. Pyle, J. R. Huizenga, A. Hirsch, W. M. Manning, C. I. Browne, H. L. Smith and R. W. Spence 1952	B. B. Cunningham, J. C. Wallmann, L. Phillips and R. C. Gatti 1961
100	fermium	Fm	A. Ghiorso, S. G. Thompson, G. H. Higgins, G. T. Seaborg, M. H. Studier, P. R. Fields, S. M. Fried, H. Diamond, J. F. Mech, G. L. Pyle, J. R. Huizenga, A. Hirsch, W. M. Manning, C. I. Browne, H. L. Smith and R. W. Spence 1953	

Table 17 (Continued)

Atomic Number	Element	Symbol	Discoverers and Date of Discovery	First Isolation in Weighable Amount
101	mendelevium	Md	A. Ghiorso, B. G. Harvey, G. R. Choppin, S. G. Thompson and G. T. Seaborg 1955	
102	nobelium	No	A. Ghiorso, T. Sikkeland, J. R. Walton and G. T. Seaborg 1958	
103	lawrencium	Lr	A. Ghiorso, T. Sikkeland, A. E. Larsh and R. M. Latimer 1961	
104	rutherfordium (U.S.A.)	Rf	A. Ghiorso, M. Nurmia, J. Harris, K. Eskola and P. Eskola 1969	
104	kurchatovium (U.S.S.R.)	Ku	G. N. Flerov, Yu. Ts. Oganesyan, Yu. V. Lobanov, V. I. Kuznetsov, V. A. Druin, V. P. Perelygin, K. A. Gavrilov, S. P. Tretiakova, and V. M. Plotko, 1964. Also I. Zvara et al. 1966	
105	hahnium (U.S.A.)	Ha	A. Ghiorso, M. Nurmia, K. Eskola, J. Harris and P. Eskola 1970	
105	(After Niels Bohr) (U.S.S.R.)		G. N. Flerov, V. A. Druin, A. G. Demin, Yu. V. Lobanov, N. K. Skobelev, G. N. Akap'ev, B. V. Fefilov, I. V. Kolesov, K. A. Gavrilov, Yu. P. Kharitonov and L. P. Chelnokov 1967	

atomic number 101, mendelevium, the key problem in the discovery of the transuranium elements was chemical identification—that is, the classical test of chemically separating the new element from all previously known elements. Beyond element 101, on the other hand, the radioactive isotopes of the elements decay away so fast that the discoveries were first based on nuclear evidence alone. In most instances this involved the identification of daughter elements. However, these elements have been subsequently identified chemically.

Beginning with element 101, the discoveries of transuranium elements have been based on experiments in which the elements have been produced and identified literally one atom at a time. This atom-by-atom identification of new elements has been one of the outstanding achievements in the history of science. When one realizes that it takes about a million billion atoms to make a speck barely large enough to see, one gets some glimmering of the difficulties involved in identifying a new element one atom at a time.

The first element to be synthesized and discovered on this one-atom-at-a-time basis was mendelevium. The fundamental methods worked out for this successful experiment have served as the basis for the production and discovery of the elements heavier than mendelevium, and will doubtless continue to serve as a basis for the discovery of still further elements.

The mendelevium experiment was conducted in 1955 by a group of scientists at the Lawrence Radiation Laboratory of the University of California at Berkeley. The investigators thought that techniques had advanced to a point where it might be possible to create and identify element 101 through the bombardment of a target containing an unweighable and invisible amount of material—a feat never attempted before. The plan of attack involved use of the Berkeley 60-inch cyclotron to bombard the isotope einsteinium-253 with helium ions. The net effect of this reaction was to add two protons and one neutron to the target nucleus to make mendelevium-256.

A small amount of einsteinium was deposited in an invisibly thin layer on a gold foil. The helium ion beam of the cyclotron struck the einsteinium on the back of the foil, rather than on the front as in previous transmutation experiments. This enabled each newly formed atom of mendelevium to recoil in such a manner that it could be caught on a second thin gold foil. Thus, the second gold foil contained the recoiled atoms but was relatively free of einsteinium. The foil was then dissolved for purposes of chemical identification.

The next element, element 102, was also produced and identified one atom at a time, using the recoil technique. However, the difficulties that had to be surmounted here were even greater, and they show what the future holds as research progresses at the far-out frontiers of the transuranium elements. In this case, the half-life of the product nucleus was so short that direct chemical identification was not possible in the original experiment. Consequently, a "double recoil technique," involving the chemical identification of a daughter isotope, was used. This ingenious "double recoil technique," introduced by Albert Ghiorso, at Berkeley, formed the basis for subsequent experiments which have identified additional transuranium isotopes and elements.

Let us push on further into the transuranium region. Those seeking to discover element 103 found themselves facing new difficulties. Isotopes of the undiscovered element were also expected to have half-lives too short for chemical identification in the first experiments. To make the situation doubly difficult, the isotopes of element 103 did not appear to have daughter isotopes suitable for the application of Ghiorso's double recoil method of identification.

In the spring of 1961, despite these obstacles, the Berkeley group succeeded in producing an isotope which they identified as having the atomic number 103. Using a modification of the simple recoil method and heavy (boron) ions as projectiles in a particle accelerator, they created element 103, now named lawrencium.

The story of the discovery of the elements with atomic numbers 104 and 105 is more complicated. Soviet scientists at the Dubna Laboratory, near Moscow, in 1964 employed the method pioneered by the Berkeley group in their discovery of element 103, augmented by some chemical experiments in subsequent years by I. Zvara and associates. Using heavy ion projectiles and the simple recoil technique, the Soviet scientists (G. Flerov and co-workers) based their first claim for identification on the relative yields as they varied the target materials and the types and energies of projectiles used. They have suggested the name kurchatovium, after the Soviet nuclear physicist, Igor Kurchatov, for element 104.

The Berkeley group led by Ghiorso attempted to confirm the Soviet experiments but could not do so. They concluded the Soviet scientists were in error. The Berkeley group used the recoil technique with different target materials and different heavy ion projectiles. In 1969, they succeeded in identifying, beyond any doubt, several isotopes of element 104. Chemical identification was also made. The Berkeley scientists proposed that element 104 be called rutherfordium, after the British nuclear physicist, Lord Ernest Rutherford.

The Dubna scientists performed heavy ion bombardments in 1967, creating isotopes which they suggested should be identified as element 105. Again the Berkeley group failed to confirm the Dubna results. In 1970, using different heavy ions and targets, the Berkeley group positively identified an isotope of element 105. At nearly the same time, the Dubna scientists reported results generally agreeing with some of those found at Berkeley, and the results of some chemical identification experiments. Thus, the discovery of element 105, like that of element 104, is the subject of controversy. The Americans have suggested that element 105 be called hahnium, after Otto Hahn, the German chemist and co-discoverer of nuclear fission. The Soviet scientists have said that they would like to name element 105 after Niels Bohr, the Danish physicist.

Accelerators have been a key factor in the synthesis of the transuranium elements because they can hurl nuclei at other nuclei with sufficient energies to cause them to fuse into the new elements. There is another method of building heavy artificial elements. This method uses the neutrons from underground nuclear explosions to transmute target materials to heavier and heavier isotopes by means of neutron-capture reactions. Brief (on the order of a millionth of a second) but very intense, this source of neutrons makes it possible to add very large numbers of neutrons to the nuclei of the target element. This type of heavy element synthesis occurs in supernovas, and it is remarkable that man has succeeded in reproducing it here on earth. Special schemes and devices have been devised for the very rapid recovery of samples after an underground explosion to identify new isotopes and elements by chemical procedures. Larger samples are recovered by drilling down into the ground and bringing the material to the surface.

A possible sequence of nuclear reactions has been formulated as follows: The starting material, uranium-238, might, in less than a millionth of a second, capture successively as many as thirty-seven neutrons to create uranium-275. The uranium-275 might then undergo fourteen successive beta particle (electron) emissions, converting some of its excess neutrons to protons. The final result might be an isotope of the undiscovered element 106. Of course, the transuranium elements, when they become available in quantity, might be used as the target material instead of uranium so that synthesis can begin at a higher level.

Numerous experiments utilizing underground nuclear explosions in Nevada have been carried on by scientists of Lawrence Radiation Laboratory at Livermore and Los Alamos Scientific Laboratory in New Mexico. These scientists collaborated with scientists at Argonne National Laboratory and at Berkeley in chemically and physically identifying the products.

The production of heavy transuranium elements in quantity utilizes another method. This involves the bombardment of lighter heavy elements in high-flux nuclear reactors. The successive capture of neutrons, interspersed with beta particle decays, leads to the production of all the transuranium elements up to and including fermium (element 100). The quantities produced decrease with increasing atomic number of the product. These amount to tons of plutonium, kilograms of curium, grams of californium and milligrams of einsteinium.

It is logical to question how far it is possible to go in making heavier and heavier elements. Up until a few years ago, the prospects did not appear bright for going much beyond the known synthetic elements because there appeared to be a generally decreasing trend in stability as the elements became heavier. Then, a remarkable shift from pessimism to optimism occurred when nuclear physicists came up with a new theory of stability which predicted that the trend toward instability would become reversed as the atomic number (number of protons) of the nucleus increased. In other words, certain "superheavy" elements far beyond element 105 might be more stable than it and the elements immediately below and above it, creating, in effect, "islands" of stability. This theory is now generally accepted by scientists, but it has not been possible to prove it experimentally because of technical limitations of existing accelerators.

Some scientists believe so strongly that the superheavy elements will be exceptionally stable that they have begun to examine minerals, ores, meteorites, cosmic rays, samples brought back from the moon, and many other substances in the hope that these elements were created in the cosmic upheaval which led to the formation of our galaxy. They theorize that there is a possibility, albeit remote, that traces of superheavy elements may be found in nature if their radioactive half-lives are long enough to have permitted them to survive the passage of eons of time. The search, employing the most sophisticated techniques available, has not given any conclusive results, but neither can the theory be discarded yet.

It will not be long before new or modified accelerators will be available to test the theory. If successful, the experiments will produce new elements in an "island of stability" which is predicted to exist between about 110 and 125. What is so fascinating about this concept is that these new superheavy elements may be much more stable than elements 99 to 105. Although they will be made only in extremely small quantities—one atom at a time initially—it may be possible to study their chemical properties and to determine how they fit into the periodic system of the chemical elements.

Cosmology and Astronomy

Small as atoms are, the stars and the rest of the universe seem to be made of them. We assume that by knowing how atoms behave "in the small," that is, here on Earth, we also can guess how they behave "in the large"—on the brink of the universe. In this way we apply terrestrial nuclear physics to astronomy and cosmology.

The logical place to begin such a discussion is with the formation of the universe, assuming that it really did begin and has not existed forever. The so-called Big Bang Theory was introduced by the Belgian priest Georges E. Lemaître during the 1920s. A bulwark of the Big Bang Theory is its recounting of the birth of the universe—that thirty-minute inferno that supposedly flared some ten billion years ago. According to this theory, the universe winked into existence as a seething mass of elementary particles traveling at speeds close to that of light. The immense fireball expanded rapidly, cooling in minutes to temperatures comparable with those in the center of the sun. The cooler the mass of primitive stuff became, the more elementary particles coalesced into protons, neutrons, and nuclei. Some protons fused to become deuterons. Similar fusion reactions and neutron captures gave birth to helium and the heavier elements up to uranium and even beyond.

How can we even guess what happened ten billion years ago? George Gamow, a noted physicist and cosmologist, pointed out that the present distribution of elements we see in the universe is really the oldest "archeological" evidence in existence. In addition, experiments in terrestrial laboratories have shown us how such alchemy might have occurred. The hot core of a hydrogen bomb, for example, has much in common with the Big-Bang fireball. In other words, given the laws of nuclear physics, we should be able to guess a beginning, work out the nuclear consequences, and see if we end up with the universe we observe today.

Instead of simmering for billions of years, the universe may have been flash-cooked in about half an hour, according to terrestrial nuclear physics principles. The short half-life of the free neutron—only twelve minutes—required rapid synthesis of the heavy elements. In just thirty-six minutes (three neutron half-lives following the Big Bang) element building must have slowed to almost nothing because of the shortage of bricks (neutrons). In addition, the temperature of the primeval holocaust should have dropped to below fusion temperatures in about a half hour.

Cosmologists and nuclear physicists both feel quite happy with this scenario because they still see pieces of the universe apparently flying apart from the effects of the Big Bang and can account fairly well for the observed abundances of the elements. It is interesting to note that today's human nuclear alchemists build transuranium elements through the use of nuclear explosives by adding neutrons to lighter elements after the fashion of the Big Bang.

Nuclear physics has also come to the aid of those astrophysicists trying to account for the huge energy source tapped by stars, particularly the closest star, the sun. The prodigious quantities of energy emitted by stars are easy to account for today. Before the discovery of nuclear energy, astrophysicists knew of nothing that would sustain solar fires for the billions of years that geologists required to explain the Earth's accumulation of sedimentary rocks. Just before World War II, Hans Bethe, in the United States, and Carl von Weizsacker, working independently in Germany, found that the carbon nucleus could act as a high-temperature catalyst to help "cook" hydrogen into helium in stellar interiors. The so-called "carbon cycle" was soon challenged by the application of the H-H reaction by Charles Critchfield, a physicist at George Washington University. In the H-H reaction, hydrogen nuclei (ordinary protons) collide and fuse, directly without the intervention of carbon as a catalyst, and release energy, much as is done in a hydrogen bomb and just as they supposedly did during the Big Bang. Today, astrophysicists believe that the carbon cycle is predominant in hot stars while the H-H direct reaction prevails in cooler stars such as our sun. Even though the sun is usually deprecated as "only an average star," it "burns" some 600,000,000 tons of hydrogen each second and has a fuel reserve for 5 billion years at that rate of consumption.

Chemical elements are continually being synthesized by transmutation reactions in the stars. The large variety of nuclear processes involves long chains of neutron-capture reactions, which include the so-called s-process (s for slow time scale) and the r-process (for rapid time scale). It appears that transuranium elements, even the super-heavy elements, are among the products of such ceaseless stellar manufacturing processes.

Nuclear physics has been instrumental in constructing theories about how the universe began, how it continues to synthesize elements, and how it is sustained energywise. Through radioactive dating (explained in Chapter 3), the atom also provides hints about how long it has been since the curtain was raised to reveal the postulated Big Bang.

Before the universe or any part of it can be dated by radioactive methods, one must have a sample in hand. For the Earth this is easy, but is the maximum age of the Earth's solid rock representative of the universe as a whole? Probably not, because the Earth's crust may have been molten billions of years before it froze, trapping the radioactive elements and their by-products we need for dating purposes. At long last we have samples of the moon's surface, but the moon has proven rather enigmatic—we don't really know where our celestial neighbor came from—and thus we are not sure just what we are measuring when we analyze the samples brought back by the astronauts. To illustrate, the moon may have been born the same time as the Earth, or it might have been torn from the Earth by some later catastrophe, or possibly the moon is a solar system interloper, captured eons ago by the Earth's gravitational field. Finally, pieces of the universe fall to Earth every day in the form of meteorites. We date these and generally assume that they are fragments left over after the planets were formed by accretion; that is, the gravitational coalescing of cold matter (the latest theory); or that they are pieces from some ancient planet that once plied between Mars and Jupiter only to explode and fill the solar system with rocky debris. In sum, then, radioactive dating only places lower limits on the ages of the Earth, moon, and meteorites. But this kind of information is useful in testing hypotheses about these denizens of the solar system as well as the universe as a whole.

Some of the elements have isotopes with extremely long half-lives: uranium-238, 4.5 billion years; rubidium-87, 47 billion years; potassium-40, 1.3 billion years, to name some of the more important of the radioisotopes used in dating ancient fragments of the universe. The oldest rocks on the Earth come from the ancient continental "shields" in Canada and Asia and a few unusual islands, such as the Rocks of Saint Paul in the Atlantic Ocean. The oldest terrestrial rocks are about 3.4 billion years old. But meteorites seem to be roughly 4.5 billion years old at the most, indicating perhaps that they are remnants of the original debris that coalesced into the Earth-like planets; i.e., Mercury, Venus, Earth, Mars, and possibly Pluto. Potassium-40/argon-40 dating of the lunar rocks brought back by the Apollo-11 astronauts in July 1969 yielded ages between 3 billion and 4 billion years. Some of the Apollo-12 rock samples seem to be as old as 4.6 billion years, the same as the age of the meteorites.

If the Earth and moon were formed by the accretion of cold matter simultaneously, it may have been that the moon did not melt initially as the Earth apparently did. One possibility is that the internal heat generated by the decay of potassium-40 could not escape as easily from the bulkier Earth and melting ensued. Geophysicists believe the Earth still retains a fluid core.

This short discussion makes it obvious that planetology is in a state of flux as space technology and nuclear technology unite to piece together new theories.

The age of the universe—from Big Bang to the present—must be greater than the oldest solid matter we can acquire. How much older it is than the 4.5 billion years we have measured is the subject of much controversy. With the real nature of those strange new astronomical objects, the quasars (quasistellar objects) and pulsars (pulsating stellar objects), still in doubt, cosmology remains a most exciting subject. Based on red-shift or Doppler measurements of the velocities of distant galaxies, one can calculate how long it has taken these members of the expanding universe to reach their present positions. The ages computed range between 7 and 20 billion years.

Cosmology and relativity are almost inseparable; each depends upon the other for support. For instance, the curvature of space is a relativistic concept of great importance in cosmology. Ever since Einstein proposed his Special Theory of Relativity in 1905, scientists have been trying to confirm it in various ways. One of the two basic assumptions of the Special Theory is that the velocity of electromagnetic waves in a vacuum is constant. This assumption has been verified for microwaves and visible light, but it would strengthen the Special Theory of Relativity if it could be verified all along the spectrum from long radio waves to short gamma rays. Edward Teller suggests that a nuclear explosion in space (prohibited by the Limited Test Ban Treaty) would be an ideal source of electromagnetic radiation across the spectrum. By placing radio receivers, optical instruments, X-ray detectors, gamma-ray detectors, and various other instruments together somewhere in space, perhaps on a spacecraft or the moon's surface, one could measure the times of arrival for radiation from the bomb in each part of the spectrum. The radiation should all arrive simultaneously (assuming no free electrons are in the way) if Einstein's assumption is correct.

Robot Geologists and Biologists

Astronauts have made it safely to the moon and back, but it will be a decade or two before they make round trips to the other planets. Meanwhile, without samples in hand from these other worlds, we have no choice but to make our scientific analyses of their rocks and life (if it exists) by remote control.

We have already described in Chapter 3 a type of robot geologist used in the Surveyor alpha-scattering experiment. These artificial geologists did not walk around on the moon chipping off pieces of rock; they bombarded the lunar surface with alpha particles from a radioisotope. Telemetry from Surveyor indicated that the alpha particles reflected back by the soil and the protons created by transmutation were sufficient for preliminary analyses of the elements present on the lunar surface. The Soviet Union's moon robot, Lunokhod I, which landed on the moon in late 1970 and roamed about the lunar surface, also performed nuclear analyses of the kind performed by the Surveyor craft, but with the notable advantage of being able to move from one spot to another.

By "robot biologist" we mean a robot life detector. It would be nice to know whether extraterrestrial life is five-footed or furry, but let us start with trying to determine the mere existence or nonexistence of other life. The simplest microorganisms would suffice as proof of extraterrestrial life. The basic question is: How can one distinguish life from nonlife at distances of several hundred million miles? Sterile sand looks pretty much like sand crawling with microorganisms. This is where radioactive tracers come in.

A primary characteristic of life is metabolism; that is, the consumption of food and the expulsion of wastes. A specific instrument has been designed to detect metabolism remotely. It has been named "Gulliver" after Swift's character who discovered unusual life in far-off lands. When the modern Gulliver lands on an alien planet, it first collects a sample of the soil by retrieving a sticky string shot out to a distance of two or three feet. The recoiled string is next bathed in nutrients that are tagged with carbon-14. If the collected soil contains microorganisms that partake of the nutrients and then expel tagged carbon dioxide gas, a radiation detector mounted on Gulliver will indicate the presence of the tagged carbon dioxide. The presence of extraterrestrial life, then, will be encoded on the radio signals dispatched back in the direction of Earth.

When the first unmanned probes land on Mars, they will very likely carry some sort of instrument to detect possible life. Gulliver is a prime candidate for the passenger list.

Tracking the Drifting Continents

In 1929 Alfred L. Wegener wrote in his book *The Origin of Continents and Oceans*:

> At first I did not pay attention to the idea because I regarded it as improbable. In the fall of 1911, I came quite accidentally upon a synoptic report in which I learned for the first time of paleontological evidence for a former land bridge between Brazil and Africa. As a result, I undertook a cursory examination of relevant

research in the fields of geology and paleontology, and this provided immediately such weighty corroboration that a conviction
of the fundamental soundness of the idea took root in my mind.

Wegener was speaking of the theory of continental drift, which
holds that the continents are carried steadily apart, eon after eon, on
a moving substratum of viscous basalt. Beginning as fragments of a
primitive supercontinent, the continents we know today began to
drift apart hundreds of millions of years ago. In their wakes they
leave festoons of islands and a thin pavement of youthful ocean floor
still straining and cracking under the influence of subterranean currents of molten rock. What a grand concept! It certainly explains the
jigsaw-puzzle-like similarity between Africa and South America. The
closeness of fit is so striking that Francis Bacon suggested as far back
as 1620 that these continents had been once united. Others, such as
Alexander von Humboldt, suggested the idea again and again down
the centuries. However, it was Alfred Wegener who really marshalled
all the evidence for continental drift and put his reputation on the
line. But the whole idea was too extreme for Wegener's times—
everyone knew that the earth was too soundly constructed for continents to float apart thousands of miles. Wegener was attacked merci-
lessly. "Can we call geology a science when there exists such a
difference of opinion on fundamental matters as to make it possible
for a theory such as this to run wild?" asked one scientist.

Nonetheless, the idea persisted into present-day thinking and formal names were given to the primeval supercontinents. Wegener
called his single supercontinent Pangaea (translated "all-land"). Others believed that there were two supercontinents originally. The name
Laurasia was given to the northern one and Gondwanaland to the
southern land mass.

The theory of continental drift was saved from oblivion by modern
oceanography, which has shown that indeed the ocean floors are very
young and really are spreading apart. New fossils have also been discovered, linking all of the southern continents together at some time
in the past. Further, rocks on the opposite sides of the Atlantic have
been shown to be so similar in structure and age that they must have
been linked physically at one time. Radioactive dating has been instrumental in resurrecting the drift theory. In fact, during the short
span of the 1960s, continental drift came back to life, vanquished the

forces of conservative geology, and now stands as an adequately proven fact of nature.

Proof of continental drift involves almost all the physical sciences, but here we concentrate on those that have depended heavily upon radioactive dating.

The most impressive evidence of drift is the apparent youth of the sea floors. Not only do they seem to be only a couple of hundred million years old at the most, but they are younger and younger as one approaches the great mid-ocean ridges. The mid-ocean ridges are the sites of the famous 40,000-mile crack in the Earth's crust that bisects the Atlantic and invades most of the world's oceans with its branches. The crack, which in the Atlantic is eight to thirty miles wide, cuts through the ridge summit to a mile in depth. Adjacent to the ridge, oceanographers find the youngest rock of all. Sheets of rock seem to issue from beneath the ridges at the rate of an inch or so each year. One sheet moves toward the Americas, another toward Europe and Africa. The continents seem to "ride" on this conveyor belt of newly formed ocean floor rather than drifting like icebergs in a liquid.

The ocean-floor conveyor belts are also found to be magnetic tape recorders that preserve a record of the vicissitudes of the Earth's magnetic field over the last several million years. Every time the north and south magnetic poles switched positions (no one knows why), the event was recorded on the sheets of magnetizable rock issuing from the region of the mid-ocean ridges. Actually, rocks all over the world show evidence of magnetic flip-flops, but only the ocean floors give us a play-by-play account. By measuring the amount of potassium-40 that had decayed into argon-40 in rock samples taken from all parts of the world, geochronologists have shown that magnetic reversals occurred 0.7, 2.4, and 3.35 million years ago. In this way a time scale has been added to the magnetic data recorded on the ocean floor near the ridges. This evidence strongly supports the theory of continental drift, because it is so unusual that one cannot imagine any other mechanism that would produce it except sea-floor spreading and the associated continental drift.

Abundant evidence from other disciplines supports that from the magnetic tape recorders. Working together, scientists from M.I.T. and the University of São Paulo, in Brazil, have discovered striking similarities between African and South American geology. The similarity is so strong that pure coincidence seems out of the question.

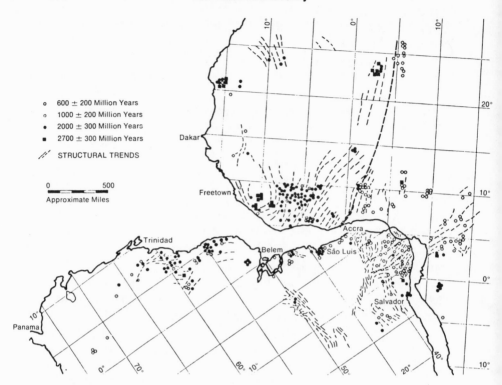

o   600 ± 200 Million Years
o   1000 ± 200 Million Years
•   2000 ± 300 Million Years
■   2700 ± 300 Million Years

/'' STRUCTURAL TRENDS

0 ——— 500
Approximate Miles

Africa and South America as fitted together prior to continental drift. Young geological formations (open circles) coincide well. (*From P. M. Hurley, et al. "Test of Continental Drift by Comparison of Radiometric Ages," Science 157 (Aug. 4, 1967):495. Reprinted with the permission of Science.*)

West Africa is divided into two major geological provinces based on rock age. In Ghana, the Ivory Coast, and regions to the west, rocks dated by the potassium-argon and rubidium-strontium methods register around 2000 million years. In sharp contrast, the rocks in Dahomey, Nigeria, and regions to the west show ages of only about 550 million years. Now, if South America and Africa are fitted together like puzzle pieces, one finds the same sort of sharp age division in South American rocks, precisely where the continents fit together. Furthermore, the types of rocks are similar and structural trends seem to run from one continent right into the other. Evidence like this is hard to deny.

Continental drift is only part of the new view we have of the universe and our planet's place in it. Continental drift raises intriguing questions: why should the Earth's magnetic field reverse in the first place? Is drift somehow related to astronomical events, such as those that cratered our close neighbor, the moon? Whatever the answers are, the atom will surely play some part in our deliberations, perhaps as a geological clock, or as a source of particles in activation analysis, or as a source of energy to take instruments to the planets and the bottom of the Earth's seas.

Archeology with a Geiger Counter

Whatever calamity engulfed the Earth and precipitated wild changes in its magnetic field during the last few million years, we are fairly sure that man was present to witness some of the later upheavals. The same type of radioactive dating that supports the theory of continental drift also tells us that man himself is a much more ancient creature than science hitherto believed. The more we study archeology with "nuclear clocks," the more intriguing human history turns out to be.

Radioactive dating of human remains has probably stimulated as many controversies as it has settled. The antiquity of man is a subject that makes almost everyone prick up his ears. The remains of the higher primates seem to go back at least 30 to 34 million years, according to nuclear clocks, but just when did man enter the picture? Before this question can be answered, "man" must be defined in terms of archeological evidence. According to the current working definition, any creature associated with the making of stone tools in a "set and regular pattern" has graduated to the human level. Before 1959, man by this definition was thought to be less than a million years old. In that year, the British anthropologist L. S. B. Leakey discovered manlike remains with such stone tools in famous, sunbaked Olduvai Gorge in Tanzania, Africa. The remains were dated by Garniss H. Curtis and Jack F. Evernden, at the University of California at Berkeley, using the radioactive potassium-argon method as the clock. The results doubled the age of man to almost two million years.

Questions were immediately raised about the applicability of the potassium-argon method for such "short" time periods; that is, short in comparison to geological time periods for which the potassium-

Uncovering a 14,000-year-old burial site in the Sudan. The ages of these remains
are determined by carbon-14 measurements of charcoal and other organic matter.
(*Anthropology Department, Southern Methodist University*)

argon technique is best suited and widely employed. Fortunately a
cross check was possible. Associated with the questioned human
remains was some pumice—a porous volcanic glass—which was
apparently contemporary with the human bones. The volcanic glass
contained a minute quantity of uranium. Now, uranium not only fissions
when hit by a neutron but it also fissions for no reason at all; this is
called *spontaneous fission*. Uranium's half-life for spontaneous fission
is about 10,000,000,000,000,000 years, which is long even compared
with the Big Bang age of the universe. When a nucleus of uranium
does fission spontaneously, the pieces fly apart with such energy that
they disturb the chemical structure of the volcanic glass along their

paths. By etching a fresh surface of a volcanic piece of glass with chemicals, the fission "tracks" become visible and can be counted through the microscope. The number of tracks seen in this way is proportional to the time that has elapsed since the glass was shot out of some volcano, solidified, and settled in Olduvai Gorge around some primitive humans frightened by the roaring mountain and shaking earth. The date of the volcanic glass also turned out to be two million years, an excellent check on the potassium-argon figure.

The discovery of new paleontological samples and improvements in dating methods continue to fill gaps in man's lineage and to extend it. Early in 1971, Professor Bryan Patterson of Harvard University announced that a fragment of primate jawbone, found in 1967 by Arnold D. Lewis of Harvard at a site in Kenya about 300 miles north of Olduvai Gorge, belonged to *Australopithecus Africanus*—a manlike creature. Potassium-argon dating of sediments in which the bone was found indicate that its age is about 5.5 million years.

So it seems that man has been wandering the Earth much longer than previously supposed. Radioactive dating has more than doubled the age of the Earth and also the age of man—and the final answers may still be to come.

Let us tackle another controversial area of archeology: the age of man in North America. Here we deal in tens of thousands of years rather than millions or billions, a range where carbon-14, with its half-life of 5,730 years, as discussed in Chapter 3, is the appropriate clock.

Archeologists of the 1950s were quite unprepared for the impact of carbon-14 upon their activities. One of them put it this way: "We stand before the threat of the atom in the form of radiocarbon dating. This may be the last chance for old-fashioned, uncontrolled guessing."

Carbon-14 did upset some cherished applecarts. Before radiocarbon techniques were introduced by Willard F. Libby in 1948, it had been dogma among archeologists to deny that man was present in the New World during the Pleistocene glaciation over 10,000 years ago. This reflex denial was due in part to previous incompetent and fraudulent work that had tried to prove that man *was* in America more than 10,000 years ago. The 1926 discovery of the famous "Folsom points" in New Mexico was a turning point. Faced with the undeniable evidence of arrowheads in association with the bones of extinct bison, the ancient-man-in-America belief gained strength once again. Based on these new discoveries and conventional archeological methods of age estimation, man-in-America was pushed back from 10,000 to 25,-000 years plus or minus 30 percent; but many conservatives would not accept these figures. This was a very wide range, even for "guesti-mates."

Radiocarbon arrived on the scene at an opportune moment. J. R. Arnold and W. F. Libby quickly narrowed the age of the Folsom point deposits to between 9000 and 10,000 years, well within the very wide range stipulated by conventional methods. Arnold and Libby released their findings around 1950. Initially, there was considerable disbelief from the archeologists—just as there had been from the ge-ologists earlier. How could unseen atoms combined with clicking ra-diation counters constitute any kind of clock? In those days, it was a "two-culture" situation in which radiochemists and archeologists each knew little of the other's methods. Radiocarbon dating soon became established, however, and is now a keystone of all sciences that delve into the recent past.

To continue the story of man in the New World, many new sites displaying evidence of old inhabitants have been uncovered in North and South America in recent years. Apparently a simple hunting cul-ture was widely diffused throughout both continents. Radiocarbon techniques date this culture as having existed between 9000 and 16,-000 years ago. And we may be sure that the oldest remains have not yet been found. There is agreement that an even earlier culture ex-isted which used cruder projectile points and rougher tools. Yet, there still exists a school of thought within archeology that opposes these theories of ancient man. The farther some scientists try to push man-in-America back in time, the harder others resist.

A number of controversial sites exist where radiocarbon dates of charcoal and other organic remains seem to indicate that men were in America 40,000 years ago and earlier. Santa Rosa Island off the Califor-nia coast is one such site. Fire sites on Santa Rosa present the ar-cheologist with the bones of mammoths and other animals that were obviously killed and barbecued by man or some closely related crea-ture. However, the charcoal in the pits is radiocarbon dated at 30,000 years—*far too old* for the conservative school. The charcoal, they suggest, came from an ancient forest fire that burned long before man occupied Santa Rosa. Of course, this is denied by the other camp.

Radiocarbon dating is still provoking argument. To the nonscientist, these controversies may seem peculiar in that science is one area of human activity where facts are supposed to be facts. Actually, the hottest arguments always come from those branches of science where the greatest advances are occurring. So it is that archeology is undergoing great changes as new scientific methods are applied.

A possible preview of coming attractions involves the antiquity of Old World invasions of the New. Just about all archeologists will agree that the Vikings did reach America long before Columbus, but did someone else beat even them across the Atlantic? Some say that the ancient Phoenicians reached American shores long before Eric the Red and his longships. Most archeologists pay little attention to the many strange stone constructions found along the East Coast of the United States. One in particular, on Mystery Hill in North Salem, New Hampshire, is true to the name of its site. On Mystery Hill is a peculiar stone edifice: a rock-walled chamber of apparently ancient origin, a grooved sacrificial stone altar, and other "things" not associated with American Indian culture. The architecture of the edifice is similar to a style used in the Mediterranean area long before the Christian era. Mystery Hill has been excavated by amateurs who are members of the New England Antiquities Research Association (NEARA).* The Association reports that charcoal taken from a level containing a stone handscraper, a hammerstone, and a broken pick has been radiocarbon dated by Geochron Laboratories at 1000 B.C. NEARA members conclude that an ancient Mediterranean people, perhaps a Phoenician ship blown off course and carried to the Americas, built the stone hut on Mystery Hill.

Many strange rock structures dot Northern Europe, with a few spilling over onto North America and the islands in the North Atlantic. Stonehenge, in England, is the classic example. Though its ancient monoliths had been thoroughly investigated by archeologists, Gerald S. Hawkins, as reported in his book *Stonehenge Decoded*, revived controversy over Stonehenge by using a computer to analyze the arrangement of the stones. Stonehenge, which is apparently 4000 years old or more, is really a brilliantly conceived astronomical observatory, according to Hawkins. The computer, radiocarbon, and other elements of modern technology are forcing science to reassess the an-

* *NEARA Newsletter*, September 1969.

tiquity and capabilities of men who lived before written records were kept.

The Atom in the Humanities

Radiocarbon continues to be a most useful dating tool right into the period of recorded history. To illustrate, the famous Dead Sea Scrolls were bound in linen, which is susceptible to radiocarbon dating. The date established, A.D. 33 plus or minus 200 years, does not seem to have the precision one would like for dates within historical times. However, 200 years is only a 10 percent error in the 2000 years that have passed since the scrolls were written. Since the scrolls are undated, the radiocarbon date serves to confirm dates arrived at on the basis of other evidence, as well as to establish that we do not have another of those carefully prepared modern hoaxes.

Neutron activation analysis, discussed in Chapter 3, has proven to be extremely useful in identifying the chemical elements present in coins, pottery, and other artifacts from the past. Before activation analysis arrived on the scene, a museum curator would have to damage one of his precious artifacts irrevocably if he wanted a chemical analysis of its constituents. Activation analysis, in contrast, is not so heavy-handed; a tiny, unnoticeable fleck of paint from an art treasure or a microscopic grain of pottery clay suffices to reveal its chemical makeup. Activation analysis is essentially nondestructive.

The versatility of neutron activation analysis is demonstrated again in tracing the trade routes and movements of ancient peoples. Most archeologists agree that the study of pottery and other ceramic objects is the most fruitful method of reconstructing the waxings, wanings, and travels of the ancient civilizations. Wherever people went, their pottery went with them. The tracing of dispersion of pottery styles has become the means for tracing migrations, conquests, trade routes, and other cultural contacts.

In the past, archeologists have relied heavily on the physical characteristics of pottery for pottery identification. Now, using neutron activation analysis, they have discovered that pottery fragments with the same physical appearance have different chemical compositions and were produced therefore from different clay sources. Tiny amounts of elements, such as cesium, lanthanum, and cobalt, can be

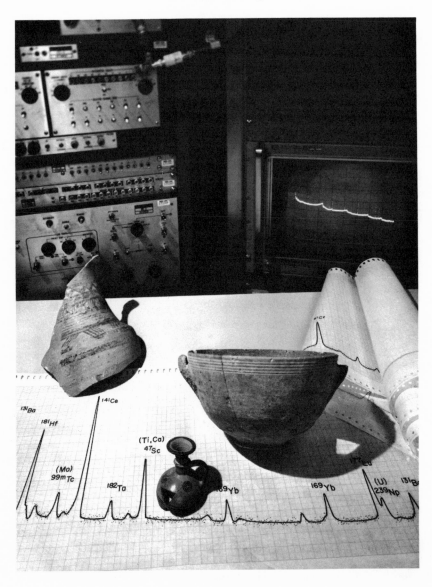

Activation analysis is now used to establish chemical profiles of pottery to help determine where artifacts were manufactured, regardless of where they are found. The peaks in the trace shown indicate isotopes in the pottery that were activated by neutron bombardment. (*Lawrence Radiation Laboratory*)

sorted out to give each source of pottery material its own distinctive fingerprint. For example, if a potsherd found in Yucatán had the same activation analysis profile as pottery originating in Peru, it would give us a more detailed picture of South American cultural contacts. Pottery styles were not always unique to a particular location.

Isadore Perlman and his associate, Frank Asaro, at the Lawrence Radiation Laboratory at Berkeley have developed a system of analysis in which about thirty-five elements can be measured in neutron-irradiated pottery, with very high accuracy in most instances. Laboratory researchers are establishing pottery chemical profiles for a number of important archeological sites. From these profiles they hope to be able to pinpoint the origin of pottery fragments no matter where they are found. One can visualize an expanding atlas of chemical profiles for archeological sites around the world—a sort of archeological Baedeker.

The world of art has always been plagued by forgers—some of them expert, indeed—and even by painters well known in their own right, who are not above "correcting" or "touching up" a masterpiece. With the technique of activation analysis at hand, art experts hope to "fingerprint" the works of famous painters. The basic assumption is that a given artist used a sequence of pigments through his career that can be identified easily and nondestructively by activation analysis. In other words, the elements in the artist's pigments should be as distinctive as his style at various points in his life. A forger, therefore, would have to match not only style and technique but also the composition of the pigments available to the original artist. In support of this new dimension of art, the AEC has made a grant to the Mellon Institute in Pittsburgh. This "Research Project in Artists' Materials" is being developed by Bernard Keisch, a nuclear chemist. His research is carried out with the treasures of famous galleries, including the National Gallery of Art in Washington, D.C.

Professional numismatists, students of coins, are delighted to be able to determine the silver contents of coins without taking chunks out of them. The millions of coins in museums represent an untapped treasure trove of information about ancient and medieval societies. The whole coin can be exposed to neutron bombardment directly. The silver will be activated and the coin's overall silver content can be measured by several silver radioisotopes that are created. The coin then goes back to the museum essentially unchanged.

Coin analysis yields interesting clues about the old civilizations. Sasanian Persia, for example, was a highly developed society between A.D. 224 and 651. Although Persian coins and art objects, some with inscriptions on them, are available to the scholar, all actual descriptions of this civilization were written by Romans, Arabs, and other peoples. Activation analysis of the coinage tells us that the silver content of the metal used at the Sasanian mint fluctuated between 70 percent and near 100 percent. To the student attuned to these things, the debasement of coinage indicates that the Sasanians were probably at war, experiencing famine, or undergoing some sort of internal strife during the periods when debasement occurred. Activation analysis has also discovered that the Sasanians had at least three distinct sources of silver. Clues such as these mean a great deal in the humanities, because unlike the situation in the physical sciences, they do not deal in reproducible experiments; the humanities have to make do with what history has preserved. Any technique that squeezes a few more facts out of scarce raw material (without harming it) is to be welcomed with open arms.

There is another way to analyze materials quickly and easily that uses atomic methods but which does not have the sensitivity of neutron activation analysis. It is called X-ray fluorescence analysis. X rays from an X-ray machine or gamma rays or X rays from radioisotope sources bombard the material being analyzed, whereupon different X rays are emitted with energies which are characteristic of the elements in the material. These X rays are measured, both for energy and intensity, and the elements which gave rise to the X rays are thus determined. This method does not work on all elements and it is not very good when one wants to determine if only a trace of an element is present, but it has the distinct advantage of employing equipment that is relatively simple and (if radioisotope sources are used) portable. One of the best ways to use this technique is in validation of the authenticity of archeological finds or valuable works of art, since the method is nondestructive.

A treasure trove of another kind consists of pictures made during the experimental days of photography. Not only do these photographs give us an intimate view of some important personages of the past, but they also tell us much about the development of the photographic art itself. As we might expect, the early experiments were less than completely successful and many of the pictures have faded and other-

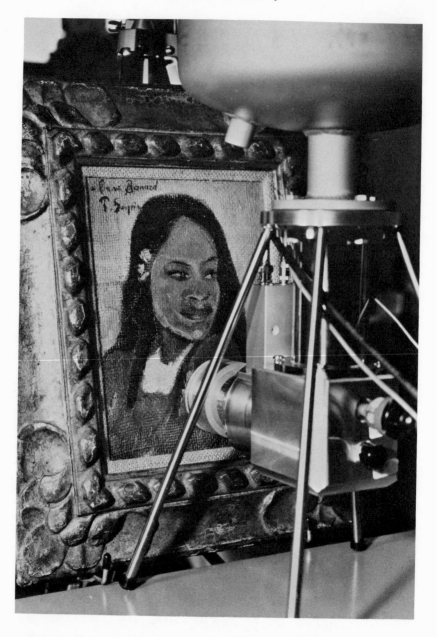

X-ray fluorescence equipment being used to study a Gauguin painting. X-ray fluorescence is nondestructive and very helpful in checking the authenticity of valuable paintings and archeological finds. (*Kevex Corporation*)

wise deteriorated. But beneath the nearly destroyed image, in the chemicals that still remain, lies the residue of what photographers call the "latent image." Part of the latent image was brought out by the developing process when the experimenters tried first this solution, then that. If the remaining latent image, which exists in the form of silver grains, can be reached chemically or physically, an old photograph might be restored for history's benefit.

Eugene Ostroff, the curator of photography at the Smithsonian Institution, has used neutron activation to "develop" the latent image in some photographs made by the English inventor W. H. F. Talbot early in the nineteenth century. Talbot's experiments centered around the light-sensitive properties of silver nitrate and silver chloride, and in this he pioneered the darkroom chemistry of today. Ostroff exposed some of Talbot's pictures, which in 130 years had deteriorated considerably, to reactor neutrons which activated the silver in the latent images. The stable silver-107 and silver-109 were transmuted to the radioisotopes silver-108 and silver-110. When the photograph was withdrawn from the reactor, it was immediately placed in contact with modern film. The beta and gamma radiation from the activated silver exposed the modern film, transferring the latent image from Talbot's photograph. Thus the photographs were restored by autoradiography; that is, they revealed themselves through their own radiation. Many old photos thought to be beyond saving have been restored to a remarkable degree.

A considerably more ancient work of man is the Pyramid of Chephren at Giza, Egypt. One would think that such an impressive monument would have been thoroughly explored long ago either by grave robbers or, if science were lucky, archeologists. There is a theory among Egyptologists, however, that the Egyptian monarchs were one step ahead of the grave robbers, and that they ingeniously planned their pyramids in a manner that would lead future generations to believe that the tombs had already been sacked. By this ruse, their bodies would be preserved in hidden chambers, ensuring their immortality. Several secret chambers in the Pyramid of Cheops were discovered accidentally a thousand years ago, but Cheops had forestalled the invasion of his privacy for 3500 years. Egyptologists have long suspected that the known substructure of the pyramid of Chephren at Giza may constitute a blind, and that rich finds are to be made in hidden chambers somewhere above. But how does one find hidden chambers in the 470 feet of stone sitting atop the lower chambers?

Take an X ray, obviously; except that no X-ray machine of man can penetrate 470 feet of stone. Nature, though, conveniently provides radiation with adequate penetrating power: cosmic rays. Scientists, under the leadership of Luis W. Alvarez, from the University of California at Berkeley, and Ein Shams University, at Cairo, have recently placed spark chambers in the pyramid's lower chambers and measured the amounts of radiation arriving from outer space in different directions. If the cosmic rays were unusually stronger in any specific direction, a cavity in the stone structure above would be a possibility. So far, only a portion of the pyramid's structure has been "cosmic-rayed," but the portion surveyed corresponds to the region where Egyptologists would expect secret chambers, based on their knowledge of other pyramids. The search has produced negative results to date. Perhaps Chephren, who built the pyramid millennia ago, had more faith than Cheops that mankind would not stoop to disturb his soul.

Chapter 12

Man to Mankind— The New Optimism

Having discussed in some detail the relationship of Man and Atom— and particularly how nuclear energy, when properly applied, will advance those who develop and use it—we would like to conclude on the broader note of the relationship of science and technology to man.

Over the next few decades—before the end of this century— the human race will have to face and resolve challenges that may well determine the shape of its life for centuries to come, if not its very survival. There is no doubt that many of these challenges are a result of the rapid growth and cumulative effect of science and technology. There is also no doubt that they are bringing into direct confrontation what men have tried to separate—fact and value. One aspect of this is that science and morality have been brought face to face. But what we believe will result from this confrontation, albeit after the period of anxiety and agony we seem to have entered, will ultimately be a united force to raise men to a new level of rationality and humanity.

What we are witnessing today in all the tension and turmoil of our times is perhaps the physical birth of "mankind" as a world community, as nations or as individual citizens—a "mankind" we have recognized only in the ideal but which we have never had to deal with so directly. In short, all the moral laws, all the religious teachings, all the poetic and philosophical writings that have exhorted us to recognize the brotherhood of man, that have urged us to understand and respect nature, to act justly and humanely toward our fellowman—all these are being made physical imperatives by the power of "neutral," "amoral" science.

Science and technology have created the conditions and have set the stage for our transition from man to mankind. They have brought men from the most distant points on Earth into close contact with each other through instant communications and rapid transportation. They have drawn people increasingly into large communities of growing interdependency—megalopolises, powerful nations, and an international community—where men have recognized that the conflicts of human and technological interfaces must be solved. And they have given many men the new perspective, the new awareness, to see humanity, all life, and the planet that supports it as the holistic system that it is.

But ironically, now that science and technology have brought us to this point, many people cannot recognize that it is only through the wisest and most humane advancement and application of these forces that we can complete this journey to mankind—*that* mankind which can live in peace and harmony with itself and its environment. Bombarded daily by the mass media's pessimistic and disheartening emphasis on such subjects as pollution, poverty, overpopulation, and the problems of explosive political and military power, they are falling victim to negativism and despair. Filled with a mixture of shocking facts and gloomy forecasts, they accept too readily the belief that we cannot or will not turn the tide of our mounting problems. They see only disaster ahead.

In his book *So Human an Animal*, Rene Dubos refers to this doomsday feeling as "the new pessimism" and says of it: "As the year 2000 approaches, an epidemic of sinister prediction is spreading all over the world, as happened among Christians during the period preceding the year 1000." Later, when speaking of the "new optimism," he notes: "Despite the foreboding of the tenth century, the world did not come to an end in the year 1000. . . ." He goes on to point out that "the new optimism finds its sustenance in the belief that science, technology, and social organization can be made to serve the fundamental needs and urges of mankind, instead of being allowed to distort human life."

We believe in the "new optimism," but to be an optimist today is a difficult burden. It is frightening and pathetic to find that any statement reflecting optimism and hope for the future inspires so many people to respond so gratefully, almost like drowning men grasping at straws. We must fight such despair and emphasize that today's problems, as big and as pressing as they are, are not insurmountable. Physically, we are better equipped than at any time in human history to resolve those problems and realize many of man's age-old dreams. And our awareness of our problems and our knowledge of their urgency are also positive factors that will work in our favor despite the current pessimism they create.

One way to overcome the paralysis of negativism and despair today is to combat the general surge of antitechnological feelings by establishing a more balanced and reasonable view of technology. The current outcry against technology has its roots to a great extent in the environmental problems that are receiving so much public attention today, particularly in the United States. Because our productivity has moved far ahead of our current ability or past desire to handle the waste products associated with it, the discomforts and dangers of the latter are overbalancing the comforts and advantages of the former. The natural, or at least simplistic, reaction to this is to "turn off" technology, to return to earlier days and simpler ways. Even if this were possible—and we know it is not—most people, after a little reflection, would not want to go back in time. The spirit of the old French proverb applies, "Ah, for the good old days—when we were so unhappy."

One of the problems we face in dealing properly with our technology today is that many of us have conditioned ourselves to a certain attitude about it. Our very concern over needed reforms has focused our attention in recent years on all that is harmful, crass, and ugly in our world, and seeking a scapegoat, we point to technology. In order to avoid blaming ourselves for our excesses, our shortsightedness, and any other human failing, we point the finger of guilt at "the machine," "the system," "the establishment." In this particular exercise of self-redemption, we conveniently overlook the many benefits of the technology of the past, but more importantly, what it might accomplish for us in the future, were we to direct it wisely.

Technology can be directed creatively so as to bring human society into close harmony with its natural environment. It can be made to create more wealth with less waste—less waste products and less waste of human and natural resources. It can be made to create beauty where we have let it spawn ugliness. It can be made to bring man both greater security and more individual freedom. To do this, however, we must stop blaming technology for our shortcomings, reassert our mastery over it and agree on what we want it to do. We must also be willing to pay for advancing those scientific and technological developments that we find necessary to meet our common goals. Particularly in the pursuit of a healthier environment there are large costs involved which, directly or indirectly, must be shared by society as a whole.

During the coming years much will have to be done in the way of "technological assessment"—in wise planning for the development of man's new tools and for their application toward the most human goals with a minimum of harmful impact on the natural environment. All in all, we must not be against technology—we must be for better technology.

Somewhat akin to the rise of the antitechnological attitude, but perhaps potentially more dangerous, has been an increase in antirationalism. Many people, seeing so much in the world that seems irrational, are doubting man's capacity to progress through reason.

And in some of the attempts to place the blame for our current plight on science and technology or the system or the establishment, we find people going beyond an attitude of antirationalism to a posture that must be described as anti-intellectualism. This attitude is tied partially to the ridiculous notion that the intellect and the emotions are divorced entities, that the thinking man cannot be a man of deep feeling and sensitivity. Quite the opposite is true. Those who develop their minds, who understand the scientific disciplines, who think most deeply and broadly about man and nature, about the physical forces and the beauty of this earth and the universe—those individuals can achieve a far greater degree of sensitivity and emotional awareness than anyone who rejects the power of his mind. We need today, more than ever, men and women who can combine great intellectual power with a new depth of feeling and a new awareness. Such a combination is both possible and necessary to human survival today. There are many sensitive people who have not turned their backs on science and technology as the means of achieving a better life but who seriously question whether these forces, supporting an ultra-rational society, might not eventually destroy all human diversity and individual freedom. While a case could be made for this, we do not believe it has to happen.

There are common values and long-range goals for mankind as a whole, the pursuit of which does not lead to a homogenized mankind or technological tyranny. In fact, there are global goals today, such as world peace, control of population and the elimination of hunger, that if not given the utmost priority could eventually produce conditions in this world that would reduce discussions of cultural values to almost metaphysical arguments. It can also be shown that a rational world based on the wise and constructive application of science and technology can be one in which human diversity and freedom are enlarged.

It is unfortunate but true that until recently most people have been convinced that the world must choose between a tyranny of total planning or the unknown that evolves from the interplay of free, competing forces. Today science and technology make a different set of courses not only possible but necessary. In the past we have exploited both man and nature. Today it is becoming possible to exploit knowledge—and in a way and to an extent that will allow us to establish new and healthier relationships between man and man and man and nature. Knowledge—particularly as advanced by today's science and technology—is our newest and by far our greatest form of capital. The general recognition of this fact would in itself be a value that could radically change our world for the better. Knowledge must be used today to create the physical conditions—the abundance and security—that will allow the more noble aspects of human nature to ascend.

This whole question of human nature has, of course, been debated through the ages. And the debate continues, even intensifies, in the light of today's conditions. There are those who see in today's violence and man's inhumanity to man the true man. And there are those who disagree with these ideas and see in the positive contribution of today's civilizations the reflection of a humanity capable of great care and concern for its members. Some feel even more hope for the future because of the values being stressed by the new generation. We are inclined to side with those who believe that the evolution of a higher man is both possible and essential, and we would agree with Joshua Lederberg's statement: "The most important ethical inference from the fact of human evolution is that we are still perfectible. It is one of the least debatable of human purposes that our posterity should be wiser than we are. . . ."

We believe they will be, and we see evidence that they are becoming such. If one searches behind the headlines of the day, if one seeks out and observes the day-to-day progress in the many worthwhile human endeavors that are under way, he will realize that we are witnessing the rise, not the fall, of man. Much more is being done that is constructive and for the good of mankind than most people imagine.

The pursuit of the peaceful atom is one such constructive step. We believe that the development of this great source of energy with its multitude of beneficial applications will become a significant turning point in human history. The release of nuclear energy offers man a hope and a challenge. Both are essential to the growth and evolution of mankind. Both could make us better men who have learned to live in harmony with our fellowman and the unique environment that sustains us on this small but precious planet.

This is the message of the nuclear age. We must hear it. We must heed it. We must help to make it a reality. There are few if any alternatives to this course of action, that of constructively merging the mighty forces that flow from man and atom.

APPENDIX I

APPLICATIONS OF RADIOISOTOPES IN MEDICINE

Table 1

Non-Scanning Applications of Radioisotopes in Medicine During 1966 [*]

Radionuclide	Compound	Procedure	Physicians licensed for procedure	Physicians performing procedure	Patient administrations Number	Patient administrations Percent
Iodine-131	Sodium iodide	Thyroid uptake	1527	1284	301,052	57.2
Iodine-131[a]	Labeled albumin	Blood volume determination	1381	908	101,994	19.4
Iodine-131[a]	Sodium iodohippurate	Renal function	817	388	33,245	6.3
Cobalt-57	Labeled vitamin B_{12}	Vitamin B_{12} absorption	952	582	24,996	4.7
Chromium-51	Sodium chromate	Blood volume determination	1064	437	22,468	4.3
Cobalt-60	Labeled vitamin B_{12}	Vitamin B_{12} absorption	1149	556	16,486	3.1
Iodine-131[a]	Labeled fats	Fat malabsorption	1054	505	7742	1.5
Chromium-51	Sodium chromate	RBC survival	1110	604	6530	1.2
Iron-59	Chloride or citrate	Iron turnover	836	287	3139	0.6
Iodine-131[a]	Labeled albumin	Cardiac output	627	79	1503	0.3
Iodine-131[a]	Rose bengal	Hepatic function	33	23	1202	0.2
		Total for principal procedures listed above			520,357	98.8
		Total for 39 other procedures with less than 1000 administrations			6119	1.2
		Total for all function procedures			526,476	100.0

[*] Taken from K. D. Williams and J. D. Sutton, *Survey of the Use of Radionuclides in Medicine*, Preliminary Report, U.S. Department of Health, Education, and Welfare, MORP-68-10, 1968.

[a] Administrations using either iodine-131 or iodine-125.

Table 2
Radioisotopes Used for Medical Scanning Purposes During 1966 *

Radionuclide	Compound	Procedure	Physicians licensed for procedure	Physicians performing procedure	Patient administrations	
					Number	Percent
Iodine-131	Sodium iodide	Thyroid scanning	1092	908	153,089	37.4
Technetium-99m	Pertechnetate	Brain scanning	528	330	63,078	15.4
Gold-198	Colloidal gold	Liver scanning	730	539	41,855	10.2
Mercury-203	Labeled mercurials	Brain scanning	329	320	37,746	9.2
Iodine-131	Labeled albumin (MAA)	Lung scanning	757	562	22,840	5.6
Iodine-131	Rose bengal	Liver scanning	823	503	19,721	4.8
Mercury-197	Labeled mercurials	Brain scanning	162	145	18,489	4.5
Mercury-203	Labeled mercurials	Kidney scanning	787	543	16,223	4.0
Mercury-197	Labeled mercurials	Kidney scanning	646	282	6972	1.7
Strontium-85	Nitrate or chloride	Bone scanning	582	398	6232	1.5
Technetium-99m	Technetium sulfur colloid	Liver scanning	202	55	4911	1.2
Iodine-131[a]	Labeled albumin	Placenta scanning	619	325	3921	1.0
Iodine-131	Labeled albumin	Brain scanning	37	34	3511	0.9

Iodine-131	Sodium iodohippurate	Kidney scanning	483	99	2091	0.5
Selenium-75	Selenomethionine	Pancreas scanning	263	121	1635	0.4
Chromium-51	Heat-treated RBC	Spleen scanning	554	287	1628	0.4
Iodine-125	Sodium iodide	Thyroid scanning	5	4	1354	0.3
Iodine-131	Labeled albumin	Heart scanning	69	65	546	0.1
Strontium-87m	Nitrate or chloride	Bone scanning	130	22	438	0.1
Mercury-197	Mercurihydroxypropane	Spleen scanning	247	57	389	0.1
Iodine-131	Labeled albumin (MAA)	Liver scanning	9	4	311	b
Technetium-99m	Pertechnetate	Thyroid scanning	5	5	259	b
Technetium-99m	Labeled albumin	Placenta scanning	12	12	254	b
Technetium-99m	Labeled albumin	Lung scanning	1	1	200	b
		Total for principal procedures listed above			407,693	99.6
		Total for 25 other procedures with less than 200 administrations			1768	0.4
		Total for all scanning procedures			409,461	100.0

ᵃ Taken from K. D. Williams and J. D. Sutton, *Survey of the Use of Radionuclides in Medicine*, Preliminary Report, U.S. Department of Health, Education, and Welfare, MORP-68-10, 1968.

ᵃ Administrations using either iodine-131 or iodine-125.

ᵇ The percentages for these entries total 0.2.

Table 3
Teletherapy Treatments Employing Radioisotopes During 1966 *

Radio-nuclide	Procedure	Physicians licensed for procedure	Physicians performing procedure	Patients	Patient administrations	
					Number	Percent
Cobalt-60	Beam therapy	524	399	153,714	1,859,845	97
Cesium-137	Beam therapy	76	25	3364	60,103	3
	Total for all teletherapy procedures			157,078	1,919,948	100

° Taken from K. D. Williams and J. D. Sutton, *Survey of the Use of Radionuclides in Medicine*, Preliminary Report, U.S. Department of Health, Education, and Welfare, MORP-68-10, 1968.

Table 4

Brachytherapy Treatments Using Radioisotopes During 1966 *

Radionuclide	Form	Procedure	Physicians licensed for procedure	Physicians performing procedure	Patient administrations Number	Percent
Radium-226	Sealed sources	Intracavitary Rx of cancer	734	543	15,700	31.6
Strontium-90	Eye applicator	Ophthalmic treatment	483	295	11,785	23.7
Radium-226	Plaques and molds	Dermatological treatment	378	109	5582	11.2
Radium-226	Needles	Interstitial Rx of cancer	604	276	4152	8.4
Radon-222	Seeds	Interstitial Rx of cancer	436	158	3069	6.2
Cobalt-60	Needles or seeds	Intracavitary Rx of cancer	283	90	2737	5.5
Radium-226	Nasopharyngeal applicator	Intracavitary Rx of cancer	306	50	2457	5.0
Strontium-90	Plaques	Dermatological treatment	8	7	1182	2.4
Strontium-90	Nasopharyngeal applicator	Nonmalignant conditions	6	6	678	1.4
Cesium-137	Sealed sources	Intracavitary Rx of cancer	22	16	548	1.1
Cobalt-60	Needles, wire, and so forth	Interstitial Rx of cancer	258	31	547	1.1
Iridium-192	Seeds—nylon ribbon	Interstitial Rx of cancer	51	32	404	0.8
Radium-226	Nasopharyngeal applicator	Nonmalignant conditions	18	12	383	0.8
Gold-198	Seeds	Interstitial Rx of cancer	20	12	106	0.2
Radium-226	Eye applicator	Ophthalmic treatment	7	6	100	0.2
		Total for principal procedures listed above			49,430	99.6
		Total for 9 other procedures with less than 100 administrations			192	0.4
		Total for all brachytherapy procedures			49,622	100.0

* Taken from **K. D. Williams** and **J. D. Sutton**, *Survey of the Use of Radionuclides in Medicine*, Preliminary Report, U.S. Department of Health, Education, and Welfare, MORP-68-10, 1968.

373

Table 5
Radiopharmaceutical Therapy During 1966 *

Radionuclide	Compound	Treatment	Physicians licensed for procedure	Physicians performing procedure	Patient administrations Number	Patient administrations Percent
Iodine-131	Sodium iodide	Hyperthyroidism	1229	1042	20,717	61.4
Phosphorus-32	Soluble phosphate	Polycythemia vera	1086	581	3405	10.1
Iodine-131	Sodium iodide	Thyroid ablation	869	295	2316	6.9
Phosphorus-32	Colloidal phosphate	Malignant effusions	724	316	2033	6.0
Iodine-131	Sodium iodide	Thyroid cancer	907	334	1768	5.2
Phosphorus-32	Soluble phosphate	Leukemia	819	132	933	2.8
Iodine-131	Sodium iodide	Cardiac dysfunction	825	224	927	2.7
Gold-198	Colloidal gold	Malignant effusions	681	152	732	2.2
Phosphorus-32	Soluble phosphate	Osseous metastases	50	41	701	2.1
Total for principal procedures listed above					33,532	99.4
Total for 8 other procedures with less than 100 administrations					211	0.6
Total for all radiopharmaceutical therapy procedures					33,743	100.0

* Taken from K. D. Williams and J. D. Sutton, *Survey of the Use of Radionuclides in Medicine*, Preliminary Report, U.S. Department of Health, Education, and Welfare, MORP-68-10, 1968.

APPENDIX II
UNIVERSITIES COOPERATING IN AEC PROGRAMS

Membership of University Groups
Involved in "University-AEC Laboratory
Cooperative Program"

Argonne Universities Association
(AUA)

University of Arizona
Carnegie-Mellon University
Case Western Reserve University
University of Chicago
University of Cincinnati
Illinois Institute of Technology
University of Illinois
Indiana University
Iowa State University
University of Iowa
Kansas State University
University of Kansas
Loyola University
Marquette University
Michigan State University
University of Michigan
University of Minnesota
University of Missouri
Northwestern University
University of Notre Dame
Ohio State University
Ohio University
Pennsylvania State University
Purdue University
St. Louis University
Southern Illinois University
The University of Texas at
   Austin
Washington University
Wayne State University
University of Wisconsin

*Associated Colleges of the
Chicago Area*
(ACCA)

Aurora College
Barat College

College of Saint Francis
Concordia College
Elmhurst College
George Williams College
Judson College
Lake Forest College
Lewis College
Mundelein College
North Central College
North Park College
Olivet Nazarene College
Rosary College
Saint Dominic College
Saint Procopius College
Saint Xavier College
Trinity Christian College
Wheaton College

Associated Colleges of the Midwest
(ACM)

Beloit College
Carleton College
Coe College
Cornell College
Grinnell College
Knox College
Lawrence University
Monmouth College
Ripon College
St. Olaf College

Associated Universities, Inc. (AUI)

Columbia University
Cornell University
Harvard University
The Johns Hopkins University
Massachusetts Institute of Technolo-
   ogy

University of Pennsylvania
Princeton University
University of Rochester
Yale University

Associated Western Universities
(AWU)

Arizona State University
University of Arizona
Brigham Young University
California State Polytechnic College
University of California (Davis)
University of California (Los Angeles)
University of California (Santa Barbara)
Colorado State University
University of Colorado
University of Denver
University of Houston
Idaho State University
University of Idaho
Montana State University
University of Montana
University of Nevada (Las Vegas)
University of Nevada (Reno)
New Mexico Institute of Mining and Technology
New Mexico State University
University of New Mexico
North Dakota State University
Rice University
San Jose State College
South Dakota State University
University of South Dakota
Texas A & M University
Utah State University
University of Utah
University of Wyoming

Central States Universities, Inc.
(CSUI)

Ball State University
Bowling Green State University
De Paul University
De Pauw University

University of Detroit
John Carroll University
Kent State University
Miami University
Northern Illinois University
Northern Michigan University
Ohio University
Southern Illinois University
State College of Iowa
University of Toledo
Western Michigan University

*Northwest College and University
Association for Science* (NORCUS)

University of Alaska
Big Bend Community College
Boise College
Central Oregon Community College
Central Washington State College
Clark College
Columbia Basin College
Eastern Oregon College
Eastern Washington State College
Everett Community College
Fort Wright College
Gonzaga University
College of Idaho
University of Idaho
Lewis-Clark College
Linfield College
Lower Columbia College
Marylhurst College
Methodist University
University of Montana
Montana College of Mineral Science and Technology
Montana State University
Northern Montana College
Northwest Nazarene College
University of Oregon
Oregon College of Education
Oregon State University
Oregon Technical Institute
Pacific University
Pacific Lutheran University
University of Portland

Portland State University
University of Puget Sound
Reed College
Rocky Mountain College
Saint Martin's College

Oak Ridge Associated Universities
(ORAU)

University of Alabama
University of Arkansas
Auburn University
Catholic University of America
Clemson University
Duke University
Emory University
Fisk University
Florida State University
University of Florida
Georgia Institute of Technology
University of Georgia
University of Kentucky
Louisiana State University
University of Louisville
University of Maryland
Meharry Medical College
University of Miami
Mississippi State University
University of Mississippi
North Carolina State University
University of North Carolina
North Texas State University
University of Oklahoma
University of Puerto Rico
Rice University
University of South Carolina
Southern Methodist University
University of Tennessee
Texas A & M University
Texas Christian University
University of Texas
Texas Women's University
Tulane University
Tuskegee Institute
Vanderbilt University
University of Virginia
Virginia Commonwealth University

Virginia Polytechnic Institute and
    State University
West Virginia University
College of William and Mary

*Savannah River Nuclear Education
Committee* (SRNEC)

(Administered under ORAU—not an
    incorporated group—membership
    from South Carolina and Georgia)

Clemson College
Emory University
Georgia Institute of Technology
Medical College of Georgia
University of Georgia
Medical University of South Carolina
University of South Carolina

Universities Research Association, Inc.
(URA)

University of Arizona
Brown University
California Institute of Technology
University of California (Berkeley)
University of California (Los Angeles)
University of California (San Diego)
Carnegie-Mellon University
Case Western Reserve University
University of Chicago
University of Colorado
Columbia University
Cornell University
Duke University
Florida State University
Harvard University
University of Illinois
Indiana University
Iowa State University
University of Iowa
The Johns Hopkins University
University of Maryland
Massachusetts Institute of Technol-
    ogy
Michigan State University

University of Michigan
University of Minnesota
State University of New York (Buffalo)
State University of New York (Stony Brook)
University of North Carolina (Chapel Hill)
Northwestern University
University of Notre Dame
Ohio State University
University of Pennsylvania
Princeton University
Purdue University
Rice University
University of Rochester

The Rockefeller University
Rutgers—The State University
Stanford University
Stevens Institute of Technology
Syracuse University
University of Texas
University of Toronto
Tulane University
Vanderbilt University
Virginia Polytechnic Institute and State University
University of Virginia
Washington University
University of Washington
University of Wisconsin
Yale University

GLOSSARY [*]

ACCELERATOR   A device for increasing the velocity and energy of charged particles, for example, electrons, protons, or heavier ions through application of electrical and/or magnetic forces. Accelerators have made particles move at velocities approaching the speed of light. Types of accelerators include betatrons, Cockcroft-Walton accelerators, cyclotrons, linear accelerators, synchrocyclotrons, synchrotrons, and Van de Graaff generators.

ACTIVATION   The process of making a material radioactive by bombardment with neutrons, protons, or other nuclear particles. Also called radioactivation.

ACTIVATION ANALYSIS   A method for identifying and measuring chemical elements in a sample of material. The sample is first made radioactive by bombardment with neutrons, charged particles, or gamma rays. The newly formed radioactive atoms in the sample then give off characteristic nuclear radiations (such as gamma rays) that tell what kinds of atoms are present and how many. Activation analysis is usually more sensitive than chemical analysis. It is used in research, industry, archeology, and criminology.

ALPHA PARTICLE   [Symbol α (alpha)] A positively charged particle emitted by certain radioactive materials. It is made up of two neutrons and two protons bound together, hence is identical with the nucleus of a helium atom. It is the least penetrating of the three common types of radiation (alpha, beta, gamma) emitted by radioactive material, being stopped by a sheet of paper. It is not dangerous to plants, animals or man unless the alpha-emitting substance has entered the body.

ATOM   A particle of matter indivisible by chemical means. It is the fundamental building block of the chemical elements. The elements, such as iron, lead, and sulfur, differ from each other because they contain different kinds of atoms. There are about six sextillion (6 followed by 21 zeros, or 6×10^{21}) atoms in an ordinary drop of water. According to present-day theory, an atom contains a dense inner core (the nucleus) and a much less dense outer domain consisting of electrons in motion around the nucleus. Atoms are electrically neutral.

AUTOMATON   [pl. automata] A machine which automatically follows a preset sequence of operations and which can correct errors or deviations during operation.

AUTORADIOGRAPH   A photographic record of radiation from radioactive material in an object, made by placing the object very close to a photographic film or emulsion. The process is called autoradiography. It is used, for instance, to locate radioactive atoms or tracers in metallic or biological samples.

---

[*] Adapted from *Nuclear Terms: A Glossary*, U.S. Atomic Energy Commission, Understanding the Atom Series.

BACKGROUND RADIATION   The radiation in man's natural environment, including cosmic rays and radiation from the naturally radioactive elements, both outside and inside the bodies of men and animals. It is also called natural radiation. The term may also mean radiation that is unrelated to a specific experiment.

BETA PARTICLE   [Symbol β (beta)] An elementary particle emitted from a nucleus during radioactive decay, with a single electrical charge and a mass equal to about $1/1837$ that of a proton. A negatively charged beta particle is identical to an electron. A positively charged beta particle is called a positron. Beta radiation may cause skin burns, and beta-emitters are harmful if they enter the body. Beta particles are easily stopped by a thin sheet of metal, however.

BRACHYTHERAPY   Radiation treatment using a solid or enclosed radioisotopic source on the surface of the body or at a short distance from the area to be treated.

BREEDER REACTOR   A reactor that produces fissionable fuel as well as consuming it, especially one that creates more than it consumes. The new fissionable material is created by capture of neutrons from fission in fertile materials. The process by which this occurs is known as breeding. A fast-breeder reactor operates with fast neutrons and a thermal breeder reactor operates with thermal neutrons.

CAPTURE   A process in which an atomic or nuclear system acquires an additional particle; for example, the capture of electrons by positive ions, or capture of electrons or neutrons by nuclei.

CHAIN REACTION   A reaction that stimulates its own repetition. In a fission chain reaction a fissionable nucleus absorbs a neutron and fissions, releasing additional neutrons. These in turn can be absorbed by other fissionable nuclei, releasing still more neutrons. A fission chain reaction is self-sustaining when the rate of neutron generation by fission equals or exceeds the rate of neutron absorption in the fissioning and nonfissioning nuclei and the rate of escape of neutrons from the system.

CONTROL ROD   A rod, plate, or tube containing a material that readily absorbs neutrons (hafnium, boron, etc.), used to control the power of a nuclear reactor. By absorbing neutrons, a control rod prevents these neutrons from causing further fission.

COOLANT   A substance circulated through a nuclear reactor to remove or transfer heat. Common coolants are water, air, helium, carbon dioxide, liquid sodium, and sodium-potassium alloy (NaK).

CORE   The central portion of a nuclear reactor containing the fuel elements and usually the moderator, but not the reflector.

COSMIC RAYS   Radiation of many sorts but mostly atomic nuclei (protons and heavier nuclei) with very high energies, originating outside the Earth's atmosphere. Cosmic radiation is part of the natural background radiation. Some cosmic rays are more energetic than any man-made forms of radiation.

CRITICAL MASS   The smallest mass of fissionable material that will support a self-sustaining chain reaction under stated conditions.

CURIE   The basic unit to describe the intensity of radioactivity in a sample of material. The curie is equal to 37 billion disintegrations per second, which is approximately the rate of decay of 1 gram of radium. A curie is also a quantity of any nuclide having 1 curie of radioactivity. Named for Marie and Pierre Curie, who discovered radium in 1898.

CYCLOTRON   A particle accelerator in which charged particles receive repeated synchronized accelerations by electrical fields as the particles spiral outward from their source. The particles are kept in the spiral by a powerful magnetic field.

DAUGHTER   A nuclide formed by the radioactive decay of another nuclide, which in this context is called the parent.

DECAY, RADIOACTIVE   The spontaneous transformation of one nuclide into a different nuclide or into a different energy state of the same nuclide. The process results in a decrease, with time, of the number of the original radioactive atoms in a sample. It involves the emission from the nucleus of alpha particles, beta particles (or electrons), or gamma rays; or the nuclear capture or ejection of orbital electrons; or fission. Also called radioactive disintegration.

DEUTERIUM   Heavy hydrogen. The nucleus of heavy hydrogen contains one neutron and one proton rather than the one proton of ordinary hydrogen. Symbol: H^2 or D^2.

DOSE RATE   The radiation dose delivered per unit time and measured, for instance, in rems per hour.

ELECTROMAGNETIC RADIATION   Radiation consisting of associated and interacting electric and magnetic waves that travel at the speed of light. Examples: light, radio waves, gamma rays, X rays. All can be transmitted through a vacuum.

ELECTRON   An elementary particle with a unit negative electrical charge and a mass about $1/1837$ that of the proton. Electrons surround the positively charged nucleus and determine the chemical properties of the atom. Positive electrons, or positrons, also exist.

ELECTRON VOLT   [Abbreviation ev or eV] The amount of kinetic energy gained by an electron or singly charged positive ion, when it is accelerated through an electric potential difference of 1 volt. It is equivalent to 1.603×10^{-12} erg. It is a unit of energy, or work, not of voltage.

ELEMENT   One of the more than 100 known chemical substances that cannot be divided into simpler substances by chemical means. A substance whose atoms all have the same atomic number. Examples: hydrogen, lead, uranium. (Not to be confused with *fuel element*.)

ELEMENTARY PARTICLES   The particles of which all matter and radiation are composed. All except protons and electrons are short-lived, and do not exist independently under normal conditions. They are of less than atomic size. Originally this term was applied to any particles which could not be further subdivided; now it is applied to nucleons (protons and neutrons), electrons, mesons, antiparticles and strange particles, but not to alpha particles or deuterons. Also called fundamental particles.

ENRICHED MATERIAL   Material in which the percentage of a given isotope present in a material has been artificially increased, so that it is higher than the percentage of that isotope naturally found in the material. Enriched uranium contains more of the fissionable isotope uranium-235 than the naturally occurring percentage (0.7 percent).

EXCITED STATE   The state of an atom or nucleus when it possesses more than its normal energy. The excess energy is usually released eventually as a gamma ray or photon.

FALLOUT   Debris (radioactive material) that resettles to earth after an atmospheric nuclear explosion. Fallout takes two forms. The first, called "local fallout," consists of the denser particles injected into the atmosphere by the explosion. They descend to Earth within 24 hours near the site of the detonation and in an area extending downwind for some distance (often hundreds of miles), depending on meteorological conditions and the yield of the detonation. The other form, called "worldwide fallout," consists of lighter particles which ascend into the upper troposphere and stratosphere and are distributed over a wide area of the Earth by atmospheric circulation. They then are brought to Earth, mainly by rain and snow, over periods ranging from months to years.

FAST BREEDER REACTOR   A nuclear reactor that operates with fast neutrons and produces more fissionable material than it consumes.

FAST NEUTRON   A neutron with energy greater than approximately 100,000 electron volts.

FAST REACTOR   A reactor in which the fission chain reaction is sustained primarily by fast neutrons rather than by thermal or intermediate neutrons. Fast reactors contain little or no moderator to slow down the neutrons from the speeds at which they are ejected from fissioning nuclei.

FERTILE MATERIAL   A material, not itself readily fissionable by thermal neutrons, which can be converted into a fissionable material by irradiation in a reactor. There are two basic fertile materials, uranium-238 and thorium-232. When these fertile materials capture neutrons, they are partially converted into fissionable plutonium-239 and uranium-233, respectively.

FISSION   The splitting of a heavy nucleus into two approximately equal parts (which are nuclei of lighter elements), accompanied by the release of a relatively large amount of energy and generally two or more neutrons. Fission can occur spontaneously, but usually is caused by nuclear absorption of gamma rays, neutrons, or other particles.

FISSION PRODUCTS   Nuclei formed by the fission of heavy elements. They are of medium atomic weight, and almost all are radioactive. Examples: strontium-90, cesium-137.

FISSIONABLE MATERIAL   Any material readily fissioned by slow neutrons, for example, uranium-233, uranium-235 and plutonium-239.

FLUX (NEUTRON)   A measure of the intensity of neutron radiation. It is the number of neutrons passing through 1 square centimeter of a given target in 1 second. Expressed as nv, where $n =$ the number of neutrons per cubic centimeter and $v =$ their velocity in centimeters per second.

FUEL CYCLE   The series of steps involved in supplying fuel for nuclear power reactors. It includes mining, refining, enrichment of fuel material, the original fabrication of fuel elements, their use in a reactor, chemical processing to recover the fissionable material remaining in the spent fuel, reenrichment of the fuel material, and refabrication into new fuel elements.

FUEL ELEMENT   A rod, tube, plate, or other mechanical shape or form into which nuclear fuel is fabricated for use in a reactor. (Not to be confused with *element*.)

FUEL REPROCESSING   The processing of reactor fuel to recover the unused fissionable material.

FUSION   The formation of a heavier nucleus from two lighter ones (such as hydrogen isotopes), with the attendant release of energy (as in a hydrogen bomb or controlled thermonuclear reaction).

GAMMA RAYS   [Symbol γ (gamma)]   High-energy, short-wavelength electromagnetic radiation. Gamma radiation frequently accompanies alpha and beta emissions and always accompanies fission. Gamma rays are very penetrating and are best stopped or shielded against by dense materials. Gamma rays are essentially similar to X rays, but are usually more energetic, and are nuclear in origin.

GAS CENTRIFUGE PROCESS   A method of isotope separation in which heavy atoms are separated from light ones by the same centrifugal forces that separate cream from milk in a dairy separator.

GAS-COOLED REACTOR   A nuclear reactor in which gas is the coolant.

GASEOUS DIFFUSION   A method of isotope separation based on the fact that atoms or molecules of different masses will diffuse through a porous barrier at different rates. The method is used to enrich uranium with the uranium-235 isotope.

GEIGER COUNTER   A radiation detector consisting of a gas-filled tube which discharges when activated by radiation such as electrons or gamma rays. Also called a Geiger-Mueller or GM counter.

GENETIC EFFECTS   Effects from radiation, chemicals, and other biological forces which change the gene structure of the exposed individual and lead to detectable physical changes in later generations. Compare *somatic effects*.

HALF-LIFE   The time in which half the atoms of a particular radioactive substance disintegrate to another nuclear form. Measured half-lives vary from millionths of a second to billions of years.

HEAT EXCHANGER   Any device that transfers heat from one fluid (liquid or gas) to another or to the environment.

HEAVY WATER   [Symbol D_2O]   Water containing significantly more than the natural proportion (1 in 6500) of heavy hydrogen (deuterium) atoms to ordinary hydrogen atoms. Heavy water is used as a moderator in some reactors because it slows down neutrons effectively and also has a low cross section for absorption of neutrons.

HYDROGEN BOMB A nuclear weapon deriving most of its energy from uncontrolled thermonuclear fusion of hydrogen isotopes, such as deuterium or tritium.

INDUCED RADIOACTIVITY Radioactivity that is created when substances are bombarded with neutrons, as from a nuclear explosion or in a reactor, or with charged particles produced by accelerators.

ION An atom or molecule that has lost or gained one or more electrons. By this ionization it becomes electrically charged. Examples: an alpha particle, which is a helium atom minus two electrons; a proton, which is a hydrogen atom minus its electron.

ISOTOPE One of two or more atoms with the same atomic number (the same chemical element) but with different atomic weights. An equivalent statement is that the nuclei of isotopes have the same number of protons but different numbers of neutrons. Thus, $^{12}_{6}C$, $^{13}_{6}C$, and $^{14}_{6}C$ are isotopes of the element carbon, the subscripts denoting their common atomic numbers, the superscripts denoting the differing mass numbers, or the sum of the protons and neutrons. Isotopes usually have very nearly the same chemical properties, but somewhat different physical properties.

MODERATOR A material, such as ordinary water, heavy water or graphite, used in a reactor to slow down high-velocity neutrons, thus increasing the likelihood of further fission.

MW Abbreviation for the megawatt, which is equal to one million watts. When applied to nuclear power plants, the unit refers to electrical power output rather than thermal power, which is much greater.

NATURAL URANIUM Uranium as found in nature, containing 0.7 percent of U-235, 99.3 percent of U-238, and a trace of U-234. It is also called normal uranium.

NERVA Nuclear Engine for Rocket Vehicle Application; a joint AEC-NASA program.

NEUTRON [Symbol n] An uncharged elementary particle with a mass slightly greater than that of the proton, and found in the nucleus of every atom heavier than hydrogen. A free neutron is unstable and decays with a half-life of about 12 minutes into an electron, proton, and neutrino. Neutrons sustain the fission chain reaction in a nuclear reactor.

NEUTRON ECONOMY The degree to which neutrons in a reactor are used for desired ends instead of being lost by leakage or nonproductive absorption. The ends may include propagation of the chain reaction, converting fertile to fissionable material, producing isotopes, or research.

NUCLEON A constituent of an atomic nucleus, that is, a proton or a neutron.

NUCLIDE A general term applicable to all atomic forms of the elements. The term is often erroneously used as a synonym for "isotope," which properly has a more limited definition. Whereas isotopes are the various forms of a single element (hence are a family of nuclides) and all have the same atomic number and number of protons, nuclides comprise all the isotopic forms of all the elements. Nuclides are distinguished by their atomic number, atomic mass, and energy state.

PHOTON   The carrier of a quantum of electromagnetic energy. Photons have an effective momentum but no mass or electrical charge.

PLASMA   An electrically neutral, gaseous mixture of positive and negative ions. Sometimes called the "fourth state of matter," since it behaves differently from solids, liquids, and gases. High-temperature plasmas are used in controlled fusion experiments.

PROTON   An elementary particle with a single positive electrical charge and a mass approximately 1837 times that of the electron. The nucleus of an ordinary or light hydrogen atom. Protons are constituents of all nuclei. The atomic number (Z) of an atom is equal to the number of protons in its nucleus.

RAD   (Acronym for radiation absorbed dose.) The basic unit of absorbed dose of ionizing radiation. A dose of one rad means the absorption of 100 ergs of radiation energy per gram of absorbing material.

RADIATION   The emission and propagation of energy through matter or space by means of electromagnetic disturbances which display both wavelike and particle-like behavior; in this context the "particles" are known as photons. Also, the energy so propagated. The term has been extended to include streams of fast-moving particles (alpha and beta particles, free neutrons, cosmic radiation, etc.). Nuclear radiation is that emitted from atomic nuclei in various nuclear reactions, including alpha, beta and gamma radiation, and neutrons.

RADIATION STANDARDS   Exposure standards, permissible concentrations, rules for safe handling, regulations for transportation, regulations for industrial control of radiation, and control of radiation exposure by legislative means.

RADIATION STERILIZATION   Use of radiation to cause a plant or animal to become sterile, that is, incapable of reproduction. Also the use of radiation to kill all forms of life (especially bacteria) in food, surgical sutures, etc.

RADIATION THERAPY   Treatment of disease with any type of radiation. Often called radiotherapy.

RADIOACTIVE TRACER   A small quantity of radioactive isotope (either with carrier or carrier-free) used to follow biological, chemical or other processes, by detection, determination or localization of the radioactivity.

RADIOACTIVITY   The spontaneous decay or disintegration of an unstable atomic nucleus, usually accompanied by the emission of ionizing radiation. (Often shortened to "activity.")

RADIOISOTOPE   A radioactive isotope. An unstable isotope of an element that decays or disintegrates spontaneously, emitting radiation. Nearly 2000 natural and artificial radioisotopes have been identified.

REM (Acronym for roentgen equivalent man.) The unit of dose of any ionizing radiation which produces the same biological effect as a unit of absorbed dose of ordinary X rays. The dose in rems = relative biological effectiveness × absorbed dose (in rads).

ROENTGEN [Abbreviation r] A unit of exposure to ionizing radiation. It is that amount of gamma or X rays required to produce ions carrying 1 electrostatic unit of electrical charge (either positive or negative) in 1 cubic centimeter of dry air under standard conditions. Named after Wilhelm Roentgen, German scientist who discovered X rays in 1895.

RTG Radioisotope Thermoelectric Generator. These small power supplies generate electricity directly from the heat of decaying radioisotopes.

SCANNER A medical tool employing radiation detectors which move in a systematic pattern over a patient to create a two-dimensional picture of radioactive tracers concentrated in various organs of the body.

SCINTILLATION DETECTOR A radiation detector consisting of a piece of crystal or plastic which flashes (scintillates) when activated by radiation. The light flashes are in turn detected by a photomultiplier tube which provides the electrical signal.

SNAP Acronym for Systems for Nuclear Auxiliary Power, a USAEC program to develop reactor and radioisotope power supplies for terrestrial and space use. Also used to designate specific power supplies, e.g., SNAP-27.

SOMATIC EFFECTS Physiological changes in individuals resulting from exposure to radiation, chemicals, or other biological forces, not transferrable to successive generations. See *genetic* effects.

SPECTRUM A visual display, a photographic record, or a plot of the distribution of the intensity of a given type of radiation as a function of its wavelength, energy, frequency, momentum, mass, or any related quantity.

SPONTANEOUS FISSION Fission that occurs without an external stimulus. Several heavy isotopes decay mainly in this manner; examples: fermium-256 and californium-254. The process occurs occasionally in all fissionable materials, including uranium-235.

SYNCHROCYCLOTRON A cyclotron in which the frequency of the accelerating voltage is decreased with time so as to match exactly the slowing revolutions of the accelerated particles. The decrease in rate of acceleration of the particles results from the increase of mass with energy as predicted by the Special Theory of Relativity.

SYNCHROTRON An accelerator in which particles are accelerated around a circular path by radio-frequency electric fields. The magnetic guiding and focusing fields are increased synchronously to match the energy gained by the particles so that the orbit radius remains constant.

TELEOPERATOR A general-purpose, dexterous, man-machine system which permits a human operator to project his manipulatory capabilities over long distances or into hostile environments. Examples: remote manipulators and man-amplifiers.

THERMAL BREEDER REACTOR   A breeder reactor in which the fission chain reaction is sustained by thermal neutrons.

THERMAL (SLOW) NEUTRON   A neutron in thermal equilibrium with its surrounding medium. Thermal neutrons are those that have been slowed down by a moderator to an average speed of about 2200 meters per second (at room temperature) from the much higher initial speeds they had at the time of fission. This velocity is similar to that of gas molecules at ordinary temperatures.

THERMONUCLEAR REACTION   A reaction in which very high temperatures bring about the fusion of two light atomic nuclei to form a heavier atom, releasing a large amount of energy.

THRESHOLD DOSE   The minimum dose of radiation that will produce a detectable biological effect.

TRACER, ISOTOPIC   An isotope of an element, a small amount of which may be incorporated into a sample of material (the carrier) in order to follow (trace) the course of that element through a chemical, biological or physical process, and thus also follow the larger sample. The tracer may be radioactive, in which case observations are made by measuring the radioactivity. If the tracer is stable, mass spectrometers, density measurement, or neutron activation analysis may be employed to determine isotopic composition. Tracers also are called labels or tags, and materials are said to be labeled or tagged when radioactive tracers are incorporated in them.

TRANSMUTATION   The transformation of one element into another by a nuclear reaction or series of reactions. Example: the transmutation of uranium-238 into plutonium-239 by absorption of a neutron.

TRANSURANIUM ELEMENT   An element above uranium in the periodic table, that is, with an atomic number greater than 92. All transuranium elements are produced artificially and are radioactive. They are (as of 1970) neptunium, plutonium, americium, curium, berkelium, californium, einsteinium, fermium, mendelevium, nobelium, lawrencium, rutherfordium (or kurchatovium), and hahnium.

TRITIUM   A radioactive isotope of hydrogen with two neutrons and one proton in the nucleus. It is man-made and is heavier than deuterium (heavy hydrogen). Tritium is used as a label in experiments in chemistry and biology. Its nucleus is a triton. It may fuel fusion reactors.

WASTE, RADIOACTIVE   Equipment and materials (from nuclear operations) which are radioactive and for which there is no further use. Wastes are generally classified as high-level (having radioactivity concentrations of hundreds to thousands of curies per gallon or cubic foot), low-level (in the range of 1 microcurie per gallon or cubic foot), or intermediate (between these extremes).

X RAY   Penetrating electromagnetic radiation emitted when orbital electrons of an excited atom return to an inner orbit. X rays are usually nonnuclear in origin and can also be generated by bombarding a metallic target with high-speed electrons.

YIELD   The total energy released in a nuclear explosion. It is usually expressed in equivalent tons of TNT (the quantity of TNT required to produce a corresponding amount of energy). Low yield is generally considered to be less than 20 kilotons; low intermediate yield from 20 to 200 kilotons; intermediate yield from 200 kilotons to 1 megaton. There is no standardized term to cover yields from 1 megaton upward.

SUGGESTED READING

CHAPTER 1. TOOLS TO BUILD A NEW WORLD

Brown, H. S., Bonner, J., and Weir, J. *The Next Hundred Years, Man's Nature and Technological Resources.* New York: The Viking Press, 1957.

Commoner, B. *Science and Survival.* New York: The Viking Press, 1966.

Elsner, H., Jr. *The Technocrats: Prophets of Automation.* Syracuse: Syracuse University Press, 1967.

Feinberg, G. *The Prometheus Project, Mankind's Search for Long-Range Goals.* Garden City, N.Y.: Doubleday, 1969.

Gabor, D. *Inventing the Future.* New York: Alfred A. Knopf, 1964.

Hillegas, M. R. *The Future as Nightmare; H. G. Wells and the Anti-Utopians.* New York: Oxford University Press, 1967.

Kahn, H., and Wiener, A. J. *The Year 2000.* New York: The Macmillan Co., 1967.

Novick, S. *The Careless Atom.* Boston: Houghton Mifflin Co., 1969.

Seaborg, G. T. "Need We Fear Our Nuclear Future?" *Bulletin of the Atomic Scientists* 24 (Jan. 1968):36.

———. *Peaceful Uses of Nuclear Energy.* Oak Ridge, Tenn.: U.S. Atomic Energy Commission, 1970.

von Neumann, J. "Can We Survive Technology?" *Fortune* 51 (June 1955): 106.

Weinberg, A. M. "Energy as an Ultimate Raw Material or—Problems of Burning the Sea and Burning the Rocks." *Physics Today* 12 (Nov. 1959):18.

CHAPTER 2. POWER AND MORE POWER

Bishop, A. S. "The Status and Outlook of the World Program in Controlled Fusion Research." Paper presented before the National Academy of Sciences, 1969.

Bond, V. "The Public and Radiation from Nuclear Power Plants." Paper presented at the AIF Public Affairs Workshop on Radiation and Man's Environment, 1970. (Available from the Atomic Industrial Forum.)

Bump, T. R. "The Third Generation of Breeder Reactors," *Scientific American* 216 (May 1967):25.

Clark, J. R. "Thermal Pollution and Aquatic Life," *Scientific American* 220 (Mar. 1969):18.

Curtis, R., and Hogan, E. *The Perils of the Peaceful Atom.* Garden City, N.Y.: Doubleday, 1969.

Eisenbud, M. "Standards of Radiation Protection and Their Implications to the Public Health." Paper presented at the University of Minnesota's Symposium on Nuclear Power and the Public, 1969. (Available from the Atomic Industrial Forum.)

Foreman, H., ed., *Nuclear Power and the Public*. Minneapolis: University of Minnesota Press, 1970.

Fox, C. H. *Radioactive Wastes*. U.S. Atomic Energy Commission, Understanding the Atom Series, 1965.

Glaser, P. E. "Power from the Sun: Its Future," *Science* 162 (1968):857.

Glasstone, S. *Controlled Nuclear Fusion*. U.S. Atomic Energy Commission, Understanding the Atom Series, 1967.

Gofman, J., and Tamplin, A. *Population Control through Nuclear Pollution*. Chicago: Nelson-Hall Co., 1970.

Gough, W. C., and Eastlund, B. J. "The Prospects of Fusion Power," *Scientific American* 224 (Feb. 1971):50.

Green, L., Jr. "Energy Needs Versus Environmental Pollution: A Reconciliation," *Science* 156 (June 16, 1967):1448.

Grigorieff, W. W., ed. *Abundant Nuclear Energy*. U.S. Atomic Energy Commission Symposium Series No. 14, 1969.

Hammond, R. P. "Low Cost Energy: A New Dimension," *Science Journal* (Jan. 1961):34.

Hewlett, R. G. "Man Harnesses the Atom," in *Technology in Western Civilization*, edited by M. Kranzberg and C. W. Pursell, Jr. New York: Oxford University Press, 1967.

Hogerton, J. F. *Atomic Power Safety*. U.S. Atomic Energy Commission, Understanding the Atom Series, 1967.

——. *Nuclear Reactors*. U.S. Atomic Energy Commission, Understanding the Atom Series, 1967.

——. "The Arrival of Nuclear Power." *Scientific American* 218 (Feb. 1968):21.

International Atomic Energy Agency. *Environmental Aspects of Nuclear Power Stations*, Proceedings of a Symposium, New York, 10–14 August 1970, Document STI/PUB/261, Vienna, 1971.

Joint Committee on Atomic Energy, U.S. Congress. *Environmental Effects of Producing Electric Power*, Hearings, Part I, October-November 1969; Part II, January-February 1970. Washington, D.C.; U.S. Government Printing Office.

Lane, J. A. "Rationale for Low-Cost Nuclear Heat and Electricity." In *Abundant Nuclear Energy*, edited by W. W. Grigorieff. U.S. Atomic Energy Commission, Symposium Series No. 14, 1969.

Lapp, R. E. *A Citizen's Guide to Nuclear Power*. Washington, D.C.: The New Republic, 1971.

Library of Congress, Legislative Reference Service. *The Economy, Energy and the Environment—A Background Study*, 1970.

Lyerly, R. L., and Mitchell, W., III. *Nuclear Power Plants*. U.S. Atomic Energy Commission, Understanding the Atom Series, 1966.

McHale, J. "World Energy Resources in the Future." *Futures* 1 (Sept. 1968):4.

Novick, S. *The Careless Atom*. Boston: Houghton Mifflin Co., 1969.

Office of Science and Technology, Energy Policy Staff. *Electric Power and the Environment*, 1970.

Sagan, L. Review of *Pollution Control Through Nuclear Pollution.* Background INFO, 1970. (Available from Atomic Industrial Forum.)

Schurr, S. H., et al. *Energy in the American Economy, 1850 to 1975. An Economic Study of Its History and Prospects.* Baltimore: Johns Hopkins Press, 1960.

Seaborg, G. T., and Bloom, J. L. "Fast Breeder Reactors," *Scientific American* 223 (Nov. 1970):13.

Shaw, M., and Whitman, M. "Nuclear Power: Suddenly Here." *Science and Technology* 75 (Mar. 1968):22.

U.S. Atomic Energy Commission. *The First Reactor,* 1968.

Note: Current literature concerning nuclear power and its environmental effects is available from the Atomic Industrial Forum, Inc., 475 Park Avenue, New York, N.Y. 10016

CHAPTER 3. LABELS, BOND BREAKERS, AND EXPLOSIVES

Corliss, W. R., and Harvey, D. G. *Radioisotopic Power Generation.* Englewood Cliffs, N.J.: Prentice-Hall, Inc., 1964.

Dewing, S. B. *Modern Radiology in Historical Perspective.* New York: C C Thomas, 1962.

Eastlund, B. J., and Gough, W. C. *The Fusion Torch—Closing the Cycle from Use to Reuse.* U.S. Atomic Energy Commission paper, 1969.

Faul, H. *Nuclear Clocks.* U.S. Atomic Energy Commission, Understanding the Atom Series, 1966.

Gerber, C. R.; Hamburger, R.; and Hull, E. W. S. *Plowshare.* U.S. Atomic Energy Commission, Understanding the Atom Series, 1967.

Lyon, W. S., Jr., ed. *Guide to Activation Analysis.* Princeton: D. Van Nostrand Co., 1964.

Mead, R. L., and Corliss, W. R. *Power from Radioisotopes.* U.S. Atomic Energy Commission, Understanding the Atom Series, 1964.

Osborne, T. S. *Atoms in Agriculture.* U.S. Atomic Energy Commission, Understanding the Atom Series, 1967.

Phelan, E. W. *Radioisotopes in Medicine.* U.S. Atomic Energy Commission, Understanding the Atom Series, 1966.

Sanders, R. *Project Plowshare—The Development of the Peaceful Uses of Nuclear Explosives.* Washington, D.C.: Public Affairs Press, 1962.

Teller, E., et al. *Constructive Uses of Nuclear Explosives.* New York: McGraw-Hill Book Co., 1968.

U.S. Atomic Energy Commission. *Engineering with Nuclear Explosives.* TID-7695, 1964.

U.S. Atomic Energy Commission. *Safety of Underground Nuclear Testing.* TID-24996, 1969.

Wahl, W. H., and Kramer, H. H. "Neutron-Activation Analysis." *Scientific American* 216 (Apr. 1967):68.

CHAPTER 4. MORE FOOD AND WATER

Bardach, J. E. "Aquaculture." *Science* 161 (Sept. 13, 1968):1098.

Gerber, C. R.; Hamburger, R.; and Hull, E. W. S. *Plowshare.* U.S. Atomic Energy Commission, Understanding the Atom Series, 1967.

Hickman, K. "Oases for the Future." *Science* 154 (Nov. 4, 1966):612.

Leopold, L. B., and Tilson, S. "The Water Resource." *International Science and Technology* 55 (July 1966):24.

MacQueen, K. F. "The Worldwide Outlook for Food Irradiation." In *International Conference on Constructive Uses of Atomic Energy.* Hinsdale, Ill.: American Nuclear Society, 1969.

Osborne, T. S. *Atoms in Agriculture.* U.S. Atomic Energy Commission, Understanding the Atom Series, 1967.

Post, R. G., and Seale, R. L., eds. *Water Production Using Nuclear Energy.* Tucson: University of Arizona Press, 1967.

Sewell, W. R. D. "NAWAPA: A Continental Water System—Pipedream or Practical Possibility?" *Bulletin of the Atomic Scientists* 23 (Sept. 1967):8.

Sigurbjörnsson, B. "Induced Mutations in Plants." *Scientific American* 224 (Jan. 1971):86.

Starmer, R., and Lowes, F. "Nuclear Desalting—Future Trends and Today's Costs." *Chemical Engineering* 75 (Sept. 9, 1968):127.

Teller, E., et al. *Constructive Uses of Nuclear Explosives.* New York: McGraw-Hill Book Co., 1968.

Urrows, G. M. *Nuclear Energy for Desalting.* U.S. Atomic Energy Commission, Understanding the Atom Series, 1966.

———. *Food Preservation by Irradiation.* U.S. Atomic Energy Commission, Understanding the Atom Series, 1964.

CHAPTER 5. OLD CITIES/NEW CITIES/NO CITIES

Baker, P. S., et al. *Radioisotopes in Industry.* U.S. Atomic Energy Commission, Understanding the Atom Series, 1967.

Decker, G. L.; Wilson, W. B.; and Bigge, W. B. "Nuclear Energy for Industrial Heat and Power." *Chemical Engineering Progress* 64 (Mar. 1968):61.

Drobny, N. L.; Hull, H. E.; and Testin, R. F. *Recovery and Utilization of Municipal Solid Waste.* U.S. Public Health Service Publication 1908, 1969.

Dyckman, J. W. "Transportation in Cities." *Scientific American* 213 (Sept. 1965):162.

Eastlund, B. J., and Gough, W. C. "Thermonuclear Plasma as a Universal Solvent and Industrial Processor." *Bulletin of the American Physical Society* 11 (Nov. 1968): 1564.

Green, L., Jr. "Energy for an Inland Agro-Industrial Community." American Society of Mechanical Engineers. Paper WA/ENER-12, 1968.

Holmes, J. M. "Energy-Intensive and Heat-Intensive Processes for a Nuclear Energy Center." In *Abundant Nuclear Energy,* edited by W. W. Grigorieff. U.S. Atomic Energy Commission Symposium Series 14, 1969.

Kohl, J.; Zentner, R. D.; and Lukens, H. R. *Radioisotope Applications Engineering*. Princeton, N.J.: D. Van Nostrand Co., 1961.

Meier, R. L. *Science and Economic Development: New Patterns of Living*. Cambridge, Mass.: M.I.T. Press, 1966.

———. "The Social Impact of a Nuplex." *Bulletin of the Atomic Scientists* 25 (Mar. 1969):16.

National Academy of Science/National Research Council. *Waste Management and Control*. NAS/NRC Publication 1400. Washington, D.C., 1966.

Newell, R. E. "The Global Circulation of Atmospheric Pollutants." *Scientific American* 224 (Jan. 1971):32.

Oak Ridge National Laboratory. *Nuclear Energy Centers, Industrial and Agro-Industrial Complexes, Summary Report*. ORNL-4291, 1968.

Spilhaus, A. "The Experimental City." *Science* 159 (Feb. 16, 1968):710.

Steinberg, M. "Chemonuclear and Radiation Chemical Process Applications." *Nuclear Applications* 6 (May 1969):425.

Stanford Research Institute. *Future Urban Transportation Systems*. PB-178 260. Menlo Park, Calif., 1967.

U.S. Congress, Committee on Science and Astronautics. *Science and Technology and the Cities*. Washington, D.C.: Government Printing Office, 1969.

U.S. Government, Department of Housing and Urban Development. *Tomorrow's Transportation, New Systems for the Urban Future*. Washington, D.C.: Government Printing Office, 1968.

CHAPTER 6. PLANETARY ENGINEERING

Emiliani, C., et al. "Underground Nuclear Explosions and the Control of Earthquakes." *Science* 165 (Sept. 19, 1969):1255.

Fleagle, R. G., ed. *Weather Modification: Science and Public Policy*. Seattle: University of Washington Press, 1969.

Gerber, C. R.; Hamburger, R.; and Hull, E. W. S. *Plowshare*. U.S. Atomic Energy Commission, Understanding the Atom Series, 1967.

Halacy, D. S., Jr. *The Weather Changers*. New York: Harper & Row, 1968.

National Academy of Sciences/National Research Council. *Weather and Climate Modification, Problems and Prospects*. Report 1350. Washington, D.C., 1966.

Teller, E., et al. *Constructive Uses of Nuclear Explosives*. New York: McGraw-Hill Book Co., 1968.

U.S. Atomic Energy Commission. *Engineering with Nuclear Explosives—Proceedings of the Third Plowshare Symposium*. TID-7695, 1964.

U.S. Congress, Hearings before the Subcommittee on Legislation, Joint Committee on Atomic Energy. *Commercial Plowshare Services*. Washington, D.C.: Government Printing Office, 1968.

CHAPTER 7. NEW WORLDS ABOVE AND BELOW

Clarke, A. C., ed. *The Coming of the Space Age*. New York: Meredith Press, 1967.

Cole, D. M., and Scarfo, R. G. *Beyond Tomorrow: The Next Fifty Years in Space.* Amherst, Wis.: Amherst Press, 1965.

Corliss, W. R. *SNAP: Nuclear Space Reactors.* U.S. Atomic Energy Commission, Understanding the Atom Series, 1966.

——. *Nuclear Propulsion for Space.* U.S. Atomic Energy Commission, Understanding the Atom Series, 1967.

Croke, E. J. "Ocean Technology and Nuclear Power," *Power Reactor Technology* 9 (Winter 1965–1966):1.

Klein, M. "Nuclear Energy in Space." *Nuclear News* (Dec. 1968):31.

Mead, R. L., and Corliss, W. R. *Power from Radioisotopes.* U.S. Atomic Energy Commission, Understanding the Atom Series, 1964.

Schmidt, H. R. "Power for Undersea Applications." *1967 AIF Annual Conference.* New York: Atomic Industrial Forum, 1968.

Schulman, F., and Smith, A. H. "Lunar Power Systems—A Long View," *Astronautics & Aeronautics* 6 (Nov. 1968):62.

Seaborg, G. T. *Nuclear Energy in Space Exploration.* U.S. Atomic Energy Commission, 1966.

Wetch, J. R. "Reactors: Key to Large-Scale Underwater Operations." *Nucleonics* 24 (June 1966): 33.

CHAPTER 8. SUSTAINING AND AUGMENTING MAN

Asimov, I., and Dobzhansky, T. *The Genetic Effects of Radiation.* U.S. Atomic Energy Commission, Understanding the Atom Series, 1966.

Blahd, W. *Nuclear Medicine.* New York: McGraw-Hill Book Co., 1965.

Clynes, M., and Milsum, J. H. *Biomedical Engineering Systems.* New York: McGraw-Hill Book Co., 1970.

Dewing, S. B. *Modern Radiology in Historical Perspective.* Springfield, Ill.: C C Thomas, 1962.

Halacy, D. S., Jr. *Cyborg: Evolution of the Superman.* New York: Harper & Row, 1965.

Kisieleski, W. E., and Baserga, R. *Radioisotopes and Life Processes.* U.S. Atomic Energy Commission, Understanding the Atom Series, 1966

Longmore, D. *Spare-Part Surgery: The Surgical Practice of the Future.* Garden City, N.Y.: Doubleday & Co., 1968.

Phelan, E. W. *Radioisotopes in Medicine.* U.S. Atomic Energy Commission, Understanding the Atom Series, 1966.

Sonneborn, T. M., ed. *Control of Human Heredity and Evolution.* New York: The Macmillan Co., 1965.

CHAPTER 9. THE ATOM AS A MOVING FORCE IN SOCIETY

Corliss, W. R. *Neutron Activation Analysis.* U.S. Atomic Energy Commission, Understanding the Atom Series, 1963.

Dean, G. *Report on the Atom.* New York: Alfred A. Knopf, 1953.

Hewlett, R. G., and Anderson, O. E., Jr. *The New World*. University Park, Pa.: The Pennsylvania State University Press, 1962.

——, and Duncan, F. *Atomic Shield*. University Park, Pa.: The Pennsylvania State University Press, 1969.

Lawrence, J. H.; Manowitz, B.; and Loeb, B. S. *Radioisotopes and Radiation*. New York: McGraw-Hill Book Co., 1964.

Lenihan, J. M. A., and Thompson, S. J., eds. *Activation Analysis*. New York: Academic Press, 1965.

Lilienthal, D. E. *The Journals of David E. Lilienthal*. Vol. II, *The Atomic Energy Years, 1945–1950*. New York: Harper & Row, 1964.

Seaborg, G. T., and Wilkes, D. M. *Education and the Atom*. New York: McGraw-Hill Book Co., 1964.

Spinrad, B. I. "Why Not National Laboratories?" *Bulletin of the Atomic Scientists* 22 (Apr. 20, 1966):20.

Strauss, L. L. *Men and Decisions*. Garden City, N.Y.: Doubleday & Co., 1962.

U.S. Atomic Energy Commission. *The Future Role of the Atomic Energy Commission Laboratories*. Washington, D.C., 1960.

U.S. Government. *Utilization of Federal Laboratories*, Hearings, Subcommittee on Science, Research, and Development, U.S. House of Representatives, Government Printing Office, Washington, D.C., 1966.

CHAPTER 10. THE INTERNATIONAL ATOM

Bechhoefer, B. G. *Postwar Negotiations for Arms Control*. Washington: Brookings Institute, 1961.

Boskey, B., and Willrich, M., eds. *Nuclear Proliferation: Prospects for Control*. New York: Dunellen, 1970.

Brennan, D. *Arms Control, Disarmament, and National Security*. New York: Pergamon Press, 1969.

Daedalus. Special Issue on Arms Control, Fall, 1960.

Dean, A. H. *Test Ban and Disarmament*. New York: Harper & Row, 1966.

Dougherty, J. E., and Lehman, J. F., Jr. *The Prospects for Arms Controls*. New York: Macfadden Books, 1965.

Feld, B. T. "After the Nonproliferation Treaty—What Next?" *Bulletin of the Atomic Scientists* 24 (Sept. 1968):2.

Grodzins, M., and Rabinowitch, E., eds. *Atomic Age: Forty-Five Scientists and Scholars Speak*. New York: Simon & Schuster, 1965.

Halperin, M. H. *China and the Bomb*. New York: Frederick A. Praeger, 1965.

Jacobson, H. K., and Stein, E. *Diplomats, Scientists and Politicians: The United States and the Nuclear Test Ban Negotiations*. Ann Arbor: University of Michigan Press, 1966.

Kahn, H. *On Thermonuclear War*. Princeton, N.J.: Princeton University Press, 1960.

Kissinger, H. A. *Nuclear Weapons and Foreign Policy*. New York: Harper and Brothers, 1957.

Kolkowicz, R., et al. *The Soviet Union and Arms Control: A Superpower Dilemma*. Baltimore: The Johns Hopkins University Press, 1970.

Kramish, A. *Atomic Energy in the Soviet Union.* Palo Alto, Calif.: Stanford University Press, 1959.

———. *The Peaceful Atom in Foreign Policy.* New York: Harper & Row, 1963.

Lefever, E. W. *Arms and Arms Control.* New York: Frederick A. Praeger, 1962.

McBride, J. H. *The Test Ban Treaty: Military, Technological, and Political Implications.* Chicago: Henry Regnery Co., 1967.

Melman, S., ed. *Disarmament—Its Politics and Economics.* American Academy of Arts and Sciences, 1962.

Polach, J. G. *Euratom: Its Background, Issues and Economic Implications.* Dobbs Ferry, N.Y.: Oceana Publications, 1964.

Quester, G. H. *Nuclear Diplomacy, The First Twenty-Five Years.* New York: Dunellen, 1970.

Roberts, C. *The Nuclear Years: The Arms Race and Arms Control, 1945–1970.* New York: Frederick A. Praeger, 1970.

Rotblat, J. *Pugwash—The First Ten Years.* New York: Humanities Press, 1969.

Scoville, H., Jr. *Missile Madness.* Boston: Houghton-Mifflin Co., 1970.

———. "The Limitation of Offensive Weapons." *Scientific American* 224 (Jan. 1971):15.

Smith, A. K. *A Peril and a Hope.* Chicago: University of Chicago Press, 1965.

Stockholm International Peace Research Institute. *SIPRI Yearbook of World Armaments and Disarmament 1968/69* and *1969/70.* New York: Humanities Press, Inc.

U.S. Congress, Committee on Foreign Relations, Senate. *Non-proliferation Treaty.* Washington, D.C.: Government Printing Office, 1969.

U.S. Congress, Senate. *Atoms for Peace Manual.* Document No. 55. Washington, D.C.: Government Printing Office, 1955.

York, H. *Race to Oblivion.* New York: Simon and Schuster, 1970.

CHAPTER 11. NEW UNDERSTANDING

Bernstein, J. *The Elusive Neutrino.* U.S. Atomic Energy Commission, Understanding the Atom Series, 1969.

Corliss, W. R. *Neutron Activation Analysis.* U.S. Atomic Energy Commission, Understanding the Atom Series, 1963.

Faul, H. *Nuclear Clocks.* U.S. Atomic Energy Commission, Understanding the Atom Series, 1966.

Ginzburg, V. L. "The Astrophysics of Cosmic Radiation." *Scientific American* 220 (Feb. 1969): 50.

Gordus, A. "Activational Analysis: Artifacts and Art." *New Scientist* 40 (Oct. 17, 1968):128.

Johnson, F. "Radiocarbon Dating and Archeology in North America." *Science* 155 (Jan. 13, 1967):165.

Keisch, B. *The Mysterious Box: Nuclear Science and Art.* U.S. Atomic Energy Commission, Understanding the Atom Series, 1970.

———, et al. "Dating and Authenticating Works of Art by Measurement of Natural Alpha Emitters," *Science* 155 (Mar. 10, 1967):1238.

Leakey, L. S. B. "Exploring 1,750,000 Years into Man's Past." *National Geographic Magazine* 120 (Oct. 1961):564.

Libby, W. F. *Radiocarbon Dating.* Chicago: University of Chicago Press, 1955.

Ostroff, E. "Restoration of Photographs by Neutron Activation." *Science* 154 (Oct. 7, 1966):119.

Poole, L., and Poole, G. *Carbon 14 and Other Science Methods That Date the Past.* New York: McGraw-Hill Book Co., 1961.

Rose, P. H., and Wittkower, A. B. "Tandem Van De Graaff Accelerators." *Scientific American* 223 (Aug. 1970):24.

Seaborg, G. T. *Man-Made Transuranium Elements.* Englewood Cliffs, N.J.: Prentice-Hall, Inc., 1963.

————, and Valens, E. G. *Elements of the Universe.* New York: E. P. Dutton & Co., 1965.

————, and Bloom, J. L. "The Synthetic Elements; IV." *Scientific American* 220 (Apr. 1969):56.

Simons, E. L. "Unraveling the Age of the Earth and Man." *Natural History* 76 (Feb. 1967):52.

Swartz, C. E. *Microstructure of Matter.* U.S. Atomic Energy Commission, Understanding the Atom Series, 1965.

CHAPTER 12. MAN TO MANKIND, THE NEW OPTIMISM

de Chardin, P. T. *The Future of Man.* New York: Harper & Row, 1969.

Esfandiary, F. M. *Optimism One.* New York: W. W. Norton & Co., 1970.

Ferkiss, V. C. *Technological Man: The Myth and the Reality.* New York: George Braziller, 1969.

McHale, J. *The Future of the Future.* New York: George Braziller, 1969.

Seaborg, G. T. "Science, Technology and the Citizen." Speech delivered at Nobel Symposium on "The Place of Value in a World of Facts." Stockholm, September 17, 1969. Teselius, A., and Nilsson, S., eds. *Nobel Symposium 14.* Stockholm: Almqvist & Wiksell, 1971. New York: John Wiley & Sons, 1971.

Toffler, A. *Future Shock.* New York: Random House, 1970.

INDEX

(Figures in italics indicate pages upon which illustrations occur)

Demin, A. G., 335
deoxyribonucleic acid (DNA), 259, 261–62
desalting of water, techniques of, 23, 117, 121–30, 168, 233, 305
DESY accelerator, Germany, 329, 331
detectors, radiation, 289–90
detergents, biodegradable, 98, 164, 290
deuterium, 43–45, 49, 121, 237
deuteron, 328–29, 340
Development of Substitute Materials (DSM) Project, 268
diagnosis, nuclear, 237–44
Diamond, H., 334
disarmament policies, 323–24
distillation processes, 121–23, 127, 164, 169
Donald W. Douglas Laboratories, 252
Donaldson, Lauren R., 120
Donner Laboratory, 239
Doppler measurements, 344
Dow Chemical Company, 270
Doxiadis, Constantine, 83
Dragon High Temperature Gas-Cooled Reactor Program, 306
Dragon Tail gas stimulation project, 177
Druin, V. A., 335
Dubna Laboratory, 328, 337
Dubos, Rene, 362
Dunning, J. R., 268
Durham, Carl T., 26

Early Apollo Scientific Experiments Package (EASEP), 203
Earth, age of, 96–97
earthquake control, 195
ecology and ecologists, 55–56, 59, 64, 116, 137, 185–88, 233. See also environment
Edison, Thomas, 57–58, 154
E. I. du Pont de Nemours & Co., 271
Ein Shams University, 360
Einstein, Albert, 25, 267, 344
einsteinium, 295, 334, 336, 339
Eisenhower, Dwight D., and administration of, 26, 295–97, 304–305, 313
Eklund, Sigvard, 305
electric power consumption, 51–52
electric rockets, 219–20
electric transmission, 156–62
electromagnetic fields, 105–106, 219, 344

electromyographic (EMG) controls, 258–59
electron linac, 328–31
El Paso Natural Gas Company, 176, 178
Emelyanov, V. S., 298
energy centers, 128, 131–32, 167–72 (see also Nuplex); nuclear: see power, nuclear; solar, 55–56; thermal, 28–29, 56, 218 (see also thermal enrichment, thermal pollution)
Eniwetok Atoll, 100, 188
Entomological Research Center, 142
environment: deterioration of, 59, 150; impact of nuclear power, 28–29, 48, 62–65, 68–73, 76, 83–85; management of, 19, 54–55, 99, 112, 162, 221, 223, 364. See also ecology and ecologists, thermal pollution
Environmental Protection Agency (EPA), 60, 63
E. O. Lawrence Radiation Laboratory. See Lawrence Radiation Laboratory
Eric XIV, King of Sweden, 294
Eric the Red, 353
Eskola, K. and P., 335
Eugene Water and Electric Board, 119
Euratom Joint Power Reactor Program, 306
European Atomic Energy Community (Euratom), 297, 306, 321–22
European Chemical Reprocessing Company, Mol, Belgium, 307
European Nuclear Energy Agency (ENEA), 306–307
Evernden, Jack F., 349
evolution, theory of, 19, 96
excavation, nuclear, 98–104, 183–89, 206
Experimental Breeder Reactor (EBR-I), 37
Explorer satellites, 199, 228
explosives, nuclear, 98–104, 115–16, 129, 175–96, 234, 239, 244, 344. See also Plowshare program
Export-Import Bank, 302
extraterrestrial bases, 198–99, 206

fallout problems, 103, 116
farms and farming, 128, 130–31, 133, 164; sea, 117–18, 132; thermal enrichment, 117–20